Springer Climate

Series Editor
John Dodson, Menai, Australia

Springer Climate is an interdisciplinary book series dedicated on all climate research. This includes climatology, climate change impacts, climate change management, climate change policy, regional climate, climate monitoring and modeling, palaeoclimatology etc. The series hosts high quality research monographs and edited volumes on Climate, and is crucial reading material for Researchers and students in the field, but also policy makers, and industries dealing with climatic issues. Springer Climate books are all peer-reviewed by specialists (see Editorial Advisory board). If you wish to submit a book project to this series, please contact your Publisher (elodie.tronche@springer.com).

More information about this series at http://www.springer.com/series/11741

Lutz Meyer-Ohlendorf

Drivers of Climate Change in Urban India

Social Values, Lifestyles, and Consumer Dynamics in an Emerging Megacity

Lutz Meyer-Ohlendorf
Member of the Leibniz Association
Potsdam Institute for Climate Impact Research (PIK)
Potsdam, Germany

ISSN 2352-0698   ISSN 2352-0701   (electronic)
Springer Climate
ISBN 978-3-319-96669-4   ISBN 978-3-319-96670-0   (eBook)
https://doi.org/10.1007/978-3-319-96670-0

Library of Congress Control Number: 2018950230

© Springer International Publishing AG, part of Springer Nature 2019
This work is subject to copyright. All rights are reserved by the Publisher, whether the whole or part of the material is concerned, specifically the rights of translation, reprinting, reuse of illustrations, recitation, broadcasting, reproduction on microfilms or in any other physical way, and transmission or information storage and retrieval, electronic adaptation, computer software, or by similar or dissimilar methodology now known or hereafter developed.
The use of general descriptive names, registered names, trademarks, service marks, etc. in this publication does not imply, even in the absence of a specific statement, that such names are exempt from the relevant protective laws and regulations and therefore free for general use.
The publisher, the authors and the editors are safe to assume that the advice and information in this book are believed to be true and accurate at the date of publication. Neither the publisher nor the authors or the editors give a warranty, express or implied, with respect to the material contained herein or for any errors or omissions that may have been made. The publisher remains neutral with regard to jurisdictional claims in published maps and institutional affiliations.

This Springer imprint is published by the registered company Springer Nature Switzerland AG
The registered company address is: Gewerbestrasse 11, 6330 Cham, Switzerland

Abstract

Anthropogenic climate change jeopardises nearly all human life-support systems, and its mitigation represents one of the most eminent challenges to humanity. To abate climate change, it requires long-term, globally oriented, and far-reaching changes of the economy, of our ways of life and coexistence, and of our consumption patterns. Such a transformation pertains to the state and the economy, but it also concerns society in general and the individual consumer in particular. In order to address this challenge and examine the interlinkages between environment, state, economy, and society with a problem-oriented focus on the individual human being, new and transdisciplinary approaches are urgently needed. A rather young social-ecological research perspective combines the concept of lifestyle with issues of sustainability and climate change.

Lifestyles are group-specific, value-, attitude-, and preference-driven patterns of everyday life that unfold within an economically, social-culturally, and environmentally prestructured field of social interaction. Such a typology-oriented approach considers a multiplicity of driving factors and their interactions in order to get hold of and understand group-specific differences in conduct of life and their underlying causes, motives, and impacts. It can also highlight vantage points for targeted climate protection policies. Moreover, the study presented here developed a simple procedure to measure the specific climate impact of single consumption practices and their levels of diffusion. This approach reveals the most relevant areas of consumption (key points of climate policy intervention). Only a few studies have coupled the lifestyle concept with an approach to analyse and explain differences in personal-level carbon footprints. This PhD thesis contributes to the theoretical and methodological development of this approach and applies it to a context that has not been examined in this way before – urban India.

Over the last two to three decades, India has faced unprecedented dynamics of development and urbanisation that involve processes such as changing incomes, a growing and increasingly globally oriented choice of goods and services, and growing social disparities. As in many other countries of the Global South, these changes particularly concentrate in urban areas and it is often argued that a great transformation to sustainability will largely be decided in cities. This study therefore applies the theoretical framework of lifestyle to the city of Hyderabad, a city which saw very rapid development dynamics and which attracted attention through global and technology-oriented urban and economic policies.

For the conceptualisation of the standardised lifestyle survey, an explorative qualitative study was conducted with 26 semi-structured interviews. In a following step, the resulting questionnaire was pretested, analysed, and modified accordingly. For the main survey (n = 600), a three-stage proportionate geographical cluster sampling approach was chosen. The most relevant methodical steps of the data analysis were to develop a carbon calculator, the application of dimension reduction methods, and cluster analysis.

The analysis of income-group-specific effects on climate change revealed significant results with higher incomes leading to higher carbon footprints, especially with respect to household electricity consumption, individual motorised transport (IMT), and air travel. Surprisingly, emissions from meat consumption showed negative effects with rising income.

The analysis of consumption-practice-specific effects on carbon footprints also delivered definite results. Key points of climate policy intervention in particular are those consumption practices which show high-carbon intensities. The analysis reveals that such carbon-intensive practices are far less prevalent in Hyderabad, but with a potential to spread vastly (e.g. air travel, use of air-conditioning systems). Other practices having low emission intensities, but being extensively used, are identified as relevant due to the potential scale effects associated with addressing them. The analysis therefore allows for a precise and targeted assessment of different consumption practices and their emission reduction potential.

The development of the lifestyle typology brought about meaningful and internally consistent groups of lifestyle. The analysis of the groups revealed interesting and relevant insights with respect to the interrelated character of the incorporated dimensions (values, practices, and social-demographic factors). Group-specific differences concerning impacts on climate change were found among three out of six lifestyle groups. The analysis of these differences allowed for conclusions in respect to the underlying behavioural and motivational drivers, which tend to remain hidden for linear models of analysis and especially for purely economic analysis. In sum, this study was able to make an important contribution to the analysis of lifestyle-related impacts on climate change. Although a big challenge, the application of the lifestyle concept to the urban Indian context succeeded and delivered valuable insights.

Acknowledgements

This thesis is the outcome of a long journey of thought and interaction. Many people and institutions have crossed this path by discussing the challenges, ideas, and results. This support is invaluable for the success of such thesis. I would like to express my sincere gratitude to all of the people, who have supported this endeavour in some or the other way.

In particular, I am very grateful to Prof. Dr. Frauke Kraas for giving me invaluable insights into the subject of geography and the understanding of getting closer to those things and processes that are invisible at the first sight. She inspired me from the first day I learned about geography and patiently guided me through to the point I reached now. I would like to express my gratitude also to Prof. Dr. Josef Nipper for his supervision of my PhD thesis and his supportive comments and insights on methodology. I am also very grateful to Dr. Fritz Reusswig for inviting me to become part of such an inspiring project and learn from him about research on lifestyle, social-ecology, and climate change.

Some parts of this study were only possible to be realised with the great support of Vivek Gilani. He and his colleagues developed the first Indian-specific web-based carbon calculator, and he has developed a database for Indian-specific emission factors. Vivek was open to share his great knowledge and expertise in carbon footprinting and sustainability in India. He supported the project wherever possible, and he has become one of my best friends at the same time.

Special gratitude also goes to my direct colleagues and friends Jahid Hassan, Sonja Hassan, Eva Eichenauer, Vera Peters, Corinna Altenburg, and Florian Winter for patiently sharing their knowledge and greatly supporting this research for a long time.

I am obliged to Philip Kumar and Vamsi Krishna for their great support in administering and organising the survey in Hyderabad and for being very patient and supportive in conducting this rather unusual research project. I am also very thankful to the students who conducted the standardised interviews in Hyderabad, especially Sardhar, Mahesh J., Sreenu L., Venkat, Raju Gadepaka, Anil Kumar K., Ramesh C. H., Christy Mathews, Mahender, Shabir Hussain, Ravinder, Tony, M. G. Devi Prassanna, Saraswathi, Sujatha, Ali, Sri Kalyan, Rani, Rizwana, and Renuka.

Furthermore, I would also like to thank Oleksandr Kit, Matthias Lüdeke, Diana Reckien, Ulrike Sylla, Mirjam Neebe, Stefanie Leder, Mareike Kroll, Saskia Ellenbeck, Nadine Kuhla, Andreas Beneking, Anselmo García Cantú Ros, Susanne Schulz, Anne Dahmen, Julian Sagebiel, Jens Rommel, Kiran Anumandla, Thomas Fibian, Andreas Stadler, Karsten Schulz, Helen Jakobsen, Jalal Mando, Angelique Lustig, and Karsten Reiter for being wonderful colleagues and friends and for giving invaluable comments, ideas, and support.

I would like to thank Alison Schlums for doing a wonderful work on proofreading this PhD thesis. I want to thank the Federal Ministry for Education and Research (BMBF) for funding the project, and I want to give my regards also to Karin Gotzmann, who was particularly supportive.

Lastly, but the most, I would like to thank Seema and Ameya for their endless patience and support. This also goes to my mother, my sisters Ute and Birte, and my brother Jörn.

Contents

1 Introduction: Climate Change and Lifestyle – The Relevance of New Concepts for Social-Ecological Research 1

2 Approaches of Measuring Human Impacts on Climate Change 9

3 The Research Context: India and the Megacity of Hyderabad 49

4 Conceptualisation and Operationalisation – A Social Geography of Climate Change: Social-Cultural Mentalities, Lifestyle, and Related GHG Emission Effects in Indian Cities 81

5 Results Part I: Descriptive Analysis of Manifest Variables and Preparation of Latent Components for the Lifestyle Analysis 141

6 Results Part II: Income, Practice, and Lifestyle-Oriented Analysis of Personal-Level GHG Emissions 175

7 Discussion 219

8 Final Conclusions: Understanding Inequalities in Consumption-Based, Personal-Level GHG Emissions 245

References 253

Index 265

List of Abbreviations

AC	Air-conditioning
ANOVA	One-way analysis of variance
AP	Andhra Pradesh
AR5	Fifth Assessment Report
BfN	German Federal Agency for Nature Conservation
CAGR	Compound annual growth rate
CBF	Consumption-based footprint
cf.	Compare
CFL	Compact fluorescent lamps
CH_4	Methane
CO_2/CO_2e	Carbon dioxide/carbon dioxide equivalents
COP	Conference of the Parties
CSE	Centre for Science and Environment
DHS	Demographic and Health Surveys
e.g.	For example
EEIOA	Environmentally extended input-output analysis
EEP	Energy Efficiency Programme
EF	Emission factor
FI	Field investigators
GDP	Gross domestic product
GHG	Greenhouse gases
GHMC	Greater Hyderabad Municipal Corporation
GPC	Global Protocol for Community-Scale Greenhouse Gas Emission Inventories
GWP	Global warming potential
HAD RM3	Hadley Centre Regional Model Version 3
HFCs	Hydrofluorocarbons
HITEC City	Hyderabad Information Technology and Engineering Consultancy City
HMC	Hyderabad Municipal Corporation
HMP	Hyderabad Metro Rail Project

HUDA	Hyderabad Urban Development Authority
IIM	Indian Institute for Management
IMF	International Monetary Fund
IMT	Individual motorised transport
INCCA	Indian Network on Climate Change Assessment
INDC	Intended Nationally Determined Contributions
IO	Input-output
IPCC	Intergovernmental Panel on Climate Change
ISSC	International Social Science Council
IT	Information technology
ITES	IT-enabled services
KMO	Kaiser-Meyer-Olkin
KWh	Kilowatt hours
LCA	Life-cycle assessment
LPG	Liquefied petroleum gas
LULUCF	Land use, land-use change and forestry
MCH	Municipal Corporation of Hyderabad
MGI	McKinsey Global Institute
MIV	Motorisierter Individualverkehr
MMTS	Multimodal Transport System
MoEF	Ministry of Environment and Forests
MPI	Multidimensional Poverty Index
MSA	Measure-of-Sampling-Adequacy
MSC	Multimedia Super Corridor
N_2O	Nitrous oxide
NCAER	National Council of Applied Economic Research
NEP	New Economic Policy
NFHS	National Family Health Survey
NMT	Non-motorised transport
NSHIE	National Survey of Household Income and Expenditure
NSSO	National Sample Survey Office
O_3	Ozone
OECD	Organisation for Economic Co-operation and Development
PCA	Principal component analysis
PCF	Product carbon footprint
ppm	Parts per million
PPP	Purchasing power parity
PVQ	Portrait Values Questionnaire
REC	Renewable Energy Certificate
RWA	Residential welfare associations
SAM	Social Accounting Matrix
SAP	Structural Adjustment Program
SD	Standard deviation
SDG	Sustainable Development Goals
SEP	Social-economic position

SPRG	Sustainable Practices Research Group
T&D	Transmission and distribution
UNFCCC	United Nations Framework Convention on Climate Change
WBGU	Wissenschaftlicher Beirat der Bundesregierung Globaler Wandel/ German Advisory Council on Global Change

List of Figures

Fig. 2.1	Model of lifestyles operating within an individual scope of action, which is determined by social-positional and external factors, effects, and their feedback mechanisms	30
Fig. 3.1	Distribution of the world's urban population by size class of urban settlement and number of cities, 1970, 1990, 2014, and 2030	53
Fig. 3.2	Urban population percent distribution in cities of different size classes and absolute distribution of population between urban and rural areas over time between 1950 and 2030	54
Fig. 4.1	Main components and structure of the concept for the analysis of social-cultural differentials in personal-level GHG emissions	82
Fig. 4.2	Concept cloud of all targeted value orientations included in the survey. The sizes of the words *coarsely* point to the number of items that aim to measure the respective targeted value. The relative positions of the concepts *roughly* indicate proximity or distance in respect to content	92
Fig. 4.3	Adapted model of Schwartz' theory on universal aspects in the structure and contents of human values, their relations, higher-order value types, and bipolar value dimensions	111
Fig. 4.4	Simplified model of consumer decision-making with regard to investive expenditures (investive consumption)	117
Fig. 4.5	Research design: outline of the research process	124
Fig. 4.6	Methodological and analytical steps for the lifestyle segmentation and lifestyle-specific GHG emission accounting	129

List of Figures

Fig. 5.1	Break-up of age groups of overall sample compared with data from Census 2011 for the Greater Hyderabad Municipal Corporation (GHMC) (The Census data have been categorised into bins of 5 years. While the author's study has only respondents starting from 18 years, the respective Census data bin contains people from 15 to 20 years of age. For this reason, respondents below 20 years of age (11 respondents) have been removed from this figure))	142
Fig. 5.2	Break-up of religious groups of overall sample compared with data from Census 2011 for the Greater Hyderabad Municipal Corporation (GHMC)	143
Fig. 5.3	Bar chart of educational degrees of respondent and her/his father in percent	144
Fig. 5.4	Comparison of income distribution in Hyderabad (author's survey, combined household and equivalised categories) and urban India. The groups in the yellow box (seekers and strivers) represent the Indian middle class according to McKinsey Global Institute	147
Fig. 5.5	Histogram and descriptive statistical overview of wealth index	149
Fig. 5.6	Criterion of explained variance (ETA^2)	167
Fig. 5.7	Criterion for the relative improvement of explained variance (PRE)	167
Fig. 5.8	Criterion for the best variance proportion (F-max)	168
Fig. 5.9	Grouped bar chart of cluster centres (mean) for each cluster. Error bars represent 95% confidence interval levels	171
Fig. 6.1	Stacked bar chart of sector-specific distribution of personal GHG emissions (CO_2e/cap/year) per equivalised income class	180
Fig. 6.2	Bar chart with mean levels of overall personal GHG emissions (CO_2e/cap/year) per equivalised income class; table with median levels for overall emissions, as well as results of H-test after Kruskal-Wallis. Error bars show 95% confidence intervals	180
Fig. 6.3	Bar chart with mean levels of GHG emissions (CO_2e/cap/year) from electricity consumption per equivalised income class; table with median levels for electricity emissions, as well as results of H-test after Kruskal-Wallis. Error bars show 95% confidence intervals	181
Fig. 6.4	Bar chart with mean levels of personal GHG emissions (CO_2e/cap/year) from cooking fuel consumption per equivalised income class; table with median levels for cooking emissions, as well as results of H-test after Kruskal-Wallis. Error bars show 95% confidence intervals	181

Fig. 6.5	Bar chart with mean levels of personal GHG emissions (CO_2e/cap/year) from food consumption per equivalised income class; table with median levels for food emissions, as well as results of H-test after Kruskal-Wallis. Error bars show 95% confidence intervals	182
Fig. 6.6	Grouped bar chart with mean levels of personal GHG emissions specified for meat and dairy consumption per equivalised income class; table with median levels for meat and dairy consumption, as well as results of H-test after Kruskal-Wallis. Error bars show 95% confidence intervals	182
Fig. 6.7	Bar chart with mean levels of personal GHG emissions (CO_2e/cap/year) from long-distance bus and train travel per equivalised income class; table with median levels for long-distance bus and train travel emissions, as well as results of H-test after Kruskal-Wallis. Error bars show 95% confidence intervals	183
Fig. 6.8	Bar chart with mean levels of personal GHG emissions (CO_2e/cap/year) from air travel per equivalised income class; table with median levels for air travel emissions, as well as results of H-test after Kruskal-Wallis. Error bars show 95% confidence intervals	183
Fig. 6.9	Bar chart with mean levels of personal GHG emissions (CO_2e/cap/year) from public transport per equivalised income class; table with median levels for public transport emissions, as well as results of H-test after Kruskal-Wallis. Error bars show 95% confidence intervals	184
Fig. 6.10	Bar chart with mean levels of personal GHG emissions (CO_2e/cap/year) from individual motorised transport (IMT) per equivalised income class; table with median levels for IMT emissions, as well as results of H-test after Kruskal-Wallis. Error bars show 95% confidence intervals	184
Fig. 6.11	Grouped bar chart with mean levels of personal GHG emissions specified for both two-wheeler and four-wheeler per equivalised income class; table with median levels for both use of two-wheeler and four-wheeler, as well as results of H-test after Kruskal-Wallis. Error bars show 95% confidence intervals	185
Fig. 6.12	Bar chart with the cluster-specific (cluster 1) agreement levels on the underlying "active" attitudinal and value dimensions. Error bars show 95% confidence intervals	191

Fig. 6.13 Bar chart with the cluster-specific (cluster 2) agreement levels on the underlying "active" attitudinal and value dimensions. Error bars show 95% confidence intervals.. 196

Fig. 6.14 Bar chart with the cluster-specific (cluster 3) agreement levels on the underlying "active" attitudinal and value dimensions. Error bars show 95% confidence intervals.. 200

Fig. 6.15 Bar chart with the cluster-specific (cluster 4) agreement levels on the underlying "active" attitudinal and value dimensions. Error bars show 95% confidence intervals.. 204

Fig. 6.16 Bar chart with the cluster-specific (cluster 5) agreement levels on the underlying "active" attitudinal and value dimensions. Error bars show 95% confidence intervals.. 208

Fig. 6.17 Bar chart with the cluster-specific (cluster 6) agreement levels on the underlying "active" attitudinal and value dimensions. Error bars show 95% confidence intervals.. 211

Fig. 6.18 Stacked bar chart of sector-specific distribution of personal GHG emissions (kg CO_2e/cap/year) for each lifestyle cluster.......... 214

Fig. 6.19 Bar chart with mean levels of overall personal GHG emissions (kg CO_2e/cap/year) per lifestyle cluster; table with median levels for overall emissions, as well as results of H-test after Kruskal-Wallis. Error bars show 95% confidence intervals.............. 215

Fig. 6.20 Combined effects of cluster membership, income, and investive consumption on personal carbon footprints. The size of the "bubbles" indicates the mean of CO_2e footprints (also given in numbers – in tonnes of CO_2e/cap/year). Percentages in brackets designate the share of the cluster in the overall sample 216

List of Tables

Table 3.1	Comparative trends in population below poverty line, with 2005 and 2015 PPP revisions	59
Table 3.2	MPI results at the national level	62
Table 3.3	Estimates and projections of percent distribution of income classes and the middle class (highlighted in grey) for all India and urban India	64
Table 3.4	Size of India's middle-class, CGD, and NCAER estimates (2009/2010)	66
Table 3.5	Summary of findings from India's "4 × 4 assessment of the impact of climate change on key sectors and regions of India in the 2030s"	69
Table 4.1	Overview of questionnaire items measuring targeted values towards consumption	93
Table 4.2	Overview of questionnaire items measuring the targeted value paradigm of "culture of necessity"	99
Table 4.3	Overview of questionnaire items measuring the targeted value paradigm of "social-ecological conscious consumption"	101
Table 4.4	Overview of questionnaire items measuring the targeted value paradigm of 'religious tradition'	106
Table 4.5	Overview of questionnaire items measuring the targeted value paradigm of "family tradition"	110
Table 4.6	Overview of questionnaire items measuring the targeted more general values based on Portrait Value Questionnaire (PVQ) and others	112
Table 4.7	Key aspects and foci, objectives, tools of the qualitative survey	126
Table 4.8	Underlying emission factors (EF) and their sources	133
Table 4.9	Overview of included amenities building the wealth index and respective measurement levels	137
Table 5.1	Break-up of major employment category, past and present	145

Table 5.2	Original and equivalised income brackets based on McKinsey Global Institute	146
Table 5.3	Descriptive statistical overview of combined household and equivalised net income	147
Table 5.4	Ownership of durable assets and housing characteristics by wealth index quintile	150
Table 5.5	Rotated component matrix of PCA of media usage for information gathering	151
Table 5.6	Rotated component matrix of PCA of shopping preferences and practices	152
Table 5.7	Rotated component matrix of PCA of leisure activities	152
Table 5.8	Rotated component matrix of PCA of holiday preferences and practices	153
Table 5.9	Parameters for the Measure of Sampling Adequacy (MSA)	153
Table 5.10	Matrix of rotated components from principal component analysis (PCA)	154
Table 5.11	Rotated component matrix: factor 1 – religious tradition	156
Table 5.12	Rotated component matrix: factor 2 – frugality	158
Table 5.13	Rotated component matrix: factor 3 – social-ecological orientation	161
Table 5.14	Rotated component matrix: factor 4 – hedonist conspicuous consumption	162
Table 5.15	Rotated component matrix: factor 5 – family tradition	164
Table 5.16	Distribution of agreement levels on the five-factor solution	166
Table 5.17	F-values for each variable of the six-cluster solution	169
Table 5.18	Frequency distribution of the six-cluster solution	169
Table 5.19	Final cluster centres of the six-cluster solution (mean)	170
Table 5.20	Overview of medians and results of H-test after Kruskal-Wallis for all "passive" metric variables of social demography	172
Table 5.21	Overview of medians and results of H-test after Kruskal-Wallis for all "passive" metric variables of social demography	172
Table 5.22	Overview of medians and results of H-test after Kruskal-Wallis for all GHG emission variables (metric)	173
Table 5.23	Overview of results from cross tabulation for all "passive" categorical variables	174
Table 6.1	Sector-specific distribution of individual GHG emissions in Hyderabad	176
Table 6.2	Sector- and mode-specific structure of the overall carbon footprint of Hyderabad ($n = 590$)	178
Table 6.3	Overview of domain- and sector-specific personal average emission effects (based on mean and median) and percent shares of respondents associated with domain	188

Chapter 1
Introduction: Climate Change and Lifestyle – The Relevance of New Concepts for Social-Ecological Research

Keywords Great transformation · Sustainable development goals · Path dependencies · Sustainable urban development · Transdisciplinarity · Problem-oriented approach · People-oriented approach · New middle classes

Anthropogenic climate change (in the following climate change) can be seen as one of the greatest "world risks" (Beck 1998) that human society has ever brought forth. With technological development and industrialisation, which largely builds on the utilisation of fossil fuel-based energy, human society has – initially unknowingly – started treading a path that leads to an unprecedented global challenge to humanity. This development pathway has not only created the problem but also slowly made us able to see it and understand it, and possibly and hopefully, it will enable us to solve it.

Probably even more than other world risks, climate change both in terms of its direct physical impacts but also in terms of its encompassing social, cultural, and political implications for human society transcends all national borders. Moreover, the causes of climate change are immanently rooted in our highly "developed" ways of living. Through their ways of consumption, humans are to a varying degree closely connected with this systemic challenge, and it is hardly possible for any individual to dispose of this individual responsibility. Solutions to it need to deal with society and the ways we conduct and evaluate our lives. Behaviour and consumption patterns are therefore at the core of the problem and should be seen as starting point for a system-wide sustainability transition. This transformation is a societal process carried out by people who share a huge potential of creativity. In a context of economic growth-driven rapid social-cultural change as found in India, the chances of success of such a transformation are higher, because it does not need to tackle existing path dependencies that have been set back in the past. This aspect is especially relevant in the context of urban areas.

For a number of reasons, it is often stated that the wsuccess of the "great transformation" will be decided in cities (WBGU – Wissenschaftlicher Beirat der Bundesregierung Globale Umweltveränderungen 2016, p. 6) and this holds true not only for transitional countries like India. Some authors view cities as "key pathways in every aspect of sustainable development" (Parnell 2016, p. 539; WBGU –

Wissenschaftlicher Beirat der Bundesregierung Globale Umweltveränderungen 2016, p. 3). This decisive role of cities for the "great transformation" is underlined not only in the report published by the German Advisory Council on Global Change, WBGU (2016). Susan Parnell (2016, p. 529) observes an "emergence of a global policy focus on cities from the vantage point of the [Sustainable Development Goals] SDGs' approval and the shift from the [Millennium Development Goals] MDGs to a post 2015 sustainable development agenda". And in fact, the signs bode well that a paradigm shift in development policy has been put on track (e.g. principle of universalism for SDGs, more weight on absolute ecological limits, better monitoring and reporting through geospatial science, complex modelling, and big data analysis) (Koehler 2015, p. 733; for a more detailed argument, see Parnell 2016, p. 529).

Most central hereby is that the Sustainable Development Goals (SDGs) that were affirmed in September 2015 endorsed the role of cities through the 11th stand-alone urban goal (henceforth SDG #11) *to make cities safe, inclusive, resilient, and sustainable* (Parnell 2016, p. 529). It is still not clear how effective SDGs will be implemented in different city contexts or how the defined targets and indicators can be measured and compared (Simon et al. 2016). Concern has also been expressed as to what extent SDGs will be able to meet the commitment of *inclusive* development, where in many city contexts elite interests and coalitions, a growth-first paradigm, and disregard for the environment are very likely to challenge or oppose this key objective of SDG#11 (McGranahan et al. 2016).

Against this background, the United Nations Conference on Housing and Sustainable Urban Development "Habitat III", which was held in Quito, Ecuador, in October 2016, was a milestone. It raised huge expectations in respect to utilising the chances that cities offer for transformation, and it was on the basis of this conference that appropriate post-2015 actions were defined for sustainable urban development (Parnell 2016, p. 539).

The efficacy of the above-outlined United Nations processes may be modest in terms of concrete and projectable achievements. However, such a multilateral process is likely to realign the public and political discourses on urban development across different levels. The paradigm shift towards recognising cities as anchors for the "great transformation" has the potential to initiate shifts of power and funding from central to local governments and boost civil society participation. And in fact, city governments increasingly recognise the opportunities and co-benefits associated with addressing social and environmental issues that closely relate to improvements of infrastructure, quality of life, health, and related comparative economic advantages (Chan et al. 2015; Krause 2013; Reusswig et al. 2014, p. 41).

A first and very comprehensive strategic approach contributing to the Habitat III conference has been put forward by the German Advisory Council on Global Change, WBGU (2016). It characterises an outstanding response both to the declaration of the SDGs (especially SDG#11) and to the Paris Agreement adopted at the 21st Conference of the Parties (COP21) of the United Nations Framework Convention on Climate Change (UNFCCC). The report provides a systematic definition of the most relevant action fields, and it suggests broader principles and guidelines, which aim to be universally applicable. The report advocates a "people-oriented approach", which emphasises three basic dimensions of a "normative compass" towards sustainable urban development (WBGU – Wissenschaftlicher Beirat der Bundesregierung Globale Umweltveränderungen 2016, p. 137). *First*, the "compass" considers the importance of the planetary-limited natural resource system and the local environment as the foundations for sustaining our lives. *Second*, it underlines the basic principle of political and economic inclusion for all city dwellers. And *third*, it introduces a new concept to the urban agenda, the dimension of "Eigenart" or the specific own character of a city (WBGU – Wissenschaftlicher Beirat der Bundesregierung Globale Umweltveränderungen 2016, p. 137ff). The compass provides orientation for shaping an agenda for an urban transformation towards sustainability while at the same time offering space and flexibility for a context-specific translation and reinterpretation of the framework. The report also highlights the importance of considering the dynamic interactions and the risk of associated trade-offs between the three dimensions. It suggests an integrative and systemic approach that takes into account all three dimensions (WBGU – Wissenschaftlicher Beirat der Bundesregierung Globale Umweltveränderungen 2016, p. 161f).

In a globalised world, cities and megacities concentrate social-cultural diversity and processes of rapid change. Every city is unique with its differing character of the city's drivers of global environmental change and its differing challenges associated with the local and global changes. To be effective *and* inclusive, urban environmental policies, planning, and design need to take into account the specific conditions of a city. This, in particular, includes social-cultural aspects by taking a people-oriented perspective. Such a perspective is more likely to consider and respond to potential trade-offs that may lead to conflicts and lack of acceptance of policies. Being aware and understanding the "Eigenart" of a city should therefore be seen as a resource for the objective of bringing forward sustainable development. The study at hand emphasises the important role of environment-related social sciences to contribute to this understanding. While environment-related research has long been dominated by natural sciences and more recently by economic approaches, the social sciences have been slow to recognise their important role in informing environmental policies. This has also led to a bias in scientific environmental policy advisory. However, the signs bode well for a change.

For instance, the WBGU report constitutes a big difference in this respect by emphasising social-cultural aspects as an important foundation for improved understanding of the drivers and causes of global environmental change (WBGU – Wissenschaftlicher Beirat der Bundesregierung Globale Umweltveränderungen 2016, p. 99). The World Bank has also recently made quite a statement by taking a focus on psychological aspects of behaviour in their World Development Report 2015 "Mind Society and Behaviour" (World Bank 2015a). The authors of this report argue that development economics and policy required a redesign away from the assumption that human decision-making can largely be explained based on rational choice models. This assumption, say the authors, has been challenged by social and psychological research of the last few decades. According to the report, new policies based on a more accurate understanding of social and psychological factors have shown promising effects, especially in the context of difficult development challenges, including action on climate change. Recognising these insights and experiences, it suggests greater use is made of insights from behavioural and social sciences (World Bank 2015a, p. 1). Moreover, the International Social Science Council (ISSC) has dedicated one of its recent triannual World Social Science Reports (ISSC et al. 2013) co-published with UNESCO and the Organisation for Economic Co-operation and Development (OECD) to the topic of global environmental change.

These three examples of multilateral policy actors exploring and dealing with the question of how social sciences can contribute to a better understanding of global challenges indicate a reorientation of international development policies towards transdisciplinarity. Global environmental change is driven by our consumption patterns and the way people live. Similarly, cities are far from just being technical systems that can be mechanically transformed through design and planning, technological change, and gains in efficiency. A technocratic view blinds out the fundamental role of society and social practices. Such a perspective shift encourages and invites the social sciences to adapt to this new understanding and develop new approaches to adequately address and examine these questions and issues.

This study aims to contribute to an anchoring of the topic in the social sciences. Moreover, it deliberately takes a transdisciplinary perspective. That is, it includes engaging with the physical science base in terms of taking into account the immediate physical drivers of climate change (i.e. GHG emissions), but from a perspective that puts individual humans at the centre stage of analysis in their character as social, cultural, and political beings. It addresses a problem (problem-oriented approach) that touches upon several (globally) relevant discourses, as, e.g., on the question of how we want to live or how we humans perceive ourselves in relation to nature. Hence, it attempts to shed light on a topic with the aim to bring about practical and probably applicable ideas for a solution of the climate change problem. For sustainability policies on multiple levels, such transdisciplinary social-ecological research is important, and sustainability policies increasingly accept the relevance of this understanding.

The author applies this broader perspective by taking a focus on the specific drivers of climate change in urban India. India is a unique example in respect to climate change. India ranges among those countries that are most vulnerable to climate change. Most of the projected impacts of climate change are expected to pertain to this country. However, its current and future economic trajectory will turn the country from a pure victim of future climate change into an important actor for climate change mitigation. In this light and in the context of international negotiations on climate change, the country will have to answer the question on the direction of its social-economic development. And in particular, India will have to consider the question on its future urbanisation pathway and infrastructure development.

So far, India's position in international negotiations on climate change has been tough. With average per capita GHG emissions of below 2.5 tonnes of CO_2e per capita per year, the country still ranges close to the sustainability level (WRI 2015) and has since long argued on this basis against any legally binding commitment to the international agenda. For this reason, India has been blamed for "hiding behind the poor" (Ananthapadmanabhan et al. 2007), in reference to the substantial income dynamics in the country, the emergence of a new "consumer" middle class, and huge differences of carbon footprints between higher income classes and the huge share of the poor. These differences are expected to increase considerably in the future. Moreover, a large share of the population who are poor today will make a move upward from a social-economic position that merely fulfils the basic needs to a higher level of moderate income and consumption. These moderate shifts, however, will have a huge effect on the overall GHG emissions of the country. Cities represent the focal points of these differences and dynamics.

This issue is also highlighted in the WBGU report (2016). Besides taking a focus on cities as drivers and victims of global change, the authors emphasise the important role of city dwellers, their quality of life, and their long-standing future prospects (WBGU – Wissenschaftlicher Beirat der Bundesregierung Globale Umweltveränderungen 2016, p. 140). As one of the three dimensions of the "normative compass", the concept of the "Eigenart" of a city indicates this direction. It underlines the importance of recognising the "sociocultural and spatial diversity of cities" and the importance of allowing a "plurality of urban transformation pathways". The authors argue that this social-cultural diversity is a prerequisite for the city's unique potential for creativity and innovation (WBGU – Wissenschaftlicher Beirat der Bundesregierung Globale Umweltveränderungen 2016, p. 3f).

This recognition of the social-cultural diversity of cities touches upon the most central issue of the study at hand. The "Eigenart" of a city largely builds on the "Eigenart" of its people. Cities differ from each other culturally and socially, but cities are – as a constitutive element of the urban character – also diverse in themselves. Different people assign very different meanings and functions to the cities (and their surrounding environment) in which they live, and they differently perceive, evaluate, and make use of the social-cultural, ecological, technical, spatial, and built environment and elements (cf. WBGU – Wissenschaftlicher Beirat der Bundesregierung Globale Umweltveränderungen 2016, p. 89). With its specific features and requisites, a city can figuratively be seen as a stage on which different ways and ideas of living dynamically unfold and transform in interaction with the social, cultural, and physical environment. These external factors interact with the way people interpret, evaluate, and adapt their own and others' ways of thinking and living, certainly, in accordance with their respective social-economic position (e.g. income, education, occupation, caste, etc.).

While analysing differences in human behaviour and consumption is relatively straightforward, an extended view that includes addressing the underlying factors and motivations makes it a challenge. As in most of the social-economically based approaches to analyse differences in personal GHG emissions, for a long time, social sciences have emphasised the role of vertical, class, or social-strata-related determinants to explain social inequality and differences in social behaviour. With Pierre Bourdieu's (1987) study on the state of the French culture, which was first published in French in 1979, this has changed, and sociology gained new insights in the role of social-cultural-group-specific factors for aesthetic preferences and choices people make in distinction to other social groups. With this study, Bourdieu introduced the concept of lifestyle, more or less similar to the way it is conceptualised up to now in sociology. Since then, especially in German sociology, a reorientation of social structure analysis occurred, and in particular, the lifestyle concept gained wide-reaching popularity.

Today, lifestyle is conceptualised incorporating social-economic factors, attitudes and values, as well as behaviour and social practices. With this perspective, the author aims to meaningfully integrate these different dimensions in order to reveal the ambivalences, links, and feedback mechanisms between the underlying drivers of behaviour and consumption. This study also attempts to involve a perspective on external factors that tend to prestructure the individual scope of action. Similarly and especially in the Indian context, the social-economic position plays an important role as it limits the options people have to lead their lives. In this study, lifestyles are therefore defined as group-specific and value-, attitude-, and preference-driven patterns of everyday life, which unfold within an economically, social-culturally, and physical-environmentally prestructured field of social interaction. The social-cultural, economic, and environmental contexts are differently received by different people. With their lifestyles, people differently translate these contextual aspects into individually specific patterns of meaningful behaviour and into resulting lifestyle-specific individual-level carbon footprints. This mediating character of the lifestyle concept is key to understanding the urban carbon footprint, because lifestyles conceptually combine the view on social-economic determinants and material aspects of culture with an examination of social-cultural and psychological factors.

The study therefore attempts to reveal lifestyle-specific structures or patterns of ideal typical value orientations that build the foundation of a particular lifestyle. This basic research pursues an improved understanding on whether and how certain structural configurations of values and attitudes structure behavioural and consumption patterns. These revealed patterns represent ideal types of lifestyles as a basis for an analysis of lifestyle-specific personal GHG emissions. The given objectives of this study are further detailed out as part of the definition and specification of the research questions (Sect. 2.4).

The thesis is structured as follows. The subsequent Chap. 2 will develop the two main building blocks of the underlying theory of the thesis. The first section 2.1 will delineate the emerging relevance of and different approaches to greenhouse gas (GHG) emission accounting and carbon footprinting. It highlights the different angles (consumption-based vs. production-based emissions) and functional units of analysis (from the national level over region and city-wise accounting to personal-level consumption-based accounting). The chapter will then present the different approaches to personal carbon footprinting and how individual differences can further be analysed based on different factors, mainly income and consumer expenditure. The subchapter closes with a discussion of the shortcomings of economically based analyses.

The second section 2.2 focuses on social science approaches that analyse social inequality and differences in behaviour and consumption, which are seen as the fundamental basis of differences in personal carbon footprints. It will first give an overview of classical more vertically oriented approaches to social inequality, such as class and social strata. It will then delineate the historic reasons for a reorientation in social structure analysis, which were foundational for the developments in theorising the concept of lifestyle. The subchapter will then present the theory of lifestyle and the underlying dimensions and components of the concept. The subchapter will close with a critical review of the theory, and it discusses how these limitations can be addressed through improvements in the conceptualisation.

The third section 2.3 synthesises the collected insights by stating the relevance, objectives, and challenges in linking personal carbon footprinting with the concept of lifestyle. Chapter 2 concludes by defining and specifying the research questions in consideration of the aforementioned theoretical foundations (Sect. 2.4).

Chapter 3 sets out the geographical research context of this study, which has been conducted in the megacity of Hyderabad, India. The chapter first draws on broader aspects of economic development and dynamics of urbanisation in India (Sect. 3.1), and it will then explain the important role of poverty as well as the projected dynamics of social mobility in India (Sect. 3.2). The third part of this chapter gives an overview on the relevance of climate change in India with reference to the climate change-related impacts, issues of national and international climate change policy, and the relevance of income dynamics, the emerging middle class, and associated effects on GHG emissions in India. The fourth subchapter focuses on the city of Hyderabad by drawing on historical, economic development-related aspects and administrative and urban governance-related features of the city. It will close with a discussion of symbolical representations and infrastructure-related aspects of the city, which are relevant for the conceptualisation of lifestyle in this context.

The fourth chapter explains how lifestyle was conceptualised and operationalised with the objective to analyse and explain differences in personal carbon footprints. The chapter is structured into two parts, with the first part setting out the theoretical framework and the second part delineating the methodology of the study. Especially the first part of this chapter is important to develop an understanding of how the lifestyle concept allows integrating the dimensions of social-economic factors, mentalities, and behaviour. It will also explain the relevance of other external factors such as culture, policies and institutions, level of economic development, and especially commodities and infrastructure.

The results of this study were set out in two main chapters – Chaps. 5 and 6. Chapter 5 presents the results of the descriptive analysis of the data, and it gives reference to the outcomes of constructing the latent variables, such as investive consumption (wealth index), the latent dimensions of preferences, behaviour and consumption patterns, and the latent dimensions of value orientation and attitudes. Chapter 6 presents and explains the main results of the study giving answer to the research questions as laid out in Chap. 2. The first two subchapters focus on the general outcomes of the GHG emissions analysis referring to the structure of consumption-based personal-level GHG emissions (carbon footprints) in Hyderabad (Sect. 6.1) and an income class-specific analysis of carbon footprints (Sect. 6.2). The following subchapter (Sect. 6.3) takes a different perspective detailing out common consumption practices and estimating practice-specific personal GHG emissions. This analysis highlights all those practices that are most relevant for climate policy intervention (key points of intervention). Section 6.4 focuses on the core element of this study – the interpretive analysis of the value orientation clusters and the development of a typology of lifestyles for the city of Hyderabad. It identifies six different lifestyles and describes these based on different passive variables. The final section 6.5 delivers the results of analysing the lifestyle groups with respect to differences in personal carbon footprints.

Chapter 7 discusses the results following the structure as given with the research questions. Based on these findings and against the review of the overall research process, the author will then give insight to the challenges of environment-related lifestyle research. The chapter closes with a critical reflection on methodology and applied methods and indicates the implications for future lifestyle research in India. Chapter 8 presents the final conclusions to this thesis.

Chapter 2
Approaches of Measuring Human Impacts on Climate Change

Keywords GHG emissions accounting · Carbon footprint · Production-based emissions · Consumption-based emissions · Social structure analysis · Lifestyle concept · Social inequalities · Conspicuous consumption · Consumption of necessity

2.1 Measuring Greenhouse Gas Emissions from Human Activity

The IPCC defines greenhouse gases (GHGs) as "gaseous constituents of the atmosphere, both natural and anthropogenic, that absorb and emit radiation at specific wavelengths within the spectrum of terrestrial radiation emitted by the Earth's surface, the atmosphere itself, and by clouds. This property causes the greenhouse effect" (Planton 2013, p. 1455).

There are different approaches to measuring the impact of human activities on the Earth's climate. Up to now, these approaches aim around assessing the amount of GHGs being emitted over a certain period.

The concentration of CO_2 in the atmosphere is measured in parts per million (ppm). By analysing the air enclosed in Antarctic ice and firn, Etheridge et al. have measured preindustrial CO_2 mixing ratios in the range of 275–284 ppm (1996, p. 4115). Following this long period – after 1800 A.D. – an unprecedented growth of CO_2 concentrations has been measured. In 2005, concentration levels reached the threshold of 380 ppm, which is the highest mark in the reconstructed record of values over the last 700,000 years (Rahmstorf and Schellnhuber 2007, p. 33). In March 2015, another historic record was documented with the average CO_2 levels crossing the 400 ppm mark globally for a whole month (UNFCC 2015).

The increases in concentration levels can be traced back to the amount of fossil fuels being burned and the amount of CO_2 being emitted from these processes. Other GHGs than CO_2 have also accumulated in the atmosphere as a result of human activities, such as methane, FCKW, and N_2O (Rahmstorf and Schellnhuber 2007, p. 33). For reasons of comparability and analysis, the global warming potential (GWP) of GHGs has been introduced as a common basis for all GHGs. The GWP expresses the climate impact of GHGs in terms of carbon dioxide equivalents, i.e. CO_2e (Kennedy et al. 2010, p. 4829).

2.1.1 GHG Emission Accounting

Accounting for and analysing GHGs being released in consequence of human activities are essential prerequisites for curbing further accumulation of GHGs in the atmosphere and thereby mitigating global climate change. GHG emission accounting refers to "the calculation of GHG emissions associated with economic activities at a given scale or with respect to a given functional unit – including products, households, firms, cities, and nations" (Fleurbaey et al. 2014, p. 305). Over the last two decades, a diverse set of methodologies has emerged with quite varying definitions of assessment boundaries. The definition of these boundaries determines which processes, goods, and services are to be considered in the analysis. Depending on the objectives, accounting approaches vary in terms of the underlying functional unit (e.g. city or state) or in terms of responsibility (production-based vs. consumption-based accounting). In the following, a short overview is given of the most relevant approaches to GHG accounting.

During the last 10 years, there has been a shift from *production-based approaches* (production-based or territorial framework) to accounting methods that focus on assessing the amount of emissions associated with the consumption of goods and services (*consumption-based framework*). Obviously, in terms of analysing inequalities in the contribution to the problem of climate change, the selection of the framework is essential. While the territorial framework assesses the emissions that are physically produced within the territorial boundaries of a nation or any other jurisdiction, it is the consumption-based approach that accounts emissions on the consumption side, regardless of their territorial origin (Fleurbaey et al. 2014, p. 305). The difference between the two approaches is manifested in the indirect emissions that are embodied in goods and services and that move across territorial boundaries through trade. Davis and Caldeira (2010, p. 5691) provide an excellent empirical account on the issue of international carbon leakage as they disclose the imbalances caused as a result of territorial emission inventories. They show that "23% of global CO_2 emissions [...] were traded internationally, primarily as exports from China and other emerging markets to consumers in developed countries. In some wealthy countries, including Switzerland, Sweden, Austria, the United Kingdom, and France, >30% of consumption-based emissions were imported, with net imports to many Europeans of >4 tonnes CO_2 per person in 2004" (Davis and Caldeira 2010, p. 5687).

As will be shown in the following, the problem of carbon leakage weighs even more in the context of subnational inventories. In this context, it is crucial to identify and categorise the sources of emissions and to consider both direct and indirect emissions (Williams et al. 2012, p. 65). An analytical instrument for categorising emissions in regard to the sources is provided by the internationally applied concept of scopes, which originally was developed by the World Resources Institute, WRI, and the World Business Council for Sustainable Development, WBCSD (2004). It defines three different levels of emission calculation, namely, *Scopes 1, 2*, and *3*.

Scope 1 is all direct emissions that derive from activities controlled and owned by the functional unit (depending on the defined analysis boundaries, e.g. a nation, city, organisation, or household) and taking place within its boundaries. Examples include emissions from the direct combustion of fuels or emissions released from waste that is stored or burned within the boundaries. *Scope 2* emissions are released in association with the "consumption of purchased electricity, heat, steam, and cooling that embody emissions being released elsewhere (out-of-boundary)" (Kennedy et al. 2010, p. 4829). All other indirect and embodied emissions which are not included in Scope 2 are classed as *Scope 3*, such as embodied emissions of such products, e.g. food items, paper, and electronic items (out-of-boundary up- and downstream emissions) (Kennedy et al. 2010, p. 4829). The concept of scopes has been recognised as a standard approach in all international accounting guidelines, such as the 2006 IPCC Guidelines for National Greenhouse Gas Inventories (IPCC 2006), the PAS 2070: 2013 specification for the assessment of greenhouse gas emissions of a city (BSI 2013), and the Global Protocol for Community-Scale Greenhouse Gas Emission Inventories (GPC) (WRI et al. 2014).

2.1.2 Carbon Footprinting

The result of consumption-based emission accounting is often referred to as a "carbon footprint" (Fleurbaey et al. 2014, p. 305; Minx et al. 2009, p. 187), which is actually a misnomer, as it refers to the *mass* of cumulated carbon emissions, not to a measure of the *area* (Hammond 2007, p. 256; Hertwich and Peters 2009, p. 6414). Moreover, it is also used and applied for balances that include other, not carbon-based, GHGs. Given the great variety of objectives and motivations in GHG accounting, it is obvious that methodologies for carbon footprint calculations are still evolving, and in consequence, there is still little coherence in definitions and approaches among the studies (Pandey et al. 2011, p. 135; cf. Fleurbaey et al. 2014). A broader conceptualisation that suits most of the approaches was proposed by Peters (2010, p. 245). He defines it as follows:

> The 'carbon footprint' of a functional unit is the climate impact under a specific metric that considers all relevant emission sources, sinks and storage in both consumption and production within the specified spatial and temporal system boundary.

This very open notion of what a carbon footprint refers to is a good starting point for examining the various possibilities of application. This is with regard to functional units as well as scales. The carbon footprint clearly assigns emissions to a certain functional unit, and this is fundamental for the question of responsibility. A carbon footprint can be taken and calculated from goods and services (product-level and life cycle analysis), from a defined spatial unit (regional-, national-, or city-level footprint; territorial approach), or from individuals, households, and organisations.

2.1.2.1 The Product Carbon Footprint (PCF)

The most precise way of GHG emission analysis offers the life cycle assessment (LCA), also known as product carbon footprint (PCF). It allows all GHG emissions along the way from cradle to grave to be taken into account, i.e. from the extraction of all used raw materials to processing, manufacture, distribution, use, repair and maintenance, and disposal or recycling. The result of an LCA provides information about the exact amount of GHGs being released as a consequence of demand for and consumption of a product. This information empowers and helps consumers to engage[1] in and find ways to more climate-friendly behaviours. It can also serve to inform governments about lifestyle policies (taxing, campaigns, and (dis-)incentives), and it allows industries to design more climate-friendly products. However, the devil is in the details, and there are still considerable weaknesses in the PCF approach, as discussed, e.g. in Grießhammer et al. (2009, p. 23ff) and PCF Pilotprojekt Deutschland (2009, p. 20ff).

2.1.2.2 The Territorial Approach to Carbon Footprinting

The regional or territorial approach to carbon footprinting analyses GHG emissions associated with the consumption of goods and services within clearly defined territorial functional units, such as cities, neighbourhoods, districts, regions, or nation states. On national level, a consumption-based approach is highly instrumental as it provides a detailed emission structure analysis based on the relevant consumption categories that cause the carbon footprint. Hence, it helps us understand the drivers of GHG emissions as well as the varying contributions of different consumption activities across regions and stages of development (Hertwich and Peters 2009, p. 6414).

Besides national inventories, there are an increasing number of subnational studies. Most of these subnational studies have focused on urban areas as the relevant unit of reference. As mentioned in the introductory chapter, cities bear huge potential and actually are crucial players in climate change mitigation. Also, cities are often reckoned to be major drivers of global GHG emissions (Kennedy et al. 2010, p. 4828; Satterthwaite 2008), however, to a substantially varying degree (Dodman 2009; Kennedy et al. 2009). Due to the important role and due to the challenges in accounting GHG emissions in cities, the number of studies has increased substantially over the last two decades. However, methods and what is included in the inventories vary markedly, and the results therefore diverge in quality, scope, and comparability (BSI 2013, p. iii; Ibrahim et al. 2012, p. 223; Wiedmann et al. 2015, p. 2).

---

[1] The concept of environmental engagement was first introduced by Lorenzoni et al. (2007, p. 446). They define it as the "state of connection comprising the three codependent spheres of cognition, affect, and behaviour".

2.1 Measuring Greenhouse Gas Emissions from Human Activity

To address this problem of coherence, a milestone has been reached with the first standard guideline for GHG emission accounting in cities, issued in the year 2009 by ICLEI, Local Governments for Sustainability (2009). Since then, the methodology in urban GHG accounting has made substantial advance, and guidelines and accounting standards have been further developed.

A broad overview of consumption-based city-level GHG inventories is given by Wright et al. (2011) and Baynes and Wiedmann (2012) up to 2011–2012. A more recent examination of studies for the period thereafter is presented by Wiedmann et al. (2015). Based on their review, Wiedmann et al. (2015, p. 5) argue that in these citywide inventories, accounting of indirect emissions is still partly incomplete or inconsistent. They suggest that more specific standards for Scope 3 emissions still need to be developed (Wiedmann et al. 2015, p. 5). In response to this shortcoming, they provide a new methodological framework for the development of "city carbon maps" based on EEIOA, exemplified and tested on the greater metropolitan area of Melbourne, Australia (Wiedmann et al. 2015). The carbon map approach promises to address quite some of the shortcomings of earlier inventory frameworks, such as the problem of double counting. Wiedmann et al. (2015) state that the city carbon map allows all emissions to be allocated from Scope 1 to Scope 3 "to clearly defined and standardised sectors, representing either industry or product groups, collectively covering all economies from the urban, via the regional and national to the global scale" (2015, p. 12). However, this approach has to be further tested on the ground in other cities.

As shown above, the analysis of the structure and quantity of GHG emissions in cities is of importance for an understanding of the major sources of emissions, and therefore it is a prerequisite for climate change mitigation in cities. City emission inventories are also extremely relevant for the understanding of the role of cities in national and international GHG emission profiles (Ahmad et al. 2015, p. 11312). City-level carbon footprints inform climate policies in respect to future urban infrastructure planning. They might indicate, in which sectors of consumption and investment mitigation policies can be most productive, e.g. through economic incentives or through carbon taxing. Moreover, a consumption-based inventory might be instrumental in raising awareness among consumers with regard to the impacts of lifestyle and consumption decisions (Lin et al. 2015, p. 3).

2.1.2.3 Individual- and Household-Based Carbon Footprinting

Besides the above-mentioned strengths of and arguments for the city-level territorial accounting approach, there are very important aspects that cannot be addressed with this approach: How far do GHG emission inventories in cities reach out in terms of differentiation and structural analysis? Are such inventories sufficiently informative in providing the required knowledge and perspective on the relevant drivers of GHG emissions in order to design mitigation policies more effectively?

Aggregated data of emissions in cities do not allow for an understanding of differentials between households and individuals. Hertwich and Peters (2009, p. 6417) correctly state that eventually it is the daily consumption and production *decisions* that drive global emissions. They have stated on global scale that 72% of all greenhouse gas emissions are related to household consumption, while only 10% can be assigned to government consumption and 18% to investments. In India, the domestic share is even much higher at 95% of all emissions, with shelter (including its construction), food, and private mobility being the most important consumption categories (Hertwich and Peters 2009, p. 6414). Hertwich and Peters (2009, p. 6417) also show that consumption patterns and associated GHG emissions change substantially in structure with rising income.

Ahmad et al. (2015, p. 11312) also express concern in regard to this research gap, stating that household-level emissions have been insufficiently studied and analysed so far. They claim that understanding the household level is duly important for "finding policy solutions that respect the specific population" (Ahmad et al. 2015, p. 11312). And indeed, for the case of India, there are only a few studies that address this gap in subnational per capita-level GHG accounting, most of which use an income-based categorisation.

The earliest study was done by Parikh et al. (1997) and was updated in 2009. Parikh and colleagues have quantified "differential emission effects of consumption pattern of different income classes in India" (Parikh et al. 2009, p. 1024). Their analysis is based on an input-output (IO) table and social accounting matrix (SAM). Quite similarly, a few studies have been published by Shonali Pachauri and colleagues, who have analysed the energy consumption patterns of various income classes in India, without, however, translating household energy use into GHG emissions (Lenzen et al. 2006; Pachauri 2007, 2004; Pachauri and Jiang 2008; Pachauri and Spreng 2002). Another study was conducted by the Centre for Science and Environment in 2009 that also examined GHG emission variations between income groups based on income elasticity of emissions (CSE 2009). Politically most influential was a study by Greenpeace India in 2007 (Ananthapadmanabhan et al. 2007). It highlights the question of India's growing consumer class and analyses India's greenhouse gas emissions by different income groups, based on a market survey of 819 private households. Direct energy consumption (electricity, cooking fuels, and transport – private and public) in these households has been converted into CO_2 emissions and then assigned to seven different income classes. Food and other Scope 3 emissions have not been included in the analysis.

In the most recent study, Ahmad et al. (2015) compared India's 60 largest cities, accounting for three major direct carbon dioxide emission sources on household level: electricity, cooking fuels, and fuels for private transportation. Other domains of consumption, such as public transport, long-distance travel, food, and other Scope 3 emissions, are not included in the analysis (Ahmad et al. 2015, p. 11313). As result, they highlight the important role of household consumption expenditure, which they treat as proxy for income. Among other social-demographic determinants, consumption expenditure has by far the greatest effect, in particular on emissions from private motorised transport and the use of electricity (Ahmad et al. 2015, p. 11313). Other factors are education and access to electricity, which have a positive effect, and population density and household size, which have a reducing effect on per capita emissions (Ahmad et al. 2015, p. 11317). However, the influence is relatively small in comparison to consumption expenditure.

Also on international level, research is being done with regard to individual-level carbon footprinting. Recently, an outstanding and highly influential study has been issued by Piketty, which "presents evolutions in the global distribution of CO_2e emissions (CO_2 and other GHGs) between world individuals from 1998 and 2013" (Chancel and Piketty 2015, p. 2). The authors claim that based on global analysis of individual-level inequalities in the contribution to the climate change problem, which they call the "new geography of global emitters" (Chancel and Piketty 2015, p. 2), a more equitable and more straightforward solution towards climate action in all countries can be arrived at. This is not only in regard to climate change mitigation but also in regard to financing "a global climate adaptation fund based on efforts shared among high world emitters rather than high-income countries" (Chancel and Piketty 2015, p. 2).

The study shows that on a global-level, CO_2e emission inequalities between individuals decreased over a period of 15 years (1998–2013). This decline is explained by a rise of high- and mid-income classes in developing countries and a trend of stagnation in incomes and emissions for large shares of the population in countries of the Global North. The authors state that at the same time, income and CO_2e emission inequalities increased within the countries. Most remarkable in this context is the high level of concentration of emissions among the top 10% emitters who contribute to about 45% of overall global emissions. This is while the bottom 50% emitters contribute to only 13% of all global emissions. And interestingly, one third of the top 10% emitters do live in emerging economies (Chancel and Piketty 2015, p. 2).

This result underlines the importance of analysing individual-level GHG emissions. The analysis of differentials in per capita GHG emissions informs policies on different levels (from international down to the municipal level), thus contributing to implementing more equitable solutions that consider inequalities in response to the problem of climate change. Politically – and especially for the case of India – this is highly relevant in a context where large shares of the overall population still live below or very close to the poverty line.

In consequence of this newly deployed perspective and in particular in reaction to the influential study issued by Greenpeace India (Ananthapadmanabhan et al. 2007 see above), concern for a new debate has been raised in regard to climate change mitigation in India and the role of its emerging middle classes. In international negotiations on climate change, India has so far been reluctant to accept any legally binding commitment to reducing GHG emissions unless it was on an equity-based sharing of emission rights. The major argument behind this stand has been the very low per capita average GHG emissions as well as the low historical emissions which India has released to the atmosphere up to the present. However, by taking a more precise look into the distribution of domestic GHG emissions in India, it becomes obvious that the growing and socially upwardly mobile middle classes make a significant contribution to India's overall GHG emissions. The carbon footprints of some of these groups rank close to the average per capita CO_2 emissions found in Europe or the USA (Chakravarty et al. 2009, p. 11888). Former Environment Minister Jairam Ramesh has raised attention to this issue by speaking of the growing "Indian taste for the American lifestyle", which he called the "most unsustainable in the world today" (Burke 2010). More recently, Narendra Modi called for a much stronger focus on the issue of unsustainable lifestyles, as "the lifestyles of a few must not crowd out opportunities for the many still on the first steps of the development ladder" (Modi 2015).

2.1.2.4 Household Expenditure, Income, and Other Determinants and Their Relevance in Explaining Differentials in Individual-Level GHG Emissions

Most studies that analyse individual-level GHG emissions highlight the importance of income as a determining factor for differentials in personal carbon footprints. In almost all cases in the above-mentioned studies, however, income is not measured and surveyed directly but is approximated through the data of consumer expenditure (Ahmad et al. 2015; Chancel and Piketty 2015; Lenzen et al. 2006; Pachauri 2004, 2007; Pachauri and Jiang 2008; Pachauri and Spreng 2002; Parikh et al. 1997, 2009). It seems that it has become a standard procedure to take consumer expenditure as a valid proxy for household or personal income (see Grunewald et al. 2012, p. 9), without much further reflection. And indeed, it is often stated that expenditure data is better able to reflect long-term income than estimates of income instead (Pachauri 2007, p. 124; Rutstein and Johnson 2004, p. 2). In consequence, most of the studies use income and consumer expenditure synonymously, often without any further explanation or reflections upon potential biases.

Especially when it comes to analysing differences in individual-level GHG emissions, this is problematic. Pachauri in her study on household-based energy use patterns in India (2007, p. 124) herself states in a footnote that other authors (Jiang and O'Neil 2004; Vringer 2005 cited in Pachauri 2007) have shown evidence on the inconsistencies between consumer expenditure levels and income levels with regard to energy use. Savings are assumed not to vary among households at the same level of income, and this is simply not the case (Rutstein and Johnson 2004, p. 2), especially in India. Saving levels are high in India and have increased progressively from 8.6% in 1950–1955 to 13.8% in 1970–1975 and then to 22.8% in 1990–1998 (Athukorala and Sen 2004, p. 492). The author was not able to find studies showing variance levels within equal-income groups but based on his own calculations will show how investive consumption, which is often proposed as alternative proxy to income (see Sect. 4.2.5.5), varies within the same income categories. And, as savings rates are at such high levels, the resulting bias in approximating income through household expenditure becomes obvious.

All of the above-mentioned studies show that it is consumer expenditure, which has the most significant effect on direct and indirect energy consumption levels and related personal GHG emissions.[2] And in fact, this is not too much of a surprise, as any purchase of any goods and services involves direct and/or indirect GHG emissions that are incorporated in the product or service. This is because any product or service incorporates direct and indirect energy input as well as in some cases non-energy input-related GHG emissions such as methane from dairy or meat products.[3] This argument is supported by the fact that scholars dealing with approximated income from consumer expenditure analysed an "income" elasticity (which is based on consumption expenditure data) of between 0.7 and 1.0. Elasticity is defined as a measure of the relative change of a dependent variable as an effect of a relative change of an independent variable. That means for the case of the above-given income or consumption elasticity that a doubling of "income" leads to an increase of the associated GHG emissions of 70–100% (Chakravarty et al. 2009, p. 9 in the Annex; CSE 2009, p. 3; cf. Grunewald et al. 2012).

The expenditure elasticities on emissions vary depending on the domains of consumption, and related effects are not linear. Based on their analysis of income and energy consumption levels, Herendeen and Tanaka (1976) in their study highlight the fact that indirect energy consumption increases more directly with household expenditure than direct energy consumption. Also Pachauri (2004, p. 1726) in her study on India indicates this. Chancel and Piketty explain these differences in energy usage between different "income groups" (based on expenditure) with the saturating and limited requirements for direct energy carriers up to a certain consumption level. For instance, there is a structural limit to the amount of heat or cooling an individual can use every day, or the use of several cars is limited, as one person can only drive one car at a time. This is different for indirect energy that is incorporated in consumer goods and services, as wealthy consumers can purchase almost unlimited amounts of goods and services with the respective financial resources (Chancel and Piketty 2015, p. 19).

This is also indicated by Grunewald and her colleagues (2012, p. 2) who analysed household emissions in India based on I/O analysis and micro-level household data. They show that with rising incomes, demand for more emission-intensive consumption domains increases disproportionately in comparison to less emission-intensive consumption categories.

It has been shown above that the standard procedure of approximating income through consumer expenditure leads to serious biases in individual-level GHG emission analysis. Moreover, the foregoing chapter has also shown that income is an extremely important factor for consumption decisions and associated GHG emissions. Income can be seen as conditio sine qua non for most realms of urban consumption in India. Only with the available financial resources are certain consumption patterns possible at all. For instance, for those without a car or a motorbike and without the money to afford the ticket to travel by air, emission levels

---

[2] Depending on the underlying sources of energy, energy consumption can directly be translated into GHG emissions.

[3] Surely, one can think of renewable energy sources such as wind and solar, but there are no such production processes so far, which are completely decoupled from a release of GHG emissions.

in the domain of transport will remain on a very low level. The same is true for household energy, which is related to the endowment level of a household. Electricity consumption will remain on a rather low level if a household does not use certain appliances such as a washing machine, refrigerator, air-conditioner, etc.

On the other hand, households with high income do not necessarily need to rely on emission-intensive consumption practices. Upward social mobility due to a better job or rising levels of income can just as well lead to the decision to save money for harder times and maintain the standard of living as it was. Many households maintain a low level of GHG emissions, even though their income level is very high (see Sect. 2.1.2.4). Therefore, income is a necessary, but not a sufficient condition for higher levels of consumption and carbon footprints. Chancel and Piketty (2015, p. 21) argue that income certainly is the main driver of total CO_2e emission levels among individuals, but they admit that there are many other factors which play a determining role in energy consumption and CO_2e levels.

This research is interested in analysing those factors influencing differentials in GHG emissions apart from income, and in particular, it draws on social-cultural factors, as these are understood to be the main determinants of lifestyle beyond income. The following chapter will delineate the theoretical approach of analysing social-cultural inequalities. It will first summarise classical approaches to social structure such as class and social position. It will then outline more recent concepts of social differentiation based on social practices and mentalities, and it will give insights into environment-related lifestyle research. Building on these three sections, Sect. 2.3 will give a more specific outline of the relevance, motivations, and challenges in making use of lifestyle research in individual-level GHG accounting.

2.2 Social Structure, Lifestyle, and Consumption: Understanding Social-Cultural Inequality and Related Differences in Human Impacts on Climate Change

In Europe, sociological theory emerged as a new discipline in response to the drastic social and revolutionary political changes between the early 1800s and the early 1900s. Social theorists such as Claude Henri Saint-Simon (1760–1825), Auguste Comte (1798–1857), and Emile Durkheim (1858–1917) in France and scholars such as Georg F.W. Hegel (1770–1831), Karl Marx (1818–1883), Georg Simmel (1858–1918), and Max Weber (1864–1920) in Germany and Herbert Spencer (1820–1903) in Great Britain laid the early foundations of sociology. Political revolution and enlightenment, industrialisation and urbanisation, social disorder, and the swaying of traditional institutions called classical approaches of social theory into question.

Max Weber defined sociology as the science which aims to interpretively understand social behaviour and thereby arrive at a causal explanation of its course and effects:

> *Soziologie (im hier verstandenen Sinn dieses sehr vieldeutig gebrauchten Wortes) soll heißen: eine Wissenschaft, welche soziales Handeln deutend verstehen und dadurch in seinem Ablauf und seinen Wirkungen ursächlich erklären will.* (Weber 1922, p. 1)

Apparently, the world today similarly faces drastic social, economic, political, and ecological challenges that call for new approaches not only with regard to the analyses of these challenges but also in finding solutions to deal with this complex set of threats against human society. And it is human society, ranging at the core of the problem of global environmental change. Anthony Giddens, in his prologue of Benno Werlen's edited book on *Global Sustainability* (2015a) puts it like this:

> We live today in what I like to call a high opportunity, high risk society. The biggest risks we face, as a collective humanity, come not from nature but from ourselves. They derive from our newfound global interdependence and the fragility of the systems that are driving it. (Giddens 2015, p. vii)

This is a remarkable point as it underlines the importance of addressing global environmental change from the perspective of society. John Urry in his inspiring monograph on *Climate Change and Society* (2011, p. 2) clearly gets to the point of this problem: he states that anthropogenic climate change has long been addressed only by two groups of analysts, natural scientists and economists. He argues that this bias has led to a remarkable neglect of 'society' in analysing the related determinants, impacts, and current and future implications for humanity in general and society in particular. He summarises that 'economics' needs to be displaced from its preeminent or imperialist role in examining and explaining the 'human' causes and consequences of climate change (Urry 2011, p. 5; also see Werlen 2015b, p. 4f)". The author of this study opines that especially social and cultural geography is well arrayed to bring new and urgently required perspectives into the analysis of the ecological crisis, which is as much a social as a natural problem. Geography can and should contribute in initiating and facilitating a stronger integration of social sciences, humanities, and the natural sciences. And, as Benno Werlen (2015b, p. 11) puts it, "to overcome disciplinary blind spots, we need a perspective that specifies and solves problem complexes independent of the disciplinary interests and boundaries". Therefore, it is essential to take a more specific look at the level of the everyday human actions and practices that create the practical problem. Werlen proposes to "build a bridge between knowledge and action" and "bridge the gap between global problems and national, regional, and local behaviour, as well as decision making" by taking a transdisciplinary research perspective and by pronouncing the role of "bottom-up strategies of scientific methodologies" (Werlen 2015b, p. 12).

This study is an attempt to reveal and shed light on one of these blind spots. In Sect. 2.1.2, it has been argued that analysing the direct causes of climate change, i.e. looking at consumption-based GHG accounting, has so far been based around economic models of consumption. The economics approach assumes higher levels of income leading to higher levels of consumption and thereby to higher levels of personal GHG emissions. As stated above, this approach is problematic, because the role of other factors is downplayed in most of such studies. Factors that influence differentials in consumption behaviour across different income groups are not considered. For instance, higher-income levels may result in higher saving rates or may lead to increasing demand for more sustainable goods and services. In most cases, this effect is overlooked even more due to the quite common approach of not measuring income directly – generally due to the lack of available income data. In such cases, income instead is approximated through household consumer expenditure. In consequence, the factually differing saving rates between households are discounted even more.

This study is based on ethnographic research methods (see Sect. 4.2.3) as well as a large household survey that has been conceptualised and carried out specifically to address the research gap on GHG emission determinants beyond income. First, besides comprehensively measuring individual-level carbon footprints, the survey allowed reliable data to be collected on income; second, it delivers insights in applying alternative methods to approximate income by means of the wealth index (see Sect. 4.2.5.5); and, third and most importantly, it includes highly specified data on various social-cultural determinants that build the foundation for constructing the concept of lifestyle and thereby analyse the interconnectedness of these social-cultural determinants.

The following two sections will delineate the historical and theoretical social science foundations of this study's approach, which is based on the concept of lifestyle. The author will explain, on which early disciplinary grounds the concept emerged in response to the lack of scope offered by classical approaches to social structure analysis. This study will then outline existing applications of the lifestyle concept, which is most prominent in the German context. It will then draw on the potential and the challenges of building and using typifications based on multiple social-cultural variables and the application of the lifestyle concept for social science environmental research.

2.2.1 Classical Approaches to Analysing Social Inequality: The Foundations of Advanced Social Structure Analysis

The interest in understanding the principles of social life and social interaction is apparently as old as the human thirst for knowledge. The earliest scholars such as Aristotle were interested in social differences and inequality. He defined classes on the basis of social determinants such as birth, age, family, and related status and prestige categories (Bendix and Lipset 1967, p. 1). Social inequalities are a basic characteristic of humanity, both as a naturally given and socially constructed fact (e.g. age, gender, skills, employment, income, education, taste, etc.). Inequalities influence human behaviour and social interactions substantially and have always been a structuring element of social organisation. For instance, Stefan Hradil (2001a, p. 16ff) delivers an illustrative historical account of premodern (neuzeitlich) institutionalised social differentiation. All given examples are based on basic principles of class, rank, and social position and status. Today, social inequalities still play a major role with regard to social organisation and the principles of social interactions, but the way inequalities function, and how they are instrumentalised is substantially different. This is one of the reasons why this study is so relevant. However, before exposing the conceptual foundations and basic arguments for the lifestyle concept, the author will give an outline of classical approaches towards understanding social differences and differentials in social behaviour.

Based on the historical analysis of the development of society and its material basis (historical materialism), Karl Marx conceptualised the first comprehensive theory of classes and class conflict. For Marx, the material basis of society was foundational for economic, social, and cultural development. According to Marx, class affiliation, divided between proletarians and capitalists, was based on having property rights over the means of production. Against the light of impoverishment and the unbearable living conditions in the early stages of industrialisation in Europe, Marx prognosticated a solidarisation of the working class and class struggle against the capitalists, concluding in a communist revolution (Nollmann 2008, p. 183).

A few decades later, Max Weber conceptualised class in a far more differentiated way. He defined the position within a class as follows:

> 'Klassenlage' soll die typische Chance 1. der Güterversorgung, 2. der äußeren Lebensstellung, 3. des inneren Lebensschicksals heißen, welche aus Maß und Art der Verfügungsgewalt (oder des Fehlens solcher) über Güter oder Leistungsqualifikationen und aus der gegebenen Art ihrer Verwertbarkeit für die Erzielung von Einkommen oder Einkünften innerhalb einer gegebenen Wirtschaftsordnung folgt. 'Klasse' soll jede in einer gleichen Klassenlage befindliche Gruppe von Menschen heißen. (Weber 1922, p. 177)

From an action-theoretical perspective, Weber decomposes Marx's concept of class into its economic, social, and political dimensions. Just like Marx, he conceives classes as being constituted based on the basis of unevenly distributed economic power, leading to unequal distribution of opportunities. He differentiates the concept of class further into Besitzklassen (wealth classes) and "Erwerbsklassen" (income classes). By introducing an additional concept, he arranges the variety of class positions into broader categories, the social classes.[4] According to this

---

[4] Based on his observations, Weber demarcates four social classes in the Wilhelminian Germany of his time: (a) the working class, (b) petty bourgeoisie, (c) dispossessed intelligentsia and professionals, and (d) the propertied class and highly educated privileged (Weber 1922, p. 178).

classification, mobility within the defined *economic* classes is easily possible, while a shift into a higher *social* class is seldom achievable (Hradil 2001a, p. 58).

Remarkably, for Weber, group formation is not economically determined through class affiliation. The position within a class only designates equal or similar typical interests, shared with others from the same class:

> Auf dem Boden aller drei Klassenkategorien können Vergesellschaftungen der Klasseninteressenten (Klassenverbände) entstehen. Aber dies muß nicht der Fall sein: Klassenlage und Klasse bezeichnet an sich nur Tatbestände gleicher (oder ähnlicher) typischer Interessenlagen, in denen der Einzelne sich ebenso wie zahlreiche andere befindet. (Weber 1922, p. 177)

Based on this observation, he introduces a new term, which had far-reaching influence on the social sciences in general and sociology in particular, both in his time and until the present day. And for this study, Weber's analysis is of extraordinary importance: he highlights the social fact of stylisation of living ("Stilisierung des Lebens", Weber 1922, p. 637), which is based on his concept of "status groups" (Stände). While for Marx, peoples' conduct of life was determined based on their class affiliation, Weber discriminated economic determinants against the important role of social-cultural dimensions in devising group affiliation and social identity. Other than class position, the position within a status group (Ständische Lage) is based on characteristic and shared ways of thinking and action, the basis for social esteem and honour accorded to them by others:

> Stände sind, im Gegensatz zu den Klassen, normalerweise Gemeinschaften, wenn auch oft solche von amorpher Art. Im Gegensatz zur rein ökonomisch bestimmten 'Klassenlage' wollen wir als 'ständische Lage' bezeichnen jede typische Komponente des Lebensschicksals von Menschen, welche durch eine spezifische, positive oder negative, soziale Einschätzung der 'Ehre' bedingt ist, die sich an irgendeine gemeinsame Eigenschaft vieler knüpft. Diese Ehre kann sich auch an eine Klassenlage knüpfen [...] und der Besitz als solcher gelangt, wie schon bemerkt, nicht immer, aber doch außerordentlich regelmäßig auf die Dauer auch zu ständischer Geltung. [...] Aber die ständische Ehre muß nicht notwendig an eine 'Klassenlage' anknüpfen, sie steht normalerweise vielmehr mit den Prätensionen des nackten Besitzes als solchem in schroffem Widerspruch. Auch Besitzende und Besitzlose können dem gleichen Stande angehören und tun dies häufig und mit sehr fühlbaren Konsequenzen, so prekär diese 'Gleichheit' der sozialen Einschätzung auf die Dauer auch werden mag. (Weber 1922, p. 637)

The obvious tension between class and social status that Weber has highlighted stands to reason, when one takes a closer look at the role of conventions and social honour. According to Weber, the style of living of a person and her or his conduct of life is much more closely associated with social status than with class. Social status groups are understood as specific carriers of all conventions. All stylisations of life either originate from or are conserved through a status group:

> Denn die maßgebende Rolle der 'Lebensführung' für die ständische 'Ehre' bringt es mit sich, daß die 'Stände' die spezifischen Träger aller 'Konventionen' sind: alle,. 'Stilisierung' des Lebens, in welchen Aeußerungen es auch sei, ist entweder ständischen Ursprungs oder wird doch ständisch konserviert. (Weber 1922, p. 637)

For Weber, association with a social status group is expressed through and requires a certain way in the conduct of life (Lebensführung). Per definition, it can be open, but it also may be highly exclusive and closed. The closest and most exclusive form of a status group, however, is the caste group. Weber has studied the role of caste in the context of his studies on Hinduism and Buddhism and his scholarly engagement with the economic ethics of the world religions. According to Weber, all duties and barriers associated with the membership in a social status group are also relevant in regard to caste groups, but in its most extreme form of progression (Weber 1986, p. 41). Hindu castes, based on Weber's observations, are closed and exclusive; caste membership is based on hereditary lines, requires endogamous connubial rules, and involves a complex set of ritual barriers, in particular with regard to commensality (Weber 1986, p. 41). Weber concludes that there is no such system other than the Hindu caste system which is to such a degree based on a religious-ritualistic rank order (Weber 1986, p. 41).

The classical scholars of early sociology such as Marx, Weber, Durkheim, and Simmel were inspired by the massive societal transformation processes of their times. Essentially all of them were convinced that these changes affect the individual as well as the societal level, the material, as well as the ideal or mental basis (Rosa et al. 2007, p. 90). Their profound work on "modernisation" dealt with aspects of domestication (Karl Marx), rationalisation (Max Weber), differentiation (Emile Durkheim), and individualisation (Georg Simmel) (Rosa et al. 2007, p. 21). The analyses of these processes were foundational for a new understanding of social structure, beyond explanations based on material and economic grounds. In particular, Weber and Simmel worked very closely around the basic dimensions of more complex social positions and laid the foundations of a much more differentiated approach to social structure.

As shown above, Weber analytically separated the concept of class from his views on social status groups and related aspects of mentalities and conduct of life. Similarly, Georg Simmel (1907), in his book on *The Philosophy of Money*, analysed styles of living as an aspect of identity management in the light of industrialisation, pluralisation of possibilities, and individualisation. In this early work, Simmel has already understood lifestyle as a means of social distinction and group affiliation. Veblen (1997) too in his study *Die Theorie der feinen Leute* refers to the symbolic means of distinction through demonstrative consumption or conspicuous consumption.

As a result of these new perspectives on society, the traditional concept of class, which was largely based on birth, individual property, and assets, slowly made way for more comprehensive conceptions with a consideration of other vertical parameters, such as employment, education, place of living, family status, gender, and age. Until about the early 1980s, social structure analysis became more differentiated through concepts such as social strata and social position. Still, these approaches focused on the "objective" components and external living conditions, which were assumed to influence and to some extent determine subjective dimensions, such as social position-specific mentalities and practices (Hradil 2001a, p. 44).

However, at the latest in the 1980s, concern was raised that the classical approaches were not sufficient in the light of a more differentiated, pluralised, and even individualised society and that new and more differentiated forms of social structure analysis needed to be developed (Kleinhückelkotten 2005, p. 76f; Otte and Rössel 2011, p. 9; Rink 2002, p. 36). Increasingly, the question was raised as to how far the "objective" living conditions in fact influence values, mentalities, and attitudes as well as the "subjective" conduct of life and social practices of a person. In consequence, it became ever more important to analyse the "subjective" ways of living and the underlying values and attitudes in separation from the "objective" living conditions. Only with an understanding of both these dimensions does it become at all possible to compare and evaluate the degree of correspondence between group-specific living conditions and group-typical practices and mentalities (Hradil 2001a, p. 45). With this realisation in the 1980s, the lifestyle concept found its way into the research and analysis of social inequality and has since then been applied in many different contexts. The next section will outline the theoretical grounds and the rationale of the lifestyle concept.

2.2.2 The Lifestyle Concept

In their study on US energy use and related CO_2 emissions, Bin and Dowlatabadi (2005, p. 197) take a "consumer lifestyle approach" in which they "attempt to explore the relationship between consumer activities and environmental impacts in the US". They define lifestyle as follows: "Lifestyle is a way of living that influences and is reflected by one's consumption behaviour" (Bin and Dowlatabadi 2005, p. 198). This definition fails to specify "way of living" and therefore epitomises a tautology.

The authors then make an important remark about the complexity of consumer decision making and respond to this challenge with an "interdisciplinary framework which explicitly acknowledges the multitude of interacting factors". Besides external environmental variables (e.g. cultural influences) and household characteristics (e.g. size, income, location), the authors refer to the importance of "individual determinants, such as attitudes and beliefs, which are psychological variables influencing an individual consumer's decision making" (Bin and Dowlatabadi 2005, p. 198). With this research agenda, which is supposedly based on the concept of lifestyle and which aims to gain better understanding of consumer choices and their determinants, it is in fact of due importance to consider the individual social-cultural dimension. The authors then however surprise the interested reader as they do not further explain how they would operationalise this dimension into their analysis, nor do they mention it at all in any other part of their study. This study well exemplifies the vaguely defined and often unspecific application of the lifestyle concept.

Lifestyle is a synthetic concept, which combines a number of different components typically coinciding and playing together in quite characteristic ways. Lifestyles are understood to mediate between the objective social position of a person and her/his subjective lifeworld (*Lebenswelt*) (Reusswig 1994a, p. 127; Rink 2002, p. 36f). Moreover, the way it is conceptualised varies quite substantially in social science research: there are studies that draw on patterns of consumption and preferences of taste, as there are in the same way other studies only considering values, mentalities, and attitudes as the defining dimension of lifestyle.

The following chapter will attempt to contextualise and specify the concept of lifestyle within the broader field of analysing social inequality in sociology and social geography.

2.2.2.1 The Emergence and Further Development of the Lifestyle Concept

As stated above, the concept of lifestyle is "about as old as the discipline of sociology itself, with Weber, Simmel, and Veblen representing classics of lifestyle research studying the social conditions, forms of expression, and consequences of individual lives in the early modernity" (Reusswig 1994a, p. 40). These scholars were among the first to shift the attention from a rather economic focus on class to a much broader view on social-cultural aspects of differentiation. About six decades later, Weber's analytical view on social status groups and associated styles of living inspired Pierre Bourdieu to study French society and develop a comprehensive theory of social stratification based on class, culture, and aesthetics. The outcome of this work was issued in his book *Distinction: A Social Critique of the Judgment of Taste* first published in French in 1979 (1987). With this work, Bourdieu broke new ground in sociological research of his time – a time characterised by a spirit of critical changes and innovation in science in general.

Not only had the realisation about the vast societal changes towards societal pluralisation sparked dynamic developments in the social sciences. Also improved computation capacities, the availability of personal computers, and the further development of explorative data analysis, in particular cluster analysis and correspondence analysis, opened up new pathways in social science research (Otte and Rössel 2011, p. 9; Reusswig 1994a, p. 51). Early and representative for these innovations in statistical analysis also is Bourdieu's outstanding work, in which he critically contributed to the popularisation of correspondence analysis, a statistical approach to visualise complex relations of larger sets of variables (Blasius 2001, p. 7; Otte and Rössel 2011, p. 10).

Bourdieu's study was well received internationally, but it was also differently interpreted and adopted in different linguistic and scientific-cultural contexts. Otte and Rössel (2011, p. 10) outline the differences between the German and the Anglo-Saxon reception and international science journal discourses: in the Anglo-Saxon context, Bourdieu's lifestyle concept is more closely and systematically linked to issues of social inequality and the cultural reproduction of class structures. Also, the concept of cultural capital is given more weight, and the focus rests on the role of sophisticated culture (*Hochkultur*) and its functions with regard to distinction (Otte and Rössel 2011, p. 10). This development can be interpreted as one of the major reasons for the rather marginal role that sociological lifestyle research plays in the English-speaking world today (Stadtmüller et al. 2013, p. 264). Geisler even argues that the German reception of lifestyle is very unique in its way conceptualising lifestyle as it is almost decoupled from notions of class and social stratification (Geißler 2002, p. 141).

In the German context, Bourdieu's lifestyle concept finds a much broader interpretation, which goes far beyond analysing patterns of participation in sophisticated culture. The concept is – at least partially – decoupled from class- and strata-specific characteristics (Meyer 2001, p. 260). Bourdieu was able to spark a broadly conceptualised empirical research on the social-structural determinants of lifestyle in Germany with studies by, e.g. Hradil (1987), Vester et al. (2001), and Schulze (1992). These studies can be seen as the outcome of a first phase of lifestyle research in Germany (Stadtmüller et al. 2013, p. 264). In Germany, these scholars laid the foundations of sociologically driven lifestyle research. During this first phase, lifestyles were mainly understood as group-specific forms of organising and managing everyday life, symbolically expressed through cultural taste and leisure preferences (Stadtmüller et al. 2013, p. 264).

Building on and benefitting from this early empirical work in sociology, a second generation or second phase of lifestyle research can be identified, as claimed by Stadtmüller et al. (2013, p. 264). From an action theory perspective, scholars such as Lüdtke (1989), Otte (2004, 2005), and Rössel (2005) pointed towards an improved understanding and a further development in regard to the theories and concepts of current social structure analysis. These scholars stressed the ideals, motives, ends, and purposes of an individual with regard to the way people live and orient themselves. Lifestyle serves as a classification and evaluation system that facilitates a person to stabilise their social-cultural identity against others (cf. Reusswig 1994a, p. 69; Stadtmüller et al. 2013, p. 264).

2.2.2.2 Lifestyle in Market Research

In parallel to the conceptual development in the social sciences, lifestyle was even earlier discussed and employed in consumption and market research, later also in psephology. For an understanding of the consumer in a transforming society, the emergence of the lifestyle concept was just a logical consequence in a context in which class- and strata-specific marketing was still prevalent (Reusswig 1994a, p. 80f). Reusswig (1994a) in his inspiring book on lifestyle and ecology delineates the historical cornerstones of lifestyle research in the USA and in Germany and concludes that the concept of lifestyle had the function to find a successful marketing mix for a more differentiated consumer market (Reusswig 1994a, p. 84). The market-based approach in employing the lifestyle concept aimed to find explanations and prognoses for consumer behaviour, and it attempts to identify, describe, and address specific target groups for specific products (Reusswig 1994a, p. 84). These practical arguments for the use of the concept in marketing and consumption research are also indicative for an understanding why the lifestyle concept gains increasing relevance also in environment-related social science research.

Most outstanding in market-based lifestyle research in Germany is the SINUS Institute (based in Heidelberg). SINUS conceptualises its "milieus" based on an "everyday life analysis" of society. SINUS-Milieus "group together people who are similar in terms of their attitude to life and ways of living" (SINUS 2015:3). SINUS develops its milieus from an analysis of basic values, along with everyday attitudes towards work, family, leisure, money, and consumption, and an evaluation of the social position (Reusswig 1994a, p. 85; SINUS 2015, p. 3). Their three decades of work on lifestyle in Germany and worldwide have achieved a great resonance also from the social sciences, despite the often discussed problems associated with SINUS being a private contract research institute and being very careful with publication of or statements on results and methods (Reusswig 1994a, p. 85). Since 2009, the SINUS Institute is also involved in environment-related social science research, in particular with its contribution to the Nature Awareness Study (Umweltbewusstseinsstudie) issued biennially by the German Federal Agency for Nature Conservation (BFN).

2.2.2.3 Relevant Dimensions and Components, Theoretical Implications for Operationalisation, and Definition of the Concept

One of the greatest challenges for future lifestyle research can be seen in the definitional vagueness of the concept. *First*, it is often imprecisely defined in differentiation from other concepts, such as milieu, lifeworld (Lebenswelt), conduct of life (Lebensführung), and way of life (Lebensweise). *Second*, there seems to be dissent in the way lifestyle is conceptualised and about elements and dimensions on which it should be built. And *third*, its conceptual openness and broadness allow a great variety of different operationalisation in empirical research, which leads to the problem of lacking comparability of lifestyle segmentations (Meyer 2001; cf. e.g. Otte 2004, p. 12). For the author of this study, one major challenge was to decide on the appropriate conceptual framework that allows the concept to be employed in a so far totally new context for lifestyle research, namely, India. The following section therefore aims to shed light on the various strands of conceptualisation as a basis for understanding the rationale of the conceptual framework of this study.

Lifestyle Operating on Three Interrelated Reference Levels

First, it should be considered that lifestyle is not a specifically observable fact but a theoretical-empirical construct. It integrates internal and external, individual and collective, and substantial and relational features of persons and households (Lüdtke 1989, p. 41). Also, lifestyles cannot be understood as superficial phenomena. Reusswig (1994b, p. 41f) states that lifestyles refer to a psychological and a social identity of people. Lifestyles thereby represent forms and means of social distinction. Therefore, the concept cannot be seen in isolation from the social aspects. Lifestyle analysis rather aims to get hold of "collectively shared lifestyles" (Lüdtke 1989, p. 40; Otte 2004, p. 41). Hartmut Lüdtke (1989, p. 74) correctly states that lifestyles emerge from the mediation between personal and social identity. It can be understood as a hinge between expression of individuality and at the same time as an expression of social belonging:

> Es gibt daher gute Gründe, daran festzuhalten, dass ein Lebensstil Ergebnis der 'Vermittlung' personaler und sozialer Identität bzw. als 'Scharnier' zwischen der Darstellung von Individualität und der Darstellung sozialer Zugehörigkeit zu verstehen ist. (Lüdtke 1989, p. 74)

Lüdtke (1989) therefore suggests taking a multilevel perspective towards understanding of lifestyle. Lifestyles involve a set of mechanisms that operate differently on the respective levels of reference. Lüdtke (1989, p. 71) differentiates between distinction (micro level), social closure (meso level), and segregation (macro level). According to his theory, it is the micro level which involves mechanisms of individual comparison, distinctive behaviour, and selective interaction with others. These processes of social interaction on the micro level translate into group-based social-cultural mechanisms of role attribution, emergence of social networks and closure, as well as the exchange of goods, interests, symbols, and emotions. On the macro level, according to Lüdtke's model (1989, p. 71), these processes happen to further consolidate in terms of cross-group aggregation of similar lifestyle segments and social-spatial segregation. According to Lüdtke, the macro level of lifestyle analytically involves mechanisms of group formation and organisation. This as a result leads to processes of segregation and highly explicit forms of social closure, for instance, through differences in real estate prices and resulting residential patterns. Analytically, Lüdtke's model is very indicative, and it contributes to a better understanding of the multilevel interactions from micro to macro level.

Dimensions and Components of the Concept and Why They Require an Analytical Treatment of Independence

Which dimensions and components constitute lifestyles? Different authors put emphasis on different dimensions and components of lifestyle, and they differ in how they relate these to each other. Lüdtke (1989, p. 42), for instance, highlights four theoretical dimensions relevant for the organisation of living: the social-economic situation (economic and social resources), competence (cognitive, linguistic, and social qualifications and authority), performance (Bourdieu's "Praktiken und Werke", whole sets of behavioural and interactional expressions, their forms, features, and consequences), and motivation (acquired latent and internal dispositions of action and behaviour, based on socialisation and lifelong experience, also summarised as values, attitudes, and preferences). Reusswig highlights quite similar dimensions, but he subsumes Lüdtke's concept of "competence" within "mentalities" (attitudes, values, life goals, and world views). For him, the interrelated

2.2 Social Structure, Lifestyle, and Consumption: Understanding Social-Cultural...

dimensions "mentalities", "performance", and "social position" are constitutional to his concept of lifestyle. Most authors agree to this basic understanding of these three basic foundational elements: *first*, values, attitudes, and preferences (mentalities or motivational factors); *second*, behavioural patterns and social practices (performance); and *third*, social-economic factors (social position).

Quite some disagreement can be found in the literature about how relational aspects should be conceptualised between dimensions and components and about operationalising them. A critical point of discussion deals with the question of how to involve the relevant dimensions in the empirical analysis. In their comprehensive volume on lifestyle research, Otte and Rössel criticise that in some conceptualisations, as, e.g. in Spellerberg (1996, p. 57), lifestyle builds on both realms, behaviour and attitudes. They base their arguments on the grounds of a classic social-psychological approach, the theory of planned action (Ajzen and Fishbein 1980). In their empirical work, Ajzen and Fishbein show evidence for values and attitudes explaining the behaviour of a person in situative contexts (e.g. Ajzen 1991, p. 185ff; Fazio 1990, p. 90). Otte and Rössel rightly state and argue on this account that it is problematic to lump both dimensions together in one concept, as this procedure makes it impossible to separate cause and effect in the end (Otte and Rössel 2011, p. 12). Even though Otte and Rössel highlight the importance of values and attitudes as being constitutional in regard to lifestyle (Otte and Rössel 2011, p. 14), they suggest operationalising lifestyle exclusively on the basis of behaviour. And this is the approach that authors in their edited volume on lifestyle research follow (Otte and Rössel 2011, p. 12).

Dieter Rink (2002, p. 39) in his review article on the various approaches and applications of the lifestyle concept in sociology, in which he analyses the suitability of the concept in sustainability-oriented research, comes to a similar but at the same time contradictory conclusion: similarly, he observes that the broader notion of lifestyle and milieu sometimes leads to the conclusion that these concepts allow one to analyse attitudes, values, and preferences, in the same way as and together with behaviour. He states, however, that most of the approaches conceptually remain definitive on the attitudinal and value dimension. Based on this observation, he suggests that future lifestyle research should agree to this smallest but common denominator, by explicitly limiting the concept to the aspects of aestheticisation, stylisation, and distinction (cf. Poferl 1998, p. 310; Rink 2002, p. 39). Hence Rink argues on similar grounds as Otte and Rössel that an integration of action and behaviour together with values and attitudes would prohibit an analysis of the relationship between values and behaviour.

Other authors also refer to this problem in a quite similar way. Hradil (2001a, p. 275) recommends to separating the underlying constitutional elements of lifestyle in analysis. He argues that subjective factors in particular, such as values, attitudes, and beliefs, need to be analysed separately from objective determinants such as age, education, gender, income, etc. (Hradil 1996, p. 14f; see also Reusswig 1994a, p. 51). These objective factors are in many cases closely related to aspects of behaviour and value orientation. These dimensions also interact in various ways.

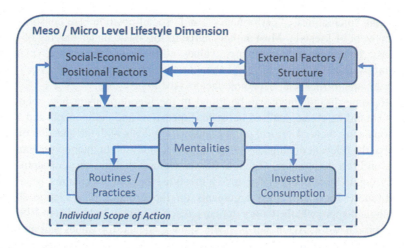

Fig. 2.1 Model of lifestyles operating within an individual scope of action, which is determined by social-positional and external factors, effects, and their feedback mechanisms. (Source: own draft)

For the understanding of lifestyle, it is therefore important to keep the possibility of tracking these linkages and points of interaction.

Especially in the context of environment-related lifestyle research, the aspect of separation and the associated possibility of analysing the interlinkages between dimensions are highly relevant, in particular regarding discrepancies between value orientations and actual behaviour. A well-known phenomenon is the value-action gap, which has raised many questions in the field of environment-related social science research. Reusswig goes as far as to say that ecologically oriented lifestyles tend to show a patchwork character with inconsistent patterns and diverging levels of environmental friendly behaviour (Reusswig 1994a, p. 113ff). According to this observation, lifestyles incorporate different behavioural and consumption aspects, while dismissing others, according to differing value orientations and preferences and according to the contextual factors.

The lifestyle concept with its integrated approach to analysing mentalities, behaviour, and social-economic position allows for analysis of a broader set of factors of individual-level rationality. Based on the underlying general values and attitudes in combination with the objective social-economic determinants, the motives, reasons, and barriers of certain behavioural patterns become clearer. The lifestyle concept therefore offers a framework for the analysis of the cognitive ambivalences and discrepancies of certain behavioural patterns as well as their underlying rationalities. The concept aims to figure out how environmental problems go back to the individual and not to a generalised way of life (Lebensweise) of a society. Figure 2.1 delineates the constitutional components of the concept and shows how they relate to each other.

Based on the relevant literature, the following paragraphs will make further reference to these interrelations and the involved feedback mechanisms and will show

how they constitute lifestyle. It should not be forgotten that these are theoretical-empirical constructs for the understanding and operationalisation of lifestyles.

Highly relevant are objective factors, i.e. social-positional and external factors, determining the space for stylisation, the individual scope of action within which lifestyle can unfold and operate. Relevant for the understanding of the role of objective factors, such as income, employment, or education, is Gunnar Otte's (1997) argumentation on low-cost and high-cost situations. Otte (1997, p. 305) defines three stages of actors maximising their benefits through lifestyle: the first stage – making sure that all basic needs are met – is a precondition for the second stage on which social distinction and recognition from others is utilised for creating a lifestyle and exhibiting it through intermediate goods ("Zwischengüter"). Intermediate goods according to Otte are all those aspects of lifestyle that are shared and recognised by larger social segments and which are societally known and well defined. In a third stage, social actors invest time and money, e.g. into investment in material assets or into certain features of leisure activities, in order to reproduce and consolidate their lifestyle (Otte 1997, p. 305).

In this context, Stadtmüller et al. (2013, p. 264) argue on the basis of Otte (2004) and Rössel (2011) that lifestyles tend to unfold only in *low-cost situations*. In situations that involve *higher costs*, it is much more the financial position that determines or requires a decision, e.g. in regard to conspicuously investing in certain goods or services. All three authors conclude in a similar direction that not all lifestyles are accessible for any social status group:

> *So wird die Entscheidung für ein bestimmtes Wohnquartier oder die Anschaffung eines bestimmten Automodells auf der ersten Entscheidungsebene durch die zur Verfügung stehenden finanziellen Ressourcen bestimmt sein und auf der zweiten Ebene durch den Lebensstil konkretisiert. Der Lebensstil ist somit, wie wir wissen, für die 'feinen Unterschiede' verantwortlich.* (Stadtmüller et al. 2013, p. 264)

In reference to the German title of one of Bourdieu's most important works (1987), this quote brings it right to the point: the lifestyle concept does not mean to replace social-economic or vertical determinants in conduct of life. In Bourdieu's conception of lifestyle and habitus, individual preferences and the ways of perceiving and thinking are closely coupled with the level of endowment with financial, cultural, and social capital, which again defines the habitus of a person. Hence, for Bourdieu, it is the position within a social class that allows for a certain character of the habitus and thereby allows for accessing a certain lifestyle. The habitus mediates between structure and scope of action (Bourdieu 1987, p. 175).

Most of the authors who work with the concept of lifestyle support this thesis that lifestyle does not empirically exist independent of external and structural factors (Hradil 2001b, p. 275). External structural boundaries (e.g. social control, power relations, cultural aspects, and infrastructure) and the limited resources of social actors determine the *objective* scope of action. Only within this objectively defined scope of action can the individual select from and decide on potential options with regard to lifestyles. Media and adverts, imitation, fashions, and social control are factors that also have a small but relevant share in structuring this space and influencing decisions on lifestyle options (cf. Schultz and Weller 1996, p. 80f). Most influential, however, for the structuring of emergent patterns of conduct of life are values, attitudes, and preferences (Ajzen 1991, p. 185ff). Reusswig argues:

> Die eigentümliche 'Vernunft' des ungleichen 'unvernünftigen' Verhaltens muß erforscht werden, damit die ökologische Vernunft am Ende bessere Realisierungschancen bekommt. (Reusswig 1994b, p. 96)

This quote refers to the fact that not all human beings are equally "irrational" in terms of, e.g. their environmentally relevant behaviour. Rationalities are socially differentiated based on constraints, structural factors, and the social context that delineate the individual scope of action on the one hand. However, values, attitudes, and preferences structure these rationalities and resulting behavioural patterns on the other hand.

This interactive relationship can be understood as a complex filter, which works on the emergence of lifestyles. Hartmut Lüdtke (1989, p. 39ff) – in reference to the "constraint-choice approach" – has introduced this view of a twofold filter: on the one hand, there are structural factors that influence or determine decisions, and on the other hand, there remains a scope of choice, which allows individuals to reflect on their own preferences and values. Therefore, lifestyles are restricted to operate within a rather limited space of individual "freedom".

Hence, values do not affect behaviour directly, but rather in a structured manner along the lines of vertical and objective determinants. And, while values are quite stable and able to withstand life events, such as birth of a child, changes in employment, or relocation, they are still not static. Constrained by limitations in their resource base and due to other external structural factors, individuals develop and reconstruct their foundational spectrum of values and attitudes. They tend to make sure that these values are sufficiently consistent with their behavioural patterns and that they accord by some means with their social-economic position. That means that values and attitudes are gradually formed and reconstructed over a lifetime in reaction to and based on social interactions and structural factors (Fig. 2.1). Within the scope of action, there is a socially specific action space for individual stylisation in tune with the underlying values and against the background of intense social interactions.

Last but not least, Lüdtke (1989, p. 39) states that the larger the resource base of an individual, the larger and richer also tends to be the space for searching for definitional aspects of lifestyle. Hence, the greater a person's individual resources are, the more it becomes an expression of personal preference that is selected, e.g. different life goals, symbols, partners, and/or behavioural patterns.

Definition of Lifestyle in This Study

In summary and in conclusion to the above theoretical considerations, lifestyle is (1) constituted based on at least three relevant dimensions: (a) the behavioural dimension; (b) the dimension of values, attitudes, and norms; and (c) social-demographic determinants. It is (2) important to keep a separation in analysing these dimensions, in order to be able to reconstruct relational and interactional aspects to gain an understanding of patterns of cause and effect between the involved dimensions (e.g. with regard to understanding group-specific reasons for the value-action gap). The lifestyle concept (3) can be figuratively understood as a hinge that links the different levels of reference, the micro or individual level, the meso or social level, and the meta- or societal level. It is (4) an expressive dimension of social inequality (Lüdtke 1989, p. 156) and involves a set of interdependent mechanisms. Structure and social-economic position of actors define their scope of action. Within this scope of action, lifestyle unfolds largely based on the underlying values and related goals. Hence, socially shared value orientations structure individual behaviour and practices. At the same time, values and their related goals can react and adapt to changed external conditions and pragmatically motivated changes of individual patterns of behaviour. Apart from these individual-level mechanisms, it is equally important to understand the social embedding of lifestyle as a means of expressing a sense of belonging and distinction. Lifestyles mediate between the social-economic position of a person and his or her lifeworld (Reusswig 1994b, p. 127).

Based on the above theoretical considerations, lifestyles are defined as group-specific, value-, attitude-, and preference-driven patterns of everyday life that unfold within an economically, social-culturally, and environmentally prestructured field of social interaction. And hence, lifestyles are limited by external and social-positional factors at a given point of time. With this definition, values, attitudes, and preferences are emphasised as important determining factor for the emergence of social-culturally differentiated patterns in conduct of life. These social-culturally-specific patterns in conduct of life are coarsely prestructured by external factors and by the resource base of the individual. These prestructured spaces of individual agency are limited differently for each individual, based on his or her own resources (income, household infrastructure, education, etc.), but also based on a variety of external factors, such as culture and historical aspects, political system, level of economic development, power relations, level and character of infrastructure development, etc.

Critical Review of the Lifestyle Concept

Critique from Practice Theory: Environmental Lifestyle Research Assumes a "Voluntaristic Theory of Action" (Warde 2014, p. 283)

The most fundamental critique of the lifestyle concept comes from practice theory. Most relevant for this study is the critique raised against the commonly used framings in social and cultural theories that are based on the assumption of individual choice (Warde 2014, p. 286). Warde (2014, p. 283) draws a revealing differentiation between sociological approaches on consumption that emerged as part of the cultural turn in comparison to approaches based on practice theory. He argues that the model of individual choice departs from a "voluntaristic theory of action" implying an "active and reflexive agent". From this viewpoint, the individual is seen as "an active, expressive, choosing consumer motivated by concerns for personal identity and a fashioned lifestyle" (Warde 2014, p. 283). Warde even goes as far as to say that the individual choice model in key terms was little different from the model of a sovereign consumer in neo-classical economics (Warde 2014, p. 283).

Elisabeth Shove follows a similar line of argument, interestingly, very specific with reference to the field of climate change policy and theories of social change. She claims that most research on the impacts of human behaviour on environment and climate change follow the "dominant paradigm of 'ABC' – attitude, behaviour, and choice" (Shove 2010, p. 1273). She argues that:

> For the most part, social change is thought to depend upon values and attitudes (the A), which are believed to drive the kinds of behaviour (the B) that individuals choose (the C) to adopt. The ABC model, derived from a strand of psychological literature grounded in theories of planned behaviour [...] and in variously rational concepts of need [...], resonates with widely shared, common sense ideas about media influence and individual agency. (Shove 2010, p. 1274)

This "individualistic understanding both of action and change" (Shove et al. 2012, p. 142) can also be found in some of the lifestyle conceptions applied in environmental sociology. Being based on the assumption that behaviours are largely motivated through beliefs, values, attitudes, and preferences, lifestyles are conceptualised as tastes and as expressions of personal choice (Shove et al. 2012, p. 3). According to Shove et al. (2012, p. 4), such conceptualisations fall short of sufficiently considering the role of structure as made out in Giddens' structuration theory.

The authors further argue that the ABC paradigm would build on the assumption that choices on behaviours and practices would reflect peoples' environmental commitments and values. Accordingly, the paradigm would underpin "two classic strategies for promoting more sustainable ways of life: one is to persuade people of the importance of climate change and thereby increase their green commitment; the second is to remove barriers obstructing the smooth translation of these values into action" (Shove et al. 2012, p. 141f).

Both lines of argument fall short of fully grasping the rationale and complexity of environment-related lifestyle research. *First*, only few conceptualisations of lifestyle exclusively focus on the role of subjective factors (e.g. Mitchell 1983; Schulze 1992). Apart from the few approaches that rather concentrate on vertical determinants, such as consumption expenditure (Sobel 1981; e.g. Weiss 1988), most of the approaches are integrative, considering to varying degree both subjective and objective factors. Most prominent in this regard are Pierre Bourdieu's (1987, p. 277ff) theory of lifestyle as a practical habitus of class or Hartmut Lüdtke's (1989) conceptualisation of lifestyle as performative and expressive inequality.

Second, environmental values represent just a small section of a more comprehensive set of general social values. A focus on such smaller sections bears the risk of ignoring values that have greater priority than the targeted value. For instance, environment-related values that greatly emphasise the importance of reducing GHG

2.2 Social Structure, Lifestyle, and Consumption: Understanding Social-Cultural…

emissions may conflict with values of probably higher priority underlining the importance of having an exciting and eventful life. Conceivable, for example, is a person who largely follows principles of sustainable consumption (e.g. cycling instead of using a car, vegetarian and organic diet, following principles of energy saving, household electricity based on renewable energy). Such a consumption pattern may cohere with other more general values (e.g. health concern), it involves rather low costs and surrender (low-cost thesis; see above), and it tends to be highly productive in terms of gains in prestige. At the same time, a person may not be willing to abstain from well-deserved once-in-a-year long-haul travel, e.g. a holiday trip to Australia, with GHG emissions, which by far outweigh emissions that were saved due to more general sustainable practices in everyday life.

Therefore, a holistic approach that combines an analysis of general social values with an analysis of everyday practices and decisions of consumption is more productive (against practice theoretical approaches) in terms of gaining better analytical understanding of social-cultural group-specific internal conflicts, ambivalences, and inconsistencies. Such an understanding may also be better able to inform environmental policies for social-culturally differentiated approaches that take into account these ambivalences and contradictions. This does not mean one has to convince people of the importance of climate change and "repair" their value inconsistencies (Shove et al. 2012, p. 141f). It rather supports a rationale that attempts to address key points of environmental policy intervention (see Sect. 4.1.4) and provide for more comprehensive solutions that take into account the conflicts and ambivalences of relevant consumers.

Nevertheless, lifestyle research can gain productive insights and a refreshing perspective by considering more strongly what practice theory delivers for environment-related social science research. The social practice approach allows for a closer look into the interdependencies between the most constitutive elements of recurring patterns of routinised behaviours (Shove et al. 2012, p. 7). According to Andreas Reckwitz (2002, p. 249), "a 'practice' (Praktik) is a routinised type of behaviour which consists of several elements, interconnected to one other: forms of bodily activities, forms of mental activities, 'things' and their use, a background knowledge in the form of understanding, know-how, states of emotion and motivational knowledge". The person or "carrier" of a social practice plays a role analytically in his or her relation to the practice only.

Especially with reference to context and the material configurations that underpin social practices, environment-related lifestyle research might gain new insights for a further development of the concept. For instance, a quite new strand in practice theoretical thought follows the argument of Schatzki et al. (2001, p. 3), who state that social practices can only be understood by apprehending the key role of things and their material configurations. Following this line of argument, Shove et al. (2012) suggest a new approach to practice theory bringing forward the three key concepts for the analysis of social practice – meanings, competences, and materials – in which they give material configurations a crucially important role. The authors underline this argument by also drawing on earlier accounts from Latour (2000) and his statement about artefacts, which are, according to him "in large part the stuff out of which socialness is made" (Latour 2000, p. 113; cited in Shove et al. 2012, p. 9). In Sect. 4.1.1, the author of this study will show how he has attempted to consider things and material configurations more strongly in relation to lifestyle. And the author will also give an outlook on how such a perspective could be further developed for environment-related social practice research.

Critique from (German) Lifestyle Research

Apart from the critique from social practice theory, there are few German authors who have taken stock of the boom times of the lifestyle concept in the 1990s. Most relevant to mention are Meyer (2001) and later Hermann (2004), and quite influential also is the review of Otte (2005). Meyer's (2001) and Hermann's (2004) accounts are criticised by Otte (Meyer's critique also by Hradil 2001b), who blames them for being too selective and too unsystematic. Both reviews issue a very negative picture of the application of lifestyle to empirical questions and even go as far as to state that lifestyle sociology has failed to comply with the promised advances in social structure analysis. Central to Meyer's (2001) critique is an assumption that lifestyle sociology seeks to replace vertical approaches to analyse social inequality. This aspect has already been addressed in Sect. 2.2.2.1: the concept of lifestyle does not set out to replace or oppose the "vertical approach" but to complement it. Vertical inequalities can only be explained in combination with horizontal differences. In the same way, lifestyle conceptualisations can only work on the basis of a combination of vertical and horizontal factors (Hradil 2001b, p. 277f). Also Hermann's critique is substantially challenged by Otte (2005, p. 3), as it fails to properly grasp the explanatory claims of lifestyle sociology.

Otte (2005) delivers an alternative critique on the lifestyle concept, in which he draws four basic weaknesses and future challenges in lifestyle conceptualisation. Based on his analysis, he also offers an alternative, refined conceptualisation, which he lays out in more detail in one of his earlier books (Otte 2004). All four problems are based on methodology and methods.

His *first* argument draws on the critique that lifestyle conceptualisations so far lack comparability. He concedes that different approaches have led to comparable types of lifestyles, but he argues that most approaches lack direct comparability in the sense of identical constructions of the respective typologies (Otte 2005, p. 24). Typifications generally are based on inductive-empirical procedures and are therefore data driven and in consequence not comparable (Otte 2005, p. 25).

Second, he argues that empirically identified lifestyle segments build on questionable real-world phenomena and therefore bear the risk of reification. The selection of lifestyle variables tends to be based on arbitrary and subjective decisions of the researcher, and only their statistical combination allows different lifestyle profiles to be defined. In consequence, typologies may touch "real life worlds", but nevertheless they often tend to be aggregates based on artefacts (Otte 2005, p. 25).

Third, Otte (2005, p. 25) argues that lifestyle conceptualisations tend to build on rather weak theoretical foundations with rather descriptive accounts on social differences with regard to leisure activities or everyday aesthetics and preferences. In many cases, the approaches fail to deliver a productive analysis of social structure with interpretational conjectures about empirically identified correlations. Theoretically stringent explanations are rarely found among the bulk of lifestyle studies, and especially explanations with regard to the logic of formation and the involved mechanisms are largely superficial and underexposed. Moreover, Otte (2005, p. 25) argues that most typologies are not reconsidered with existing higher-order theoretical models and other theoretical concepts (Otte 2005, p. 25).

Fourth, the sheer multiplicity of lifestyle-related value orientations, practices, and everyday aesthetics requires a large set of variables in order to get hold of the comprehensiveness of lifestyle. So far, there have been no successful attempts to reduce the foundational dimensions to a theoretically and content-based fixed set in order to provide for an ex ante defined simplified model that still depicts the complex lifeworld of a larger societal group (Otte 2005, p. 25). Otte (2005, p. 25) concludes that it requires between 40 and 50 variables for the construction of meaningful typology, and obviously, lifestyles are therefore markedly cost- and time-intensive instruments.

One important aspect raised both by Meyer (2001, p. 261f) and Hermann (2004, p. 166) and only shortly mentioned by Otte (2005, p. 24) is the risk of tautological explanations in theme-centred typologies. This involves especially those typologies that set out to explain behaviours on the basis of similar or even closely related behavioural patterns as definitional component of explaining lifestyle concept (Hermann 2004, p. 166; Hradil 1996, p. 27, 2001b, p. 279; Otte 2005, p. 24). For instance, a typology of tourists that pursues to explain typical patterns of touristic activities does not separate between dependent and independent variables (2005, p. 24). Surely, this issue can be avoided, but there remains a related problem, which has been raised by Hermann (2004, p. 173), who argues that the connection between lifestyle and behavioural patterns could be based on a spurious correlation. Based on the assumption that lifestyles can be seen as expressions of group-specific value orientations and because values are highly relevant for patterns of behaviour and everyday practices, spurious correlations between lifestyles and behaviour cannot be precluded. This at least suggests that both phenomena tend to have common grounds (Hermann 2004, p. 173). Indicative for this hypothesis is a study issued by Elmar Lange (1991, cited in Hermann 2004, p. 173). He shows based on multiple regressions that consumption behaviours are affected much more strongly by values and social-demographic factors than by lifestyle. In a regression that controls for values, the effects from lifestyle tend to be relatively low or even non-existent.

This issue raised above is perceived by the author to have fundamental implications for the conceptualisation of lifestyle. In Sect. 2.2.2.4, the author has shown that most authors suggest focusing on only one dimension for the purpose of lifestyle segmentation by means of cluster analysis. In this way, one makes sure that the analysis allows inferences to be drawn on cause-and-effect patterns between relevant dimensions. In reaction to this paradigm, many studies exclusively fix on behaviour as the constitutive component for the segmentation of lifestyle (e.g. Lüdtke 1989). For a lifestyle conception that pursues a general approach to social structure analysis (trans-sectoral analysis), such operationalisation of lifestyle may be constructive. However, for the purpose of selectively explaining certain behaviours and practices (theme-centred analysis), such a conceptualisation ignores the importance of values as constitutive for the structuring of human behaviour (Ajzen 1991, p. 185ff; Fazio 1990, p. 90). The author of this study therefore suggests a conceptualisation that emphasises the role of values as constitutive for lifestyle and constitutive for the structuring of behavioural patterns. Such a lifestyle analysis primarily builds on a segmentation (cluster analysis) based on values, attitudes, and preferences. These value-orientation segments or clusters are then characterised (descriptive analysis) based on social demography and typical patterns of behaviour to arrive at a lifestyle typology that builds on values and that is informed by ideal-typical patterns of behaviour and social demography. This approach is better able to shed light on the geneses of certain lifestyles based on values, and it allows for a closer understanding of directly related and typical performative and expressive behaviour.

Some of the above-mentioned more general critical aspects apply only partly to the conceptualisation of this study. In the Indian context, there have been no other attempts so far to apply the lifestyle concept in its sociological way of use, and so there was no other study to draw on or compare the results with. Therefore, this study largely draws on experiences from lifestyle research in Germany. The explorative character of this study required the author to start from scratch, i.e. identify relevant dimensions and related indicators by means of explorative qualitative research. Testing of the questionnaire was also restricted by time and money and in consequence, the author initially proposed to have a much larger set of variables, out of which only a share was actually used for the analysis. However, the author has attempted to address some of the most critical points raised, especially from the side of practice theory. These considerations will be mentioned and outlined in Sect. 4.1.

2.3 Relevance, Objectives, and Challenges in Linking Personal Greenhouse Gas Emission Accounting with the Concept of Lifestyle

The world today faces very drastic social, economic, political, and ecological challenges. The biggest risks and challenges come from ourselves, i.e. most of the challenges that we face today are related to our society and how we live. Global environmental change in general and global climate change in particular are among the most burning issues with regard to the long-term future of humanity.

Solutions to address these challenges are largely based on known paradigms. John Urry rightly states that anthropogenic climate change has long been addressed only by two groups of analysts, natural scientists and economists (Urry 2011, p. 5; also see Werlen 2015b, p. 4f). Urry (2011, p. 5) further argues that this bias has led to a remarkable neglect of "society" in analysing the related determinants, impacts,

2.3 Relevance, Objectives, and Challenges in Linking Personal Greenhouse Gas...

and current and future implications for humanity. He summarises that "economics" needs to be displaced from its preeminent or imperialist role in examining and explaining the "human" causes and consequences of climate change (Urry 2011, p. 5; also see Werlen 2015b, p. 4f)". Solutions for this societal crisis are likely to fail, if they build on already well-known paradigms and symptomatic solutions. In the light of this fundamental crisis, it is important to raise fundamental questions, such as how do we want to live as a society. The author of this study opines that especially social and cultural geography are well arrayed to bring in new and urgently required perspectives into the analysis of the ecological crisis, which is as much a social as a natural problem. Geography can and should contribute towards initiating and facilitating a stronger integration of the social sciences, humanities, and the natural sciences. And, as Benno Werlen (2015b, p. 11) puts it, "to overcome disciplinary blind spots, we need a perspective that specifies and solves problem complexes independent of the disciplinary interests and boundaries".

Dealing with issues of society, social values, human behaviour, consumption, and social-cultural change requires broad-based integrated approaches from the social sciences, which need to be well informed by the natural sciences. Such a reorientation should contribute to an improved understanding of environment-related (consumption) practices and how they are structured and determined based on external structural factors, social-economic factors, values and attitudes, as well as social interaction (imitation and social control). Moreover, environment-related social sciences need to find answers to issues related to the dynamics of changes in social practices and how people relate to these dynamics in terms of their values and attitudes. Only in this way can environmental policies be informed about how certain patterns of behaviour can be changed and how people can better be convinced of and accept certain political-structural changes, e.g. with regard to energy systems, transport systems, food production and consumption, etc.

Furthermore, there is no guarantee that the growing economic and environmental crisis can be met by technological development and by maximising efficiency, as the green growth paradigm tends to promise. There is an inherent flaw in this paradigm of continuous growth, which does not take into account the limits of resources and space. The fundamental question should be how the Earth can sustain liveable conditions for all humans and further provide the resources required to sustain the basis for life. A development path that one-sidedly follows the green growth paradigm could easily fail to meet this goal. And it is dangerous that (as is almost determined by the green growth paradigm) alternative approaches to the problem tend to be seriously under-researched.

Indicative in this regard is the often biased conception of sustainability, both in research and policy frameworks. Sufficiency one of the three main pillars of the concept is essentially neglected. Can we be sure whether the conventional growth paradigm will be able to deliver the right solutions to the crisis in time? A biased science discourse, which is partly a consequence of a biased structure of research funding, is a rather uncertain basis for informing and advising policy-making. Essential is an improved understanding of our economic system and how it is going to provide for the minimum but universal condition of a certain baseline of quality of life for all earth dwellers. This question is also related to the question of how we want to live. There are many social-culturally differentiated answers to this question, and it is a matter of fact that there is no such thing as *a* public. A better and more differentiated understanding of the aspirations, future visions, and utopias people may have will contribute considerably to finding a more equitable and more democratic solution to the current global human crisis.

2.3.1 Lifestyle as Conceptual Approach for an Improved Understanding of Differences in Human Impacts on Climate Change

Related to the question above is the question of this study that aims to shed light on the differences in social-cultural group-specific conceptions of life and how these affect the climate differently. Thereby, this study provides a new framework to measure social-cultural group-specific contributions to the climate problem. It identifies patterns of behaviour and practices and highlights those domains that are most relevant for targeted interventions. This study combines experiences from assessing personal-level GHG emissions with an approach that attempts to differentiate a population based on lifestyle-defining values and attitudes. Through a descriptive analysis of these value-based segments, the study aims to get hold of the drivers of ideal-typical patterns of behaviour. Behaviour and values are understood as closely interrelated dimensions that tend to structure one another (values structure behaviour more than the other way around). The interactive complex of values and behaviour operates within a social-economically prestructured scope of action and interpretation. The environmental effect of this three-dimensional lifestyle complex is then measured in terms of average personal GHG emissions (carbon footprints).

This approach is new, especially in its application to a context in the Global South. It is therefore a basic research and explorative in terms of conception as well as methodology. In Sect. 2.1.2, the author has shown that the understanding of consumption-based personal- and household-level GHG emission structure is very limited and almost exclusively based on economic research. Those frameworks largely analyse the structure of carbon footprints in terms of income and consumption expenditure as drivers of differences. However, there is evidence, *first*, that there are other relevant factors apart from social-economic determinants and, *second*, that an analysis of behavioural patterns and their underlying motivations and drivers allows for a much more differentiated understanding of people contributing differently to the climate problem. This understanding is fundamental for the further development of environment-related social science theory and methodology. A more differentiated understanding of consumers and their motivations could help to identify new strategies to convince consumers to engage with and on behalf of environment (e.g. on a neighbourhood level). It also provides important scientific insights to better inform environmental policies.

2.3 Relevance, Objectives, and Challenges in Linking Personal Greenhouse Gas... 41

For instance, results of the qualitative study in Hyderabad show that respondents of the highest-income segment (globals) have a footprint that ranges between 2 and 20 tonnes of CO_2e per capita per year (see Annex IX). The second highest-income group (strivers) ranges between 1.5 and 12 tonnes of CO_2e per capita per year. Girod and de Haan (2009, p. 5650) in a study in urban Switzerland deliver insights of similar significance: the annual carbon footprints of equal-income Swiss households in their study range from 5 to 17 tonnes of carbon dioxide per person (see also Sect. 2.1.2.4). What do these results show? They show that income does not determine the level of consumption-related GHG emissions. Income cannot alone serve as independent variable for analysing inequalities in carbon footprints, and its reliability in survey data tends to be questionable. Even more important is the fact that income-oriented carbon footprint estimations are often based on consumer expenditure data as a proxy for income. Consumer expenditure, however, is too directly associated with consumption-related personal carbon footprints, as has been shown in Sect. 2.1.2.4. Consumer expenditure can therefore not be used as an income proxy to explain income-driven patterns of consumption-related GHG emissions.

Moreover, assuming that income data is reliable, how well are conventional income-oriented approaches able to shed light on the actual and more direct drivers of consumption? Disposable income is a necessary condition for almost all domains of consumption, but money does not sufficiently explain consumption decisions. The rationale here is that most of the income-oriented GHG emission estimations take a neoliberal economic and rational choice-based approach of thinking. Studies assume that consumers act according to a logic that says, the more I can afford, the more I will consume. It blinds out motivations and practices that are based, e.g. on ideas of post-materialism or traditionally motivated virtues of frugality and thrift. To address consumption-related issues requires an understanding of the underlying guiding principles of behaviour, consumption, and associated GHG emissions. Already Bourdieu has characterised the relevance of the mediating role of habitus and taste in relation to income and consumption:

> *Wenn es ganz danach aussieht, als gäbe es eine direkte Beziehung zwischen Einkommen und Konsum, dann liegt das daran, dass der Geschmack fast immer aus denselben ökonomischen Bedingungen hervorgeht, in deren Rahmen er agiert, so dass sich dem Einkommen eine kausale Wirkung zuschreiben lässt, die es aber tatsächlich nur in Verbindung mit dem Habitus ausübt, der ihn hervorgebracht hat. In der Tat zeigt sich der Einfluss des Habitus deutlich, wenn denselben Einkünften verschiedene Konsumgewohnheiten entsprechen, was nur unter der Voraussetzung verständlich wird, dass andere Kriterien mitwirken.* (Bourdieu 1987, p. 590)

Bourdieu shows other factors than financial determinants that structure individual consumption. With the concept of habitus, Bourdieu delivers a new perspective on social practices and consumption, which considers economic resources but also highlights the role of cultural and social capital as well as mechanisms of social distinction. With his conception, Bourdieu was among the first scholars who conceptualised lifestyle in this comprehensive form. While Bourdieu was highly interested in highlighting mechanisms of social distinction based on consumption, it is the aim of this study to focus on identifying more general values, attitudes, and preferences, as they serve as abstract motives for certain more general patterns of behaviour and consumption. These values and attitudes are not so much based on aesthetic attitudes as Bourdieu emphasises them, but more generally on issues of everyday life and consumption. Based on this perspective, the concept of lifestyle integrates social-economic determinants, values and attitudes, as well as behavioural patterns. The author of this study has further developed the concept for analysing motivational differences in consumption patterns and related GHG emissions. In this framework, values and attitudes are conceptualised as structuring factors of behaviour and consumption, obliquely situated to income.

2.3.2 Culture, Poverty, and Stylisation: Potential and Challenges in Applying the Concept of Lifestyle to the Indian Context

India is extremely heterogeneous in terms of linguistic, religious, ethnic, social-economic, education and employment-related as well as other general culture-related differences. In *urban* India, this specificity is even more pronounced due to the high shares of well-educated, globally oriented higher social-economic segments. Especially with its colonial history and its long phase of Nehru socialism, India as it has emerged now as an economic world power brings far-reaching challenges for the application of the lifestyle concept. In addition to the heterogeneity, India also is extremely dynamic in its economic transition towards an economically oriented society, also in regard to the involved dynamics of urbanisation, environmental change, and social-cultural and social-economic changes. These features of contemporary India challenge but also invite application of a lifestyle analysis.

The rapid social-economic changes in India happening now and over the last 25 years suggest interesting analogies with the times of transition in the post-World War II period in Germany and the related emergence of the late but due reorientation of social structure analysis in the 1980s and 1990s (cf. Sect. 2.2.2.1). Especially in the urban context, the dynamics of social mobility and associated social-cultural change are significant. The proliferation of literature dealing with the new (urban) middle classes indicates at the relevance of research addressing this newly emerging heterogeneity, which is not only a result of a growing number of opportunities and increased levels of income.

Extremely relevant for lifestyle-related environmental research in the context of growing heterogeneity and rapid social mobility is the substantial share of lower social-economic segments. In terms of their overall share, in terms of absolute numbers, and in terms of the diversity and scope of aspirations, the lower social-economic segments need to be given due consideration. Increases in disposable household income allow people to raise their standard of living accordingly. Due to scale effects, even minor changes (e.g. acquisition of an air-conditioner or a motorbike) tend to have substantial GHG emission effects in absolute terms. This study will show how single steps on the "technology ladder" tend to increase the overall carbon footprint of a person or household. In terms of individual-level carbon footprints, such increases can be argued to be negligible, but if applied to broad masses of people, the scale effect is massive.

In this context, an improved understanding of the underlying motivations influencing consumption decisions is of great importance. A higher level of disposable income may enable a person or household to take a step on the "technology ladder" (e.g. from bicycle to motorbike), but it does not reflect and explain the considerations, reasons, and motivations of people in deciding for or against certain patterns of behaviour and consumption. Whether a person or household decides to take this step, and what matters most in their decision for or against it, is of major concern for this study. The complexity of the drivers and mechanisms involved creates additional challenges in terms of conceptualisation and operationalisation.

The most challenging issue, however, in applying the lifestyle concept remains poverty. Poverty does not allow people to *choose* from a larger variety of potential stylisation options, e.g. based on choices of consumption and/or investments. Poor people are restricted in their scope of action due to their limited endowment with resources – or in the words of Bourdieu (1983) "forms of capital". For instance, control over economic capital, i.e. all those resources that are directly convertible into financial means, enables people to draw on these resources and utilise them for stylisation. Economic capital is the most basic, most obvious, and most direct capital base that can be utilised in various ways, e.g. for the purchase of consumer goods, holidays, and school education or for increasing the chances of a prestigious marriage match. Bourdieu's (1983) path-breaking analysis of different forms of capital helps to understand the mechanisms behind stylisation. Different sorts of capital (economic, cultural, and social capital) can be used and translated into symbolic capital with the consequence of gaining a certain position in the social field based on a person's habitus (Bourdieu 1987, p. 175). According to Bourdieu (1983, p. 183), it is due to the differentiated distribution of different types of capital that the societal and economic interplay operates not just as a game of luck, in which everybody has the same chances and in which there is no inertia, no accumulation, and no inheritance of possessions or properties:

> *Aber die Akkumulation von Kapital, ob nun in objektivierter oder verinnerlichter Form, braucht Zeit. Dem Kapital wohnt eine Überlebenstendenz inne; es kann ebenso Profite produzieren wie sich selbst reproduzieren oder auch wachsen. Das Kapital ist eine der Objektivität der Dinge innewohnende Kraft, die dafür sorgt, daß nicht alles gleich möglich oder gleich unmöglich ist.* (Bourdieu 1983, p. 183)

This quote says that it takes time to accumulate capital, but capital also has a reproductive tendency, it can produce profits (e.g. in form of other capital) and reproduce itself, and it can grow. The individual endowment with different capitals involves a force that is immanent in all objectivities (Bourdieu 1983, p. 183). This force deeply structures and predetermines the individual scope of action, which therefore is a result of the individual level of capital endowment in interaction with the external structure (see section above 2.2.2.4). The less a person is endowed with capital, the smaller is the individual scope of action. And as lifestyle operates only within this limited scope of action, it is the poor who tend to have a very limited space for stylisation (cf. Bourdieu 1987, p. 594). Bourdieu explains that a "stylisation of living" can only unfold under conditions of a certain level of freedom of choice, i.e. an absence of or distance to objective and subjective material and time constraints. Living conditions that are significantly affected by such constraints (e.g. financial) cannot provide for enough of the space in which a relaxed and indifferent attitude towards consumption can unfold (Bourdieu 1987, p. 591).

With this observation in mind, the question may arise as to whether the lifestyle concept can really fulfil its objectives, namely, to provide a framework to analyse social-cultural differences across all social classes (cf. Dangschat 1996, p. 99f). This question is even more relevant in India, i.e. in a context in which the lower social segments still represent the majority of the society even in an urban setting and where the financial situation of these segments leaves no objective scope of action.

The author of this study argues, however, that even though the poor's scope of action may be very limited, there remains some space for evaluation and self-reflection, i.e. constructing a subjective picture of one's own situation and related practices. This subjective (e-)valuation of practices tends to be quite optimistic and positive and is framed as a conscious choice for the necessary. Bourdieu argues that the lower classes' patterns of behaviour can only seemingly be deduced from the objective conditions; rather they should be understood as resulting from the *decision* taken by members of the lower classes to adopt practices and consumption patterns of necessity ("Entscheidung für das Notwendige", Bourdieu 1987, p. 594). The lower classes opt for all those things and practices that are technically necessary, "practical", or functional:

> [...] was aus ökonomischem und sozialem Zwang die 'einfachen' und 'bescheidenen' Leute zu einem 'einfachen' und 'bescheidenen' Geschmack verurteilt. (Bourdieu 1987, p. 594, original emphasis)

To conclude, basic attitudes and values always mediate between objective conditions (e.g. financial resources) and actual social practices. Values are informed by and adapted to the objective boundary conditions, just in the same way as practices are. And these so formed value-orientation patterns provide a more or less consistent framework that allows for a subjectively positive (e-)valuation of "chosen" patterns of behaviour in differentiation to other unattainable practices. That means that economically driven "consumption of necessity" is subjectively a matter of choice and taste ("Geschmack am Notwendigen", Bourdieu 1987, p. 587) and the objective conditions play a lesser role in the evaluation.

And even with very limited economic resources, just like everyone else, poor people are bound to develop certain values, attitudes, and preferences in distinction or reference to others. This process of identification with certain forms of (e-)valuation in differentiation to other forms of value-orientation patterns takes place within a broader social field, which dynamically interacts with a continuously changing set of external, contextual determinants. Social-economic factors, the endowment with social and cultural capital (as Bourdieu understands it), external and contextual boundary conditions, all these factors prestructure the scope of action and (e-)valuation for all social segments. This structure affects both values and attitudes in interaction with behavioural and consumption patterns.

For the purpose of environmental and climate-related social science research and especially in the context of India, it is highly productive to get hold of these fundamental determinants of everyday behaviour. A restriction in the application of lifestyle to higher social-economic segments of the (urban) society would compromise the objective of the research to arrive at a more differentiated view of the social structure across all social classes. Especially in India, the lower social segments are extremely relevant in gaining an understanding of social-cultural-specific general motives and patterns of self-(e-)valuation. With the economic and income dynamics (esp. income mobility), it is essential to develop an understanding of the subjective principles of consumption in order to be better able to project the impacts of and potential political approaches to economically driven shifts in the overall social structure.

The author of this study therefore aims to arrive at a conceptualisation that allows for an application of lifestyle to urban contexts and which is able to also meaningfully include lower social segments under conditions of social-economic constraints. For this purpose, the author emphasises on general social values and attitudes and how these structure differences in everyday patterns of behaviour and consumption. Values are social-culturally specific, and they indicate how people evaluate existing patterns of behaviour and consumption and by how far they identify themselves with or distinguish themselves from it. By analytically integrating all three dimensions, values, behavioural patterns, and social-economic determinants within one concept of lifestyle, this study is able to shed light on the social-cultural-specific interactions between all three dimensions across all social segments. The focus is put on the interactions between basic value orientations, motivations, and general social-cultural-specific forms of everyday life.

With this approach to lifestyle, this study is able to analyse a multiplicity of lifestyle-related issues in a society or an urban population, such as health-related issues, issues of quality of life, affectedness to environment-related problems, education and awareness, as well as social-cultural-specific differences in personal GHG emission levels.

After defining the research questions in the following section, the author will introduce the research context, in which he will also highlight relevant drivers and boundary conditions of social-cultural-specific behaviour and (e-)valuation.

2.4 Definition and Specification of Research Questions

The first research question of this study deals with the problem of analysing personal-level carbon footprints based on consumption. Consumption data in this research framework refers to individual-level consumption of goods and services, which are characteristic in their overall GHG emission-based contribution to the greenhouse gas effect. A carbon footprint is the measured amount of individual level GHG emissions estimated in tonnes of CO_2 equivalents over a year of time. It is calculated in its sectoral structure, i.e. domains of everyday consumption, e.g. diet of meat and dairy products or mobility-related emissions. The first question reads as follows:

1. What is the sector-specific structure of personal GHG emissions being emitted in Greater Hyderabad?

The second question raises the issue of social-economic differences in personal-level carbon footprints. It is hypothesised that with higher levels of disposable income, an individual or household is able to consume greater volumes of consumer goods and services. As an effect, carbon footprints increase due to higher levels of consumption-based GHG emissions.

2. What is the influence of income on carbon footprints?	(a) Do higher levels of income lead to higher levels of personal GHG emissions?
	(b) Which domains of consumption are more affected by income variations?

Apart from income, there are other driving factors that structure consumption decisions. This study aims to shed light on these factors, but it also seeks to contribute to a better understanding of the structure of personal- and household-based GHG emissions. It moreover aims to explore and develop new methodological approaches to analyse consumption-related emissions that may become easy-to-handle tools for applied studies in this field. A focus on consumption practices and their associated social-technical systems is such an approach, which has been developed by the author. A sector-specific analysis (see above) does not allow for conclusions about specific lifestyle practices that are most relevant for climate change mitigation. A social practice approach is much better able to identify relevant key points of intervention. This theme is addressed through the third research question, which reads:

3. What are the consumption practice-related key points of intervention?	(a) What is the average contribution of specific consumption practices to the overall personal GHG emission balance of a person?
	(b) How big is the overall share of people following each of these consumption practices?
	(c) Based on the results about the contribution and share of users, which of the everyday consumption practices are most relevant for an intervention?

2.4 Definition and Specification of Research Questions

The author emanates that not all values affect behavioural and consumption patterns in a linear manner. The lifestyle concept focuses on the configurations of values and attitudes that allow delineate ideal-typical patterns of value orientations. It highlights the structural non-linear effects of value orientations on behaviour and consumption. The study therefore attempts to reveal lifestyle-specific structures or patterns of ideal-typical value orientations that build the foundation of a particular lifestyle. This basic research pursues an improved understanding on whether and how certain structural configurations of values and attitudes structure behaviour and consumption patterns. These revealed patterns represent ideal types of lifestyles as a basis for an analysis of lifestyle-specific personal GHG emissions. The fourth research question therefore is:

1. Does the concept of lifestyle contribute to an improved understanding of differences in personal carbon footprints?	(a) Does the concept of lifestyle that draws on ideal-typical patterns of value orientation apply to the urban Indian context?
	(b) Does this concept explain group-specific differences in behavioural and consumption patterns and related personal carbon footprints?

Before laying out the conceptualisation of the research framework of this study (Chap. 4), the author will contextualise this study in the next chapter. This contextualisation builds on the author's empirical research and serves for an understanding of the meso- and meta-level mechanisms of lifestyle. It will also dwell on the most critical contextual issues and challenges that the author had to face concerning transferring the concept of lifestyle to the Indian context.

Chapter 3
The Research Context: India and the Megacity of Hyderabad

Keywords Hyderabad · Megacity · Urbanisation · New economic policy · Poverty · Measuring poverty · Social mobility · New middle classes · Climate change in India

India has been facing rapid transformation processes for about the last two and a half decades. One of the most critical manifestations and drivers of these changes is economic growth, which is a consequence of external (globalisation) and internal factors (liberalisation and economic reform since the early 1990s) – both interrelated. Along with economic development, there are other dimensions of social and environmental change, such as rapid urbanisation, rising incomes, a maturing young workforce, emergence of a new and rapidly growing middle class, and associated social-cultural changes. These apparently positive developments involve a multiplicity of interrelated processes and mechanisms that directly and indirectly create social, ecological, and political risks and challenges. For instance, the social effects of economic development have to be looked at from two sides. On the one hand, the last two decades of growth have led to an escalator effect, which has lifted large parts of the poor out of poverty. However, on the other hand, social disparities have grown substantially – between different regions and between urban and rural areas. Moreover, the fast character of change and the lack of effective institutions have led to tremendous social and environmental risks and challenges. And, as a form of "umbrella effect", most of these changes take place in urban areas in a concentrated manner, with the consequence of powerful feedback mechanisms that further drive urbanisation and serve as a pull factor for rural-to-urban migration. Hyderabad is taken as a representative example for many other megacities in India. This city defines the geographic frame for an analysis of processes that transcend physical-spatial boundaries by far. Such concentration, however, allows that the manifest locational aspects anchoring these processes on the ground are not lost sight of in the analysis.

3.1 Economic Development and Dynamics of Urbanisation in India

3.1.1 Liberalisation Politics, Trends of Economic Development, and Future Visions of (Urban) Development

One of the most significant boundary conditions of India's social-economic and social-cultural development of the last two and a half decades can be seen in economic liberalisation policy, mainly initiated by P. V. Narasimha Rao taking office in June 1991. At that time, the new government had to deal with the conditions of a nearly bankrupt economy, with a massive current account deficit, imbalances in foreign exchange, and a largely inefficient public sector based industry (Rieger 1995, p. 523). International pressure, mainly led by the World Bank and the International Monetary Fund (IMF) and the absence of other options urged the relatively weak governing majority to drive forward massive and unprecedented reforms of the economic order – from a largely government-dominated to a market-oriented system. The reform, which was mainly conceptualised by Manmohan Singh in his role as finance minister, involved a new industrial policy, a partial withdrawal of the government from paternalising the economy, and a stepwise liberalisation of the market in order to incentivise foreign direct investments (Rieger 1995, p. 524).

With the success of the initial reform years and even with some setbacks, the following years involved further reorganisational steps with the consequence of a more or less stable growth rate over the last 25 years. As an effect, after a slowdown in growth rates in 2009 due to the global economic recession and another downward shift in 2012, India has recovered and gained new impetus. With a growth rate of 7.3% in the fiscal period 2014/2015 and even higher estimates for 2015/2016, India for the first time outpaced China (World Bank 2016b: 142). Today, with a gross domestic product (GDP) at market prices (current US$) of $2.049 trillion, India is the No. 4 economy in the world today (World Bank 2014). Soon, the country will climb up the global "economic ladder" surpassing Japan in the next year and the Euro area in about 20 years (Johannson et al. 2012, p. 22).

However, in spite of the vast growth and rapid urbanisation rates, more than half of the employment is still based on agriculture and its allied sectors (forestry and fishing) (GoI 2014a). This is while the agricultural sector's output is relatively low, accounting for around 18% of the GDP in 2014 (World Bank 2016a).

Other than in most other emerging economies, economic growth in India has not led from agriculture to industries but to an expansion of the service sector. The service sector gives employment to around 25% of the workforce (GoI 2014a), which is argued to be very low relative to other economies (Mukherjee 2013, p. 1). However, in terms of economic output, services account for the largest share of the overall economy, at 52% in 2014 (World Bank 2016a). Moreover, services in India depict the highest labour productivity, and in terms of services' exports and imports, India ranges among the top ten WTO members in international trade (Mukherjee

2013, p. 16). In comparison, industrial development has not kept pace with services development. In terms of employment, it ranges around 20% (GoI 2014a), and in respect to economic output, industries account for just 30% (compared to more than 42% in China) (World Bank 2016a).

To sum up, agriculture still makes out to be the most important base for livelihood in India, where a share of almost 70% of the population still lives in rural areas (GoI 2011). This substantial share of rural population can be termed as largely disadvantaged against the trends of development in urban areas. The share of agriculture in accounting for the overall GDP remains low, and its trend is continuously falling. Lack of capital goods and financial resources, fragmented small-scale acreage, poor quality or lack of infrastructure for efficient and fast supply, and stagnating crop yields are very relevant issues that call for political intervention (Bronger 1996, p. 150). Moreover, in terms of the living conditions, disadvantages substantially contribute to the push factors in rural areas that yearly drive millions of people from rural to urban areas. For instance, in 2013, more than one fourth of all rural dwellers still had no access to electricity, i.e. 237 million people (IEA 2015). And in 2012, based on the Global Hunger Index, India was even ranked behind Sub-Saharan Africa (Von Grebmer et al. 2012, p. 12).

Also with regard to the service sector, there is huge potential for progress. Observers argue that its share in providing employment in terms of number as well as quality still lags behind its actual possibilities (Mukherjee 2013, p. 16). Also in respect to access to different services, there are great disparities – socially and regionally – especially basic services such as healthcare, electricity, education, and water and sanitation (Mukherjee 2013, p. 16). Much hope is also placed in industrial development in order to create new jobs in this sector.

This strategic realm has also been taken up by the new BJP government under Narendra Modi, who has taken office in Mai 2014. For instance, his industrial policy initiative "Make in India" has raised quite some international attention (Betz et al. 2013, p. 3). In this initiative, Modi builds on the country's advantages of a huge and still growing young labour force and the large domestic market in competing with China (cf. Thite 2014, p. 290). The programme aims to ease and enhance the conditions for doing business and thereby incentivise foreign direct investment in India. It foresees investments into large infrastructure projects such as development of industrial corridors that connect important economic hubs (e.g. Delhi-Mumbai Industrial Corridor). It also involves a programme on developing selected smart cities (part of industrial corridors programme) (Ganesan 2014; GoI 2014b). Moreover, large investments are being planned and made in terms of transport system development in railways, aviation, and shipping.

According to some observers, however, the rather newly framed discourse around "smart" development is quite as much an envisioning or a "seductive projection" that aims to frame and shape urban and industrial policy (Bunnell and Das 2010, p. 277). It is argued that it follows a language of technology-led utopian "imaginings" with terms such as "leapfrogging", "smart", and "intelligent" (Bunnell and Das 2010, p. 281). This language is underlined with well-designed statistical figures, pictures, and digital simulations "to visualise the 'multimedia utopia'"

(Bunnell and Das 2010, p. 281, emphasis in original). In respect to the language and representation of policies and government-led programmes, a lot has changed since Modi took office. Newly designed government webpages mirror the recently invented digital marketing campaign. The prime minister's highly "data-driven" (Fraser 2015) and professionalised use of social media such as Facebook and Twitter is successful in terms of demonstrating "connectedness". With issues such as sanitation and hygiene, growth, digitisation, and technology, it aims directly to reach out to an increasingly relevant and growing share of young and educated people, mostly representing the new middle classes (BBC 2015; Fraser 2015).

This language and imagination of a new modernity tends to mask bottleneck issues such as local (Datta 2015, p. 14) or national (Sherwell 2015) resistance, and/or technological and finance-related challenges, especially with respect to the smart city programme. Certainly, Modi has achieved a broad-based political backing, especially among the Hindu population and from the corporate sector. But the success of many large-scale infrastructure projects depends on a variety of factors and boundary conditions.

3.1.2 Urbanisation in India: Cities as Foci of Diversity and Lifestyle

3.1.2.1 Urbanisation as Central Aspect of Global Change

The process of urbanisation is one of the most remarkable issues of global change. The year 2007 marks a silent turning point in human history, when for the first time more people lived in urban areas than in rural areas. In 2014, the world's urban population has increased to a share of 54%, and urban growth is expected to continue with estimates saying that by 2050, around two thirds of the world's population will be living in urban areas (United Nations 2014a, p. 7). Urbanisation in most countries of the Global North has already reached quite high levels, e.g. in Europe (73%) and the USA (82%) (United Nations 2014b). Latin America and the Caribbean also account for very high levels of urbanisation (80%). In many countries of the Global South, especially most countries in Sub-Saharan Africa and Asia, the level of urbanisation is still very low, while the rate of growth is remarkably high here. Almost 90% of the urban population growth in the coming 35 years will take place in Asia and Africa. In absolute terms, the world's urban population has grown from around 700,000 in 1950 to close to 3.9 billion in 2014 and is expected to reach 6.3 billion in 2050 (United Nations 2014a, p. 12). The UN projections further estimate that the urban population in Africa will triple and in Asia it will grow by 61% by 2050. As a result, most of the world's urban population will then be concentrated in Asia (52%) and Africa (21%). China, India, and Nigeria alone will account for around 37% of the future global growth in urban population between 2014 and 2050 (United Nations 2014a, p. 12).

3.1 Economic Development and Dynamics of Urbanisation in India

A quite critical issue of the world's urbanisation history of the last two centuries is seen in the unprecedented concentration of people in urban agglomerations. The largest and most concentrated agglomerations are known as *megacities*, with populations – depending on definitions – of above five million (e.g. Kraas 2007, p. 79), more than eight million (e.g. Fuchs et al. 1994, p. 1), or more than ten million inhabitants (e.g. United Nations 2014a, p. 78). In this study, a threshold of five million inhabitants has been decided upon in order to take into account all those cities, which have recently emerged as megacities and which often grow more rapidly than larger megacities. These are especially relevant in the Global South and in transitional countries (Kraas 2007, p. 82). In 2014, about 10% (758 million people) of the global population lived in only 51 megacities (9 in India) with a size of more than five million inhabitants (United Nations 2014a, p. 78). In 2030, there will be more than 100 cities with a population above the threshold of five million people (Fig. 3.1). However, it is not the megacities that grow at the fastest pace but the medium-sized cities or cities with less than one million inhabitants. Most of these fastest-growing medium-sized cities are located in Asia or Africa, and a very large share of them are found in China alone (United Nations 2014a, p. 20).

Concerning quantitative definitions of megacities, the given minimum/maximum thresholds are bound to be subjective and invite debate. In the end, all quantitative data involve such definitional problems and in addition bear the risk of statistical and reporting problems. Therefore, the given trends and data have to be seen in this light and taken with proper caution (Kraas 2007, p. 82; Kraas and Nitschke 2006, p. 19). Frauke Kraas therefore suggests a "more qualitative, process-oriented

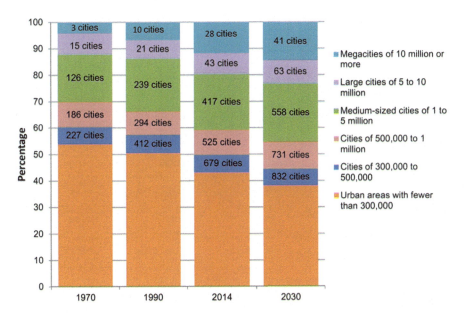

Fig. 3.1 Distribution of the world's urban population by size class of urban settlement and number of cities, 1970, 1990, 2014, and 2030. (Source: United Nations 2014a, p. 17)

perception and a more comprehensive understanding of megacities as functional mega-urban regions" (Kraas 2007, p. 82). However, it is worth taking a specific look at the urbanisation trends and data for India.

3.1.2.2 Urbanisation Dynamics in India

According to UNPD and the 2011 Census, the level of urbanisation in India is still very low at 32.4% (GoI 2011; United Nations 2014b). This rate is still below the average urbanisation level of Asia and still below that of Sub-Saharan Africa. The average rate of annual growth of the urban population in India has been quite modest since independence. From a reasonably high rate in the 1950s, it fell sharply over the 1960s and reached a peak in the 1970s (Kundu 2014, p. 197). Over the last two decades (1991–2011), the annual growth rate has ranged between 2.73% and 2.76%. According to the UNPD projections (2014 revision), the Indian urban population grows exponentially, while the growth of the rural population slowly decreases until it comes to a point of nearly zero growth (Fig. 3.2). Figure 3.2 also indicates the growth of the urban population in absolute terms. In only two and a half decades (1990–2015), the urban population in India has increased by almost 200 million people – this number exceeds the total population of Western Europe. By 2030, another 160 million people are expected to add to the existing share of the urban population, which will then reach to a total number of more than 580 million people (Fig. 3.2). Today, one out of ten city dwellers of the world lives in India. In 2050, urban India will account for about 14% of the world's urban population (more than 800 million people), and the majority of Indians will be living in urban areas.

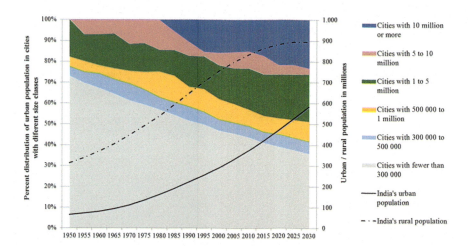

Fig. 3.2 Urban population percent distribution in cities of different size classes and absolute distribution of population between urban and rural areas over time between 1950 and 2030. (Source: Own draft based on United Nations 2014b)

It is worth looking at the patterns of urbanisation across different size classes of urban areas. A commonly used size classification of urban centres and towns in India is based on six classes: at its lowest end are Class VI towns with populations of below 5,000 inhabitants and its highest order level are Class I towns and cities of above 100,000 inhabitants. Based on this low-threshold classification, the largest population increases over the past several decades happened in Class I cities. The share of the overall urban population residing in these largest cities has reached 70.2% in 2011 (from 26% in 1901). According to Kundu (2014, p. 201), the reasons for this shift are not based on a faster pace of growth of these larger cities. Rather it is the graduation of smaller towns and cities into the highest category with a resulting rise in the number of these cities. Given the fact of the existence and high relevance of megacities in India, a more differentiated analysis of the growth patterns of larger cities is required.

The UNPD provides a more suitable classification, depicted in Fig. 3.2. It subsumes all towns and cities of below 300,000 inhabitants into the smallest size category. All larger cities are classified into five higher-order classes with an uppermost threshold of ten million, which is the actual UN megacity minimum threshold. Figure 3.2 illustrates the population distribution among different city sizes over time. It shows that the share of the overall urban population steadily shifts away from the smaller class cities towards larger cities and megacities in particular. In 2014, for the first time, more than a quarter of the urban population in India resided in megacities of above 5 million inhabitants; this is about 100 million people altogether. The overall population share of these five million-plus cities is projected to remain at this one-quarter level. However, by 2030, three more cities will reach the ten million threshold and thereby add to the share of this largest size class. In 2015, there were four of these largest cities contributing to the urban population with a share of about 17% (71 million inhabitants; Delhi, Mumbai, Kolkata, and Bangalore).[1] In 2030, it will be seven ten-million-plus cities with a share of 23% (135 million inhabitants; Delhi, Mumbai, Kolkata, Bangalore, Chennai, Hyderabad, Ahmadabad) (United Nations 2014a, p. 98).

Kundu (2014, p. 227) and other scholars underline the fact that urbanisation in India has been "top-heavy" with an orientation towards large cities. Population growth can be traced back to natural increases as well as higher net in-migration. Some authors argue that this tendency of concentration points towards the phenomenon of city primacy (Butsch 2011, p. 21; Stang 2002, p. 120). In terms of infrastructure, global integration, better healthcare, education, and employment, these megacities are outstanding with very central functions for their particular regions. Even more important is the economic relevance of these cities, making them highly efficient in generating growth and attracting investments. All these aspects are serious pull factors for migration. Also, New Economic Policy (NEP) has contributed to this trend of concentration. In a globalised context, national as well as global investments are more likely to concentrate in more developed states, regions, and cities. This in return also makes it easier for local urban bodies to initiate and invest

---

[1] At this time, Chennai ranged just marginally below the threshold of ten million inhabitants.

into public works and infrastructure development projects (Kundu 2014, p. 227). The tendency of decentralisation and the strengthening of local urban bodies over the last two and a half decades have also substantially contributed to this shift. Kundu (2014, p. 227) argues that "the resulting decline in central and state financial assistance has led to an exacerbation of inequity in the provision of basic services among states and among size categories of urban centres".

3.1.2.3 General Implications, Opportunities, and Challenges of Rapid Urbanisation Processes

The above analysis remains incomplete without an understanding of the meanings and implications that these dynamics have for the country, the regional development, and the conditions in and around the urban areas. Since the Neolithic revolution, processes of urbanisation have been a consequence as well as a driver of human development. Cities and urban areas have always served as hubs for trade, exchange of knowledge and services, innovation, creativity, and cultural development. And cities are likely to provide the critical link between the development of rural areas and the larger global economy (see also Kraas and Nitschke 2006, 21; Sánchez-Rodríguez et al. 2005, 12; World Bank 2009). Cities work as nodes and hubs under processes of globalisation and they often serve as nuclei of societal change and social-cultural innovations. The large majority of the world's future population will reside in urban areas, and this is one of the main reasons why it is of such importance to search for solutions for humanity's most urgent problems in the context of cities and urban areas.

Furthermore, urbanisation processes in the Global South tend to be fast, often outpacing adequate governance and institutional responses. Therefore, many of the development processes are not well or effectively regulated in order to steer development adequately (slums, growth of informal sector, emergence of unauthorised or badly planned areas and neighbourhoods). In consequence, there are processes involved that create and accumulate largely unconsidered and unaddressed risks and challenges.

These risks and challenges have a social dimension, as vulnerability levels increase and social disparities in regard to education, health, housing, access to basic infrastructure, and quality of life grow considerably. Closely related to this social dimension is the political dimension. As formal institutions fail to address the complexity of problems, informal mechanisms and institutions emerge that fill the social disorganisation gap. Informality however bears the risk of high transaction costs and high levels of institutional insecurity for large portions of the society. These problems of governance and the dynamics of change are likely to also shift the priorities and responsiveness away from issues of sustainability that tend to have a rather long time horizon. Environmental issues are very characteristic in this realm. Urban areas spatially concentrate a multiplicity of human activities with a concentration of related environmental impacts. Some of these impacts are localised within the boundaries of the city, while others partly transcend the boundaries across scales up to the global level, as, for instance, impacts related to the release of GHG emissions into the atmosphere. Even if some of these impacts directly lead to problems related to health and quality of life (e.g. air, water, waste, and noise pollution), often there remains the problem of attribution, and in most cases environmental issues are highly complex and not intuitively understood (e.g. plastic waste and the associated risks of vector-borne diseases). One of most complex issues relates to the causes and impacts of climate change, such as the release of GHG emissions: a local process leads to the release of GHG emissions and contributes to the accumulation of GHGs in the global atmosphere. Largely decoupled from the local process, anthropogenic climate change indirectly translates and feeds back into the urban system. And this feedback may be as "direct" as a local manifestation of a globally working climate-based mechanism (climate-change impacts), or it may work even more indirectly in the form of a political response (e.g. carbon tax).

Also with regard to these issues of sustainability and the need of a great transformation towards zero-carbon emissions, cities play a crucial role. Urban planning in this context offers huge potential for effectively reaching out to a large and growing share of population, bringing forward improved sustainability and quality of life. As cities concentrate population, consumption, infrastructure, and economic activities, so are the causes for the release of GHG emissions concentrated here. This fact and the related economies of scale offer huge potential for the establishment of sustainability strategies (Butsch 2011, p. 12). Therefore, cities and megacities can be understood as "priority areas" and "drivers of change" (Kraas and Mertins 2014, p. 4).

In this context, Satterthwaite (2003, p. 74) highlights the fact that poverty has led to the still very low levels of resource consumption in most cities of the Global South. At the same time however, given the sheer number of the urban poor, slight upward changes in the income situation of the large lower economic segment of the urban population are likely to have vast effects on levels of resource consumption and environmental degradation. Against the fact that large shares of the future urban infrastructure still needs to be built, it is duly important to develop sustainability strategies and plan for the projected demand in sectors such as transport, electricity, food, and housing. To regulate and direct this social-economic development and to effectively incentivise and stimulate social change towards sustainability, it is highly relevant to achieve a founded understanding of these future consumers, their practices, and potential demands. Hence, lifestyle research in cities of the Global South offers one important approach to better tackle climate change and environmental degradation.

Alongside with the remarkable chances for a more sustainable urban development (Kraas and Nitschke 2006: 21), rapid urbanisation also carries challenges and risks of many uncontrolled and unwanted side effects that are today found in many larger cities, particularly in the Global South. Often, governance and institutions are largely ineffective, and the implementation of rules lacks rigour and is challenged by serious problems of corruption across all levels. Moreover, responsibilities among the political and administrative actors are fragmented, and largely cross-sectoral problems are therefore often treated within rather artificially maintained sectoral administrative approaches. For instance, road infrastructure planning in many cities still lacks an integration of all modes of transport into the planning process (GIZ 2015, p. 2). Often, non-motorised transport (NMT), such as walking, cycling, and the use of cycle-rickshaws is not given adequate consideration. In many cases, facilities for pedestrians and cyclists are not planned for, or existing infrastructure is removed in order to improve the flow of the motorised traffic and to make way for construction of bridges, flyovers, and subways (Tiwari and Jain 2013, p. 45). With a share of between 40% and 50% of the overall modal split, NMT is still the most important mode of transport in megacities in India (Tiwari and Jain 2013, p. 2). However, the relevance is receding as incomes rise and conditions for NMT increasingly get worse and more dangerous (GIZ 2015, p. 1).

The example of the transport sector is illustrative of the importance of contextual factors, such as infrastructure and urban planning. The way cities function and how they are planned in regard to transport and municipal services and the way cities are outfitted with features such as recreational facilities and parks, shading greenery in streets, shopping, and other facilities are crucial factors for lifestyles. Infrastructure lays out the foundation for the choices people may have in regard to arranging their daily routines, such as commuting, shopping, recreation, leisure, and room comfort. And infrastructure and housing are often planned for based on longer time horizons. Therefore, planning decisions being made today have critical long-term implications and create path dependencies as, e.g. in the case of road infrastructure planning. And most relevant in this regard is the fact that with the low level of urbanisation in India today and the expected pace of urbanisation in the future, obviously largest parts of the urban areas, infrastructure, and housing are still to be built. Urban planning therefore needs to build on standards that work across sector boundaries and incentivise sustainable lifestyles in the widest sense of the term by a balanced consideration of efficiency *as well as* sufficiency and consistency.

3.2 Poverty and Projected Dynamics of Social Mobility

3.2.1 *Poverty and the Policy-Statistics Interface*

In the foregoing chapters, dynamics of economic development and urbanisation have been outlined. The following chapter will draw on the status of social inequality, current and projected levels of poverty, and the projected dynamics of social mobility.

With New Economic Policy in India, lots of hope were put in the associated effects concerning poverty reduction as trickle-down effect from faster economic growth. And in fact, both national and World Bank estimates of poverty in India show how the country has made impressive progress in reducing poverty since early 1990s. National estimates of poverty in India are regularly published by the National Planning Commission based on large sample surveys conducted by the National Sample Survey Office (NSSO) of the Ministry of Statistics on a quinquennial basis. The estimates are solely based on household consumer expenditure data. Income is not accounted for as the NSSO surveys do not measure household income (cf. Datt et al. 2016, p. 7f; Rutstein and Johnson 2004, p. 2ff). The calculations are based on

3.2 Poverty and Projected Dynamics of Social Mobility

Table 3.1 Comparative trends in population below poverty line, with 2005 and 2015 PPP revisions

	World Bank estimates: population below global poverty line		National estimates: population below poverty line		
	$1.25 a day (2005 PPP)	$1.90 a day (2011 PPP)	Based on Tendulkar method		
	Total	Total	Total	Rural	Urban
1993	–	46.1	–	–	–
1994	49.4	–	45.3	50.1	31.8
2004	–	38.4	–	–	–
2005	41.6	–	37.2	41.8	25.7
2009	–	31.4	–	–	–
2010	32.7	–	–	–	–
2011	23.6	21.3	21.9	25.7	13.7

Source: compiled from GoI (2013), p. 3, and World Bank (2016b)

interstate price differentials and a state-specific poverty line for rural and urban areas. For a more detailed overview of the method, see GoI (2013), and for a comprehensive critique, see Deaton and Kozel (2005) and Ferreira et al. (2015). Table 3.1 gives an overview of the share of population living below the national poverty line in total and disaggregated into rural and urban areas.

Apart from this national estimate, an international approach to measuring global poverty levels has been taken up by the World Bank. Based on a conversion of the world's poorest countries' poverty lines into a common currency, the World Bank constructed a single global poverty line. This benchmark aims to measure "extreme poverty" (World Bank 2015b) in all countries by the same standard and is made to reflect a person's minimum nutritional, clothing, and shelter needs in the respective country. The conversion is based on exchange rates in purchasing power parity (PPP) to ensure that the same quantity of goods and services are priced equivalently across countries (World Bank 2015b). Based on recurrent surveys on global price levels (PPP), the $1 a day poverty line from the first round in 1990 was revised three times, in 1993 to $1.08 a day (at 1993 PPP prices), in 2005 to $1.25 a day (at 2005 PPP prices), and in 2015 to $1.90 a day (at 2011 PPP prices) (Ferreira et al. 2015, p. 3). However, the poverty lines have been revised only because of changes in the relative price levels, not in response to economic development and related overall improvements in living standards in many parts of the world. The World Bank states that they have "sought to keep the definition of the line unchanged, and its new value as close as possible to that of the $1.25 line (in 2005 PPPs) in real terms" (Ferreira et al. 2015, p. 3f). In this sense, the new benchmark for the minimum level of well-being follows the same definition as the one dollar a day benchmark set two-and-a-half decades earlier in 1990. For each of the revised PPPs, the World Bank has backcasted the estimates for previous years in consideration of the adjusted prices.

For India, the World Bank poverty line is based on the same data as the national poverty line – consumer expenditure data from large NSSO sample surveys. Table 3.1 compares the two more recent revisions of the World Bank estimates for India with the Indian national estimates beginning in 1993 up to 2011. In each of the three backcasting cases, the same methodology has been applied to adjust the poverty lines to the respective price levels. All three figures show a significant downward trend indicating that extreme poverty levels have been more than halved over a time of less than two decades.

Estimating social inequality and levels of poverty is a highly political issue, and for the case of India alone, debates about definitions of poverty lines and methodology have been going on for many decades. Deaton and Kozel (2005) have provided an excellent review of the ongoing debate. They have shown how closely politics and statistics interact in this mainly domestic debate, and they highlight the considerable weaknesses of estimating poverty based on consumer expenditure data over time. Especially in the case of India, changes in the questionnaire design, e.g. in regard to the length of the reporting period ranging from 7 days over 30 days up to 365 days, have led to debatable inconsistencies regarding data quality issues (Deaton and Kozel 2005, p. 183; Ferreira et al. 2015, p. 14). Moreover, it is problematic to mix income data with consumer expenditure data to arrive at a common basis for well-being, especially for building a common international poverty line as followed by World Bank. As Ferreira et al. (2015, p. 12) state, measures of income and consumption are "neither conceptually nor empirically comparable measures of welfare. Conceptually, income is usually described as defining the opportunity set, while consumption defines realised outcomes" (Ferreira et al. 2015, p. 12).

Following a similar methodology, but providing for a more in-depth assessment, a recent World Bank study (2016a) has made an attempt to analyse the effects of long-term economic development over the last 60 years in India. The study puts a special focus on the post-1991 reforms and their effects on mitigating poverty. Just as the described above assessments, the study also builds on data from 51 NSSO household surveys (3rd round, 1951, up to 68th round, 2011/2012). The authors show that poverty in India follows a downward trend since 1970. This trend has accelerated in the post-1991 era. This faster post-reform reduction of poverty is shown to be even more significant in rural than in urban areas[2] (Datt et al. 2016, p. 28). And as an effect, a convergence of rural and urban poverty was observed with the share of the urban poor having significantly increased. Today, one in three of the poor live in urban areas compared to one in eight in the early 1950s (Datt et al. 2016, p. 48). In spite of the decline in absolute poverty levels, the authors admit a significant rise in levels of inequality. Much of the rise in inequality is primarily driven by growing inequalities in urban areas and especially by the increasing gap between urban and rural areas (Datt et al. 2016, p. 28; Motiram and Vakulabharanam 2012, p. 50).

The above-given overview of classic approaches to measuring poverty raises the question of how meaningful and significant these purely financial assessments actu-

---

[2] Definition of urban and rural was based on NSS standards following the India Census definition of urban areas. It includes all places with a municipality, corporation, cantonment board or notified town area committee, and places that meet a number of criteria including a population greater than 5000, a density not less than 400 persons per sq. km. and three fourths of the male workers engaged in nonagricultural pursuits as well as certain pronounced urban characteristics (Datt et al. 2016: 9f; Kundu 2014: 543).

ally are. This is even more at question against the background of the above-highlighted weaknesses and discrepancies in the methods that makes comparisons over time and between regions problematic. The authors of the study cited above admit that one-dimensional approaches lack scope and that there are various other dimensions of well-being that are worth considering, especially in regard to evaluating the social effects of the NEP (Datt et al. 2016, p. 7f). Studies such as the one issued by Datt et al. (2016) are insightful and valuable, as they deliver broad and nationally representative trends on poverty and levels of inequality (Motiram and Vakulabharanam 2012, p. 47). Such studies exemplify classic approaches to poverty which are based on the assumption that poverty is a function mainly of income and consumption.

Over the last two or three decades, the discussion on poverty has increasingly begun to recognise the multidimensional character of poverty that involves various sources of deprivation. According to this perspective, a set of multiple factors challenge and hinder the poor in trying to improve their overall livelihood situation and well-being. And these impediments and deprivations are further linked to the way people live and work, to the level of access to resources and infrastructure, and to the extent to which poor people can raise their voices politically and organise themselves collectively. Poverty therefore involves a multiplicity of mechanisms that work at the same time in creating inequalities and leading to social segregation and exclusion. Baud and colleagues call this complex of structural determinants "collective structures of constraint" (Baud et al. 2008, p. 1385f). These closely interacting mechanisms become apparent in many realms of everyday life: deprivation in one area may work as a determinant of deprivation in another area, e.g. insufficient sanitation may cause health problems that again cause problems in insecure employment relations and considerable income loss. Along the chain of these exemplary determinants, there are feedback mechanisms at work, and the factors have much broader implications for the whole household (Baud et al. 2008, p. 1386).

Meanwhile, there are a number of new initiatives and methodological approaches that aim to shed light on the diverse character of poverty and its underlying determinants. These approaches, e.g. the livelihood approach and Amartya Sen's (1999) capabilities approach, aim to take into account a multiplicity of factors contributing to poverty. Due to space restrictions, only the results of two such approaches can briefly be discussed here.

The Multidimensional Poverty Index (MPI) aims to measure acute poverty based on "a person's inability to meet minimum international standards in indicators related to the Millennium Development Goals and to core functionings" (Alkire and Santos 2014, p. 251). It has been tested and applied for over 100 developing countries including India. The MPI involves the three dimensions health (nutrition, child mortality), education (years in school and attendance), and living standard (cooking fuel, sanitation, water, electricity, floor, and assets).[3] Based on these three dimensions, a person is rated as being multidimensionally poor if they are deprived in at least one third of the factors, i.e. the cut-off for poverty (k) is 33.3% of the weighted

---

[3] For information on the MPI methodology, see Alkire and Santos (2014).

Table 3.2 MPI results at the national level

	All India (%)	Urban (%)	Rural (%)
Incidence of poverty	53.8	24.6	66.6
Vulnerable to poverty (k = 20–33.3%)	16.4	–	–
In severe poverty (k >50%)	28.6	–	–
Destitute	28.5	–	–

Source: OPHI (2015)

indicators. But the index further differentiates between "Vulnerable to Poverty" (k = 20–33.3%), "Severe Poverty" (k >50%), and "Destitute", when at least one third of more extreme indicators[4] apply (OPHI 2015, p. 1). Data for this assessment is based on the National Family Health Survey (NFHS-3) for India conducted in 2005–2006. Table 3.2 gives an overview of the poverty headcount on national level based on the MPI approach.

Compared to the above-given figures for the national poverty line (which delineates 37.2% of the population into poverty) and the global poverty line (making it 41.6% based on $1.25 in 2005 and 38.4% based on $1.90 in 2004; see Table 3.1), the MPI measurement (53.8%) turns out to be considerably higher in respect to the share of extreme and absolute poverty. This bleaker figure mainly goes back to substantially higher levels of deprivation in rural areas. Compared to the national Tendulkar poverty line and the figures for rural areas, there is a resulting difference of almost 25% (66.6% MPI vs. 41.8% of the national population based on national Tendulkar poverty line). This is while the share of the poor in urban areas is in both cases markedly lower at around 25%.

This simple comparison of the two approaches indicates the problem of contextual and structural determinants that condition poverty and its effects on well-being, vulnerability, and deprivation. It also shows the importance of more comprehensive approaches that aim to understand poverty, its interrelated determinants, and the related structural mechanisms that impede human well-being. Besides this, locating poverty and its structural determinants can play an essential role in informing policy makers and planners and thus assist a more targeted response.

For instance, Baud et al. (2008) have used GIS mapping in Delhi applying a multi-criteria index – disaggregated to the ward level – in order to analyse the spatial concentration of poverty, the diversity of deprivation, and how single aspects interact with others. Moreover, they have examined how far poverty on the ward level correlates with other measures such as prevalence of slums and number of households living below the poverty line. The authors employ a livelihoods assets framework taking into account social, human, financial, and physical capital, which they operationalise on the basis of insights from complementary surveys and available data from the Indian Census.[5]

---

[4] For example, "two or more children in the household have died (rather than one), no one in the household has at least 1 year of schooling (rather than 5 years), the household practises open defecation, and the household has no assets (rather than no more than one)" (OPHI 2015, p. 1).

[5] In this way, social capital, for instance, is measured based on ward-level percentages of households with scheduled caste background (indicator for social discrimination). Physical capital is

In result, they highlight that the hotspots of poverty are not necessarily located in slum areas and that these hotspots are diverse in terms of the underlying factors. Through this study, the authors show that multidimensional measures of poverty much better reflect the underlying barriers to improved well-being and that such an index combined with GIS is better able to inform policy makers to make targeted interventions on the ground (Baud et al. 2008, p. 1385).

3.2.2 Social Mobility and the Emergence of the New Indian Middle Class(es)

As shown in the chapter above, India is still a poor country. According to the National Council of Applied Economic Research, NCAER (Shukla 2010, p. 100), the lower classes in India account for about 85% of the population with a household income below of 200,000 INR per year in 2009/2010 ("aspirers" and "deprived"). Against these figures, the role of the so-called middle class seems to be of minor importance. However, its role in terms of both consumption and politics should not be underestimated.

An analytical assessment of the social-economic relevance of the Indian middle classes cannot be separated from more general issues, such as liberal reforms of the 1990s, the role of state policies, or India's position in a globalising world. Much of the debate has focused on the size of the middle class and the criteria to be used in drawing the boundaries. Political critics of liberalisation tend to both downplay the share of the middle class in India's social structure and to criticise its presumed "predatory consumerism", while proponents of liberalisation tend to overestimate its size and to downplay its negative impacts on society and natural resources (Reusswig et al. 2012, p. 35).

While much of the market-oriented research defines "middle class" basically via income (e.g. MGI 2004), more sociologically oriented researchers focus on structural characteristics such as occupational position or cultural capital (Béteille 2001; Deshpande 2003; Sridharan 2004). Nevertheless, the basic quantitative findings of both types of research converge (cf. Reusswig et al. 2012, p. 36).

A study published by McKinsey Global Institute (MGI) in (2007) is an example for a very optimistic and market-oriented assessment and a bold projection of the emerging middle class in India. The MGI study assumes an income classification and definition of the middle class suggested by the National Council of Applied Economic Research, NCAER (2004) (see also Sect. 5.1.4). According to this, the middle class has a household income level ranging between INR 200,000 and INR 1,000,000 per year.

---

measured on the basis of household infrastructure with "use of handpump", "no latrine", "no electricity", and "little space" being indicators for low levels of physical capital (for further details on operationalisation, see Baud et al. 2008, p. 1395).

Table 3.3 Estimates and projections of percent distribution of income classes and the middle class (highlighted in grey) for all India and urban India

MGI Income Classes	All India			Urban India		
	2005(%)	2015(%)	2025(%)	2005(%)	2015(%)	2025(%)
Deprived < 90K INR	54	35	22	21	9	5
Aspirers 90K – 200K INR	41	43	36	66	32	12
Seekers 200K – 500K INR	4	19	32	10	53	51
Strivers 500K – 1.000K INR	1	1	9	2	4	26
Globals > 1.000K INR	0	1	2	1	3	6

Compiled from McKinsey Global Institute (2007, pp. 45, 69)

Based on this definition, MGI estimated Indian middle-class households (see Table 3.3; seekers and strivers, highlighted in grey) to represent a share of 5% of total population in 2005. MGI projects this segment to grow fourfold to a share of 20% in 2015 and with an eightfold increase to more than 40% of all households in 2025. According to the authors, more than two thirds of the consumption growth between 2005 and 2025 will be concentrated in urban India, and also the largest growth of the middle classes will be found in cities (McKinsey Global Institute 2007, p. 61). In 2005, the middle class comprised about 12% of all households in urban India. It increased almost fivefold to 57% in 2015 and is expected to grow to a share of 77% in 2025, according to MGI projections. Irritating are the given figures in absolute numbers:

> *The middle class currently constitutes just 13 million households (50 million people), or 5 percent of the population. [...] [B]y 2025 India will transform itself into a nation of strivers and seekers with 128 million households (583 million people), or 41 percent of the population, in the middle class.* (McKinsey Global Institute 2007, p. 46)

Apart from the general confidence in this projection, it specifically remains unclear how the authors have arrived from the underlying number of households for the estimated number of people: 13 million households translate into 50 million people by a factor of 3.85, while it is a factor of 4.55 to compute 128 million households into 583 million people. This error certainly results in a much better figure for a projection that assumes India to "climb from its position as the 12th-largest consumer market [...] [in 2005] to [...] the world's fifth-largest consumer market by 2025" (McKinsey Global Institute 2007, p. 10).

A more recent study, based on rescaled data of the National Survey of Household Income and Expenditure (NSHIE) conducted in 2004/2005, was issued by Shukla (2010) under the aegis of NCAER. Shukla (2010) gives a very detailed picture of the income distribution, and he also takes into account a specific analysis of income in relation to consumer expenditure and savings. The study modifies the former NCAER definition of the middle class (see above; Shukla et al. 2004) by extending the upper limit to give a new range of INR 200,000 to INR ten million. Shukla (2010, p. 100) estimates the middle class have doubled in size from 5.7% in 2001/2002 to 12.8% in 2009/2010. In absolute terms, this is an increase from 10.7

million households (58 million people) to 28.4 million households (153 million people) within a decade.

The latter study also draws on aspects of social inequality, in particular regionally, both between states as well as between urban and rural areas (Shukla 2010, p. 97). With this growing gap between the rich and the poor, the question of what is middle class remains crucial. There is no consensus internationally in regard to defining a "new, income-based 'class' of the not-poor but not-rich in developing countries" (Meyer and Birdsall 2012, p. 2). Some authors who deal with an income-based definition of the middle class in the Global South suggest a bottom line just above the international poverty line: e.g. Banerjee and Duflo (2007, p. 4) set the range between $2 and $10 a day (in PPP), while Ravallion (2009, p. 5) designates as middle-class people living between $2 and $13 a day (in PPP). In Europe, this income segment would be regarded as poor, and therefore such a classification is specifically set for the Global South. And people defined as being within this segment are still very vulnerable even to an economic downturn, as their incomes are so low that they do not allow precautionary savings or assets to be accumulated (Birdsall 2010, p. 5).

A more recent strand in income-based middle-class measurement argues for a much higher benchmark starting at a minimum of $10 per capita per day (in PPP) (Birdsall 2010, p. 4; Ferreira et al. 2012, p. 2; Kharas 2010, p. 6; Meyer and Birdsall 2012, p. 2). Such a threshold is considerably higher than the World Bank's international poverty level, but it still implies a "minimum vulnerability to most economic and political shocks" (Meyer and Birdsall 2012, p. 2). Birdsall (2010, p. 6f) contends that – although it is a "round" and ad hoc number – this benchmark of around $10 a day per person demarcates a financial position that allows people to care about and save for the future and that it conveys a feeling of economic security against downturns of "the normal business cycle" (Birdsall 2010, p. 6). The basic argument here is that "a household is unlikely to need to sell household or business assets or take children out of school, and is insured through savings or formal insurance arrangements against such idiosyncratic risks as a family health catastrophe or a brief spell of personal unemployment" (Birdsall 2010, p. 6). While the upper benchmark in delineation to the rich still varies considerably, some authors predict a $10 per capita minimum threshold to be emerging as a new global standard definition for the middle class (Meyer and Birdsall 2012, p. 2).

Based on data from the NSSO Socio-Economic Survey 66th round (2009/2010), Meyer and Birdsall from the Center for Global Development (CGD) in Washington have attempted to assess the size of the middle class with an underlying middle-class definition ranging between $10 and $50 per capita per day. With this upper threshold, 0.06% of the rural population and 0.23% of the urban population have been found to have higher income than $50 per day, altogether about 1.33 million people (Meyer and Birdsall 2012, p. 6). The authors compare their findings with the results of the NCAER estimates from Shukla (2010, p. 100ff; see above) showing that less than 6% (about 70 million people) can be termed as middle class according to this definition, less than half of the NCAER estimation. Interestingly, about 60%

Table 3.4 Size of India's middle-class, CGD, and NCAER estimates (2009/2010)

	Meyer and Birdsall (2012)/CGD		Shukla (2010)/NCAER	
	Population share (%)	(Million)	Population share (%)	(Million)
Rural	3.37	27.84		
Urban	11.79	41.33		
Total	5.88	69.17	12.8	153

Source: Meyer and Birdsall (2012, p. 6)

of the Indian middle class lives in urban areas, making out a share of almost 12% (about 41 million people) of the overall urban population in India (Table 3.4).

The size and aspirations of the emerging middle class are also part of a social discourse on India's new power and future development. Even if it is just "a small segment of urban upwardly mobile people that has provided the basis for the discursive production of the image of 'the new middle class'"(Fernandes 2006: 89), this public discourse not only reveals the developmental desires but also reflects the shifting social realities of urban India. Also quite controversial appears to be the academic and market-research literature of the last two decades that turns around the issue of an emerging new middle class and the advent of a so far unseen consumer culture. Especially market-research-oriented studies follow a quite dominant narrative of new middle-class consumer lifestyle with rapidly growing levels of consumption and very optimistic projections of the growing middle class (Mathur 2010, p. 213). The German-language rather than popular-science-based literature has also joined this optimistic canon quite unequivocally, with book titles such as *Wirtschaftsmacht Indien, Weltmacht Indien, Die neue Wirtschaftsmacht am Ganges,* or *Tanz der Riesen: Indien und China prägen die Welt.*

Very importantly, the Indian state also intones into this "new middle class rhetoric" (Fernandes 2009: 219). Leela Fernandes (2009: 219) has taken a closer look at this phenomenon. She conceptualises the narrative of a new middle class in India as an aspect of "a state-led project [...] of development rather than as an expanding consumer group that has naturally been produced by economic growth" (Fernandes 2009: 219). First, the massive influx and sudden availability of a broad variety of new consumer goods and the ubiquitous visualisation of their use through advertising and marketing is an outcome of state-led liberalisation policy. Along with this highly visible market development, "new languages of development" emerged centring on the promise of a growing middle class that directly benefits from this transition towards consumer lifestyle (Fernandes 2009, p. 223). These two strands, the changed market situation on the one hand and the new middle-class narrative on the other, create a so far unknown atmosphere of new "imaginations" (Appadurai 1996, p. 10) towards future development and the question of what a good life could look like.

The above-given context and the figures on poverty in India (Table 3.1 in Sect. 3.2.1) indicate the huge potential of social mobility in the near future. The figures in Table 3.1 allow the assumption that more than half of the overall Indian population still lives under conditions that at the utmost allow them to satisfy little more than basic human needs, such as water, food, shelter, clothing, sanitation, education, and healthcare. Yet some improvement in the income situation for this bottom segment of Indian society allow them to increase their level of consumption and to change some aspects in their way of life. Against this background, growing demands for consumer goods such as cars, homes, household appliances, etc., seem inevitable. And obviously, with the liberalisation of the Indian economy, this demand is met by a virtually boundless variety of new mass consumer goods that have emerged on the Indian market.

Moreover, crucially important in transporting images of lifestyle are mass media, advertising, and marketing strategies. They play key roles in creating new imaginations and in shaping and manifesting the different world views, values, and preferences (McFarlane 2013) that guide behaviour and consumption. Urban areas and cities are in the focus and serve as projected area for marketing and advertising, as they offer an unprecedented market potential for new products. This especially holds true in emerging economies, such as India, where advertising occupies a major space in the public sphere as well as in mass media (cf. Franck 2010; Brosius 2010). Existing and newly emerging urban public spaces have therefore become a target of market-driven scenic colonisation and commercialisation that leads to an intense atmosphere of departure. These urban-specific features materialise in the form of oversized hoardings, advertising, and locations of exclusionary consumption, such as malls, cinema halls, leisure parks, cafes, and restaurants. They characterise the setting and the social space, and they are at work quite directly in stimulating and creating demand for new consumer products. These publicly celebrated and commercially staged sceneries of a newly emerging consumer culture create a set of new images and a semiotic language for conspicuous consumption and distinctive behaviour. As highly visible markers of social inequalities and exclusion, they convey comprehensive images and references for various styles of living. In this way, new lifestyles and new ways of consumption become conceivable, much of it in reference to a "Western" or "global" role model.

By considering consumption as a medium to perform and express a specific way of life, these images and references and their reception and translation into individually specific consumption patterns play an important orientational role for stylisation and construction of a social identity. Such expressive modes of behaviour and consumption, for instance, are manifested in diverging dietary patterns, modes of transport, religious practices and rituals, leisure activities, practices of vacation, and through a material culture conspicuously exhibited in the form of acquiring a multiplicity of consumer goods. Thereby, individual consumption and lifestyle tends to be based on following existing behavioural patterns and is based on shared imaginations of what a good life should look like (see Sect. 3.2.2). In this sense, cities can be seen as stages where a multiplicity of different social practices are performed and expressed.

It is well known from research on globalisation that the flow of information, images, goods, and products does not lead to a globally homogenised consumer culture. The information, images, products, sceneries, and practices are received very differently depending on the locational context as well as on the people's social-cultural and social-economic background. Income is not by far the only determinant for certain patterns of consumption, as income does not say much about spending or saving. Therefore, income-based projections of the middle class – no matter how exact they might be – are unable to reflect how consumption levels may actually change. Most observers expect the greatest dynamics will take place in urban areas. In many ways, such trends in consumption dynamics – and especially in respect to economies of scale as part of the lower economic segments – have severe implications for urban governance and climate-change mitigation. Whether and how this "dividend" of increasing incomes will be spent depends considerably on external and cultural factors such as the market, infrastructure, housing conditions, cultural dynamics, and institutions. It also builds, however, on socially shared representations that are based on the structure of individual attitudes and values.

3.3 Climate Change in India[6]

3.3.1 Impacts of Climate Change in India

Most of the impacts of climate change will affect India severely in many ways. Some of the more general environmental and climate-related risks and hazards that already exist without anthropogenic climate change are likely to be exacerbated, taking the form of extremes and increased variability. Extreme weather events such as torrential rain with flooding or failed monsoon seasons are expected to rise in number and intensity. Such events often have far-reaching and in some cases cascading effects for large parts of the population but also for whole economic sectors with related feedback mechanisms, for instance, in agriculture or the transport sector. In addition, larger and more systemic effects such as sea-level rise, glacier change in the Himalayas, and the destabilisation of the Indian monsoon are as yet unclear. Moreover, most of the climate signals translate into a multiplicity of impacts, which often interact in complex ways and which represent often long networks and pathways of effects (Reckien et al. 2009, p. 3). Given the enormous size and variation of India's physical and social geography, the possible impacts but also its adaptive capacity will vary widely, resulting in a broad range of vulnerability to climate change (O'Brien et al. 2004). In particular, major impacts are projected for agriculture. Moreover, there are important health implications, as climate change is likely to impose an additional layer on already existing, severe environmental health risks – to name just a few – risks associated with air, water and soil pollution, heat stress, flooding, waterlogging, and vector-borne diseases. This is in particular the case in urban areas within a context of rapid urbanisation.

In 2010, India released its first-ever scientific assessment of climate-change-related impacts projected for 2030 and deduced from the Hadley Centre Regional

---

[6] Some parts of this chapter have been taken from one of the author's earlier publications, namely, that authored by Reusswig and Meyer-Ohlendorf (2010: 19ff).

3.3 Climate Change in India

Table 3.5 Summary of findings from India's "4 × 4 assessment of the impact of climate change on key sectors and regions of India in the 2030s"

Impact category	Expected changes and variability by the year 2030
Temperatures	Rise in annual mean surface air temperature between 1.7 and 2.0 °C. Potential of increase in variability of seasonal mean temperature in winter months
Precipitation	Small increases in annual precipitation
Extreme events	Extreme temperatures: potential of intensification of daily temperature minimum and maximum in surface air temperature. Spatial pattern change in lowest daily minimum and highest maximum temperature suggests warming of 1–4 °C. Night temperatures likely to rise more over south peninsula and central and northern India. Central and northern India may experience increase in daytime warming
	Extreme precipitation: extreme precipitation events likely to increase by 5–10 days in all regions
	Cyclones: frequency of cyclones likely to decrease, but increase in cyclonic intensity
	Storm surges: All locations along eastern coast north of Vishakhapatnam, except at Sagar and Kolkata, show increase in storm surge levels in the 100-year return period by about 15–20% with respect to the 1970s. For Sagar and Kolkata, increase was found less than 5%
Sea-level rise	Sea level along Indian coast has been rising at rate of 1.3 mm/year; likely to rise in consonance with global sea level rise in future
Agriculture	In all regions, irrigated rice likely to gain in yields marginally from CO_2 fertilisation compared to rain-fed rice. Maize and sorghum projected to have reduced yields in all regions. With overall warming, the thermal humidity index projected to increase in all regions, especially in months of May and June, leading to stress to livestock and reduction in milk productivity
Forestry	All forest vegetation types in the four eco-sensitive regions vulnerable to projected climate change in short term, even under moderate climate change scenario (A1B). Impacts vary from region to region with changes ranging from 8% to 56% in vegetation cover
Human health	Malaria projected to spread in new areas in Jammu and Kashmir in Himalayan region. In the north-east region, opportunities for transmission likely to increase for longer period. In Western Ghats, no change observed. In coastal region, especially in eastern coast marked decrease in number of months; this increases likelihood of malaria transmission
Droughts and floods	Water yield projected to increase in Himalayan region by 5–20%; water yields likely to be variable across north-east region, Western Ghats, and coastal region. Moderate to extreme drought severity for Himalayan region, as compared to other regions. All regions likely to experience flooding with exceeding existing magnitudes by 10–30%

Source: data compiled from MoEF (2010)

Model Version 3 (HAD RM3) run for the IPCC SRES A1B scenario. The 4x4 assessment study addresses *four* of the key sectors of the Indian economy, namely, agriculture, water, natural ecosystems and biodiversity, and health. It concentrates geographically on the *four* climate-sensitive regions of India, namely, the Himalayan region, the Western Ghats, the coastal area, and the north-east region" (MoEF 2010). Table 3.5 summarises the most important conclusions from the assessment.

When the report was released, the then Environment Minister Jairam Ramesh stressed: "There is no country in the world that is as vulnerable on so many dimensions, to climate change as India is", and he emphasised the importance of building indigenous, independent research capacity for assessing the risks of climate change (MoEF 2010, p. 9).

3.3.2 India's Role in (International) Climate Policy

It is not only these climate-change impacts that provoke India's major interest in influencing the outcome of a global agreement on climate change. The pressing issue of future adaptation to climate change, growing environmental problems, the importance of energy security as a precondition for economic development, and increasing pressure from a broad, concerned, and well-informed (English speaking) public certainly provoke the interest in an effective post-Kyoto international climate policy regime – in favour of non-Annex I parties (Wagner 2010, p. 70). Most importantly, such an agreement provides the opportunity of assistance, financially as well as in the form of a technology partnership and strategic climate and energy relations with Annex I parties such as the USA. This has implications for infrastructure and non-polluting and energy self-sufficient economic development, as well as the potential of probably seising new markets globally (e.g. solar technology, electric cars). Against this background, India's position in international climate change negotiations is unique and worth looking at in more detail.

Since the Conference of Parties (COP13) in Bali, India increasingly faces rigid pressure in the international climate change negotiations towards taking legally binding commitments to a post-Kyoto agreement. In the Kyoto Protocol, binding emission-reduction targets were set only for 37 industrialised countries under the principle of "common but differentiated responsibilities". With its late economic awakening in the early 1990s, India has until now contributed relatively little to the overall cumulative GHG emissions (so-called historic GHG emissions) when compared with the early industrialisers. But given its rapid economic development of the last 25 years and the vast population numbers and area, India has now reached the fourth position with a contribution of almost 6% to overall global annual GHG emissions following the USA, China, and EU (WRI 2015). However, by taking population into account and calculating per capita emissions, India falls back to the 129th position with only 2.44 tonnes per capita per year in 2012 (WRI 2015).

Mainly holding to this argument of very low per capita GHG emissions, India has long retained a tough position in the international climate-change negotiations under the United Nations Framework Convention on Climate Change (UNFCCC). In fact, India cannot be lumped with China as a global economic player with its almost 250 million Indians surviving on less than a dollar a day (see above). Moreover, India still struggles with one third of all households still lacking access to electricity (GoI 2011), and in regard to energy security, India still faces major challenges. Hence, economic development, poverty alleviation, and energy security are among the major priority areas that India has to deal with in the coming years. A reduction of GHG emissions holds the risk of being bound to seriously compromise on these priority areas, such as building up a stronger economy, addressing poverty, and investing in broad-based infrastructure development. In addition, Indian foreign policy has – since independence – insisted on maintaining its own sovereignty, especially since the founding of the Non-Aligned Movement in 1961 (Wagner 2010, p. 67). Any legally binding commitment to an international agreement on climate change counteracts these priorities.

Against this background, it appears surprising that India has in fact taken major steps forward in the recent years to address climate change voluntarily and in a self-determined way, and it is obvious that climate change ranks high on the political agenda. Quite relevant here are technological and economic opportunities and challenges associated with strategies to reduce GHG emissions and develop sustainably. Quite groundbreaking was the constitution of the Prime Minister's Council on Climate Change in 2007, established to coordinate national action plans for assessment, adaptation, and mitigation of climate change. In this light, the visionary National Action Plan on Climate Change (NAPCC) was released, defining eight core national missions as a framework for the implementation of state action plans on climate change. Since then, quite a number of initiatives and programmes have emerged: just to name a few, the Energy Efficiency Programme (EEP), the Carbon Strategy, a coal tax to finance clean energy research and development, fuel efficiency standards, a Renewable Energy Certificate (REC) trading scheme, and the initiation of the Indian Network on Climate Change Assessment (INCCA), which represents a network of over 120 institutions and 220 scientists aiming to further improve Indian climate science. Also, the new prime minister, Narendra Modi, signalled changed priorities, as he initiated a nomenclature change of the Ministry of Environment and Forest renaming it the Ministry of Environment, Forest and Climate Change.

A breakthrough in the international negotiations was reached in 2015 at COP21 in Paris, where India has been lauded for the quite ambitious targets that it has set as part of the Intended Nationally Determined Contributions (INDCs). India has communicated its future commitment based on eight targets, out of which three are quantitatively tangible. The other targets are more inexplicit, but they are not less remarkable. The very *first* mentioned target draws on the issue of increasingly changing lifestyles. It aims to "put forward and further propagate a healthy and sustainable way of living based on traditions and values of conservation and moderation" (GoI 2015, p. 29). This reference to values of thriftiness and frugality rooted in the Gandhian ideals of simplicity and asceticism (see Sect. 4.1.2.1) contains an implicit critique of an adoption of "Western" lifestyles and values and an emerging consumer culture. This critique is further underlined by a *second* objective to "adopt a climate friendly and a cleaner path than the one followed hitherto by others at corresponding level of economic development" (GoI 2015, p. 29). These first two targets are noteworthy as they reflect one side of an ambivalent development discourse. The discourse on conservation and moderation on the one hand is considerably at variance with the dominant technology-oriented and liberal development paradigm that has been deeply embraced since the emergence of the NEP (see Sects. 3.1.1 and 3.1.2). The following three targets are more concrete and quantitatively measurable. In its *third* objective, India proposes to "reduce the emissions intensity[7] of its GDP by 33–35% by 2030 from 2005 level" (GoI 2015, p. 29). *Fourth*, it aims to "achieve about 40 percent cumulative electric power installed capacity from non-fossil-fuel-based energy resources by 2030 with the help of transfer of technology and low cost international finance including from Green Climate Fund" (GoI 2015, p. 29). And *fifth*, India aims to "create an additional carbon sink of 2.5 to 3 billion tonnes of CO_2 equivalent through additional forest and tree cover by 2030" (GoI 2015, p. 29). Also remarkable is that only the *sixth* target refers to the issue of climate change adaptation in sectors "vulnerable to climate change, particularly agriculture, water resources, Himalayan region, coastal regions, health and disaster management" (GoI 2015, p. 29). The last two goals rather draw on the issue of implementation and finance with the support of Annex I countries and the creation of a domestic framework and international architecture in order to bring forward research and development of future technological solutions (GoI 2015, p. 29).

While it is still open, whether and how the Indian government would operationalise and implement the "soft" targets, it is worth looking at the feasibility of the quantitatively tangible objectives and the proposed mitigation strategies and actions. India aims to develop an installed capacity of 175 GW based on new renewable energy sources to be built by 2022: 60 GW of grid-connected wind power, 100 GW of solar power (60 utility scale, 40 rooftop), 10 GW of biomass energy, and 5 GW of small hydro. While there is no given nuclear capacity target mentioned for 2022, the document proposes a target of 63 GW for 2032. Also large hydro plants, which currently make out the largest source of non-fossil energy, are not mentioned directly.

While some observers doubt the viability of these quite ambitious targets, others are quite positive about the proposed commitments. Chakravarty and Ahuja argue that in order to meet the 4% non-fossil target in 2030, India requires about 300–400 GW of non-fossil capacity. With an expected increase of about 5 GW of wind power and 15 GW of solar PV per year, this target is not seen as unrealistic (Chakravarty and Ahuja 2016, p. 475). For this, however, huge investment needs to be made, and it is unlikely that much of it will be drawn from the Green Climate Fund, a financial mechanism under the UNFCCC. However, this doesn't seem necessary, as the Indian market for renewable energy is highly attractive for Indian and international private-sector firms (Chakravarty and Ahuja 2016, p. 475f).

Apart from investments in generation systems, such as wind and solar power, the requirements of renewable energy systems in respect to the electricity grid are considerable. A robust, flexible, and well-integrated grid system that connects the Indian subcontinent needs to be put in place in combination with reliable storage facilities for balancing out remaining variabilities. This is quite challenging as smart

---

[7] Emission intensity is defined as tonnes of CO_2 generated per unit of GDP corrected for purchasing power parity.

and well-integrated solutions still need to be found and huge investments in the energy grid need to be made (Chakravarty and Ahuja 2016, p. 476).

Increasing the share of low-carbon electricity-generation technologies certainly does not suffice to meet the substantially growing energy demands in India. Such a scheme needs to be combined with significant reductions in the emission intensity, as rightly envisioned in the INDCs. A high and increasing share of the service sector and considerable gains in energy efficiency are among the main factors which could contribute towards realisation of the intensity targets. The former and the new government have put into place a variety of programmes that aim to raise energy efficiency. The Star appliance labelling programme, the promotion of compact fluorescent lamps (CFL), and another programme to replace all incandescent light bulbs with LEDs by 2020 have induced and are likely to further increase energy savings. This is an essential step to reduce peak power demands, especially in the evening hours, and thereby balance out the lack of solar power with the setting sun (Chakravarty and Ahuja 2016, p. 476).

Huge potential concerning both energy efficiency and in the reduction of non-CO_2 GHG emissions lies in the realm of refrigeration and air-conditioning (AC). Technological advancements can raise efficiency levels to a substantial extent. Adding to this, building design and urban planning as well as behavioural and operational adaptations can contribute to huge energy savings in room and vehicle air-conditioning, considering the sector's growth rate and the current lack of awareness (see, e.g. Noé21 2014). Some observers estimate the peak AC load alone as high as 140 GW in 2030, which represents almost the all-India peak load today (Chakravarty and Ahuja 2016, p. 476). ACs, room and water heaters, chillers, refrigerators, etc., have tremendous potential for energy savings. And in addition to the issue of efficiency increases, even can be achieved in terms of avoiding the use of refrigerants (hydrofluorocarbons, HFCs) increasingly used in air-conditioners and refrigerators (Velders et al. 2009, p. 10949), yet having a very high global warming potential (GWP) ranging between about 700 and 4,000 GWP (Xu et al. 2013, p. 6084).

However, the greatest challenge India will face in the future is based on the fact of rising incomes and large fractions of the population entering the middle class. This transition will significantly increase the consumption of manufactured goods, electricity, ACs, heating, automobiles, and air transport, and it has just started with the New Economic Policy in India and will continue for decades to come (Chakravarty and Ahuja 2016, p. 476). The next chapter will examine this factor in more detail, taking a closer look at the effects of urbanisation, rising incomes, and changing patterns of consumption.

3.3.3 The Indian Middle Class, Rising Incomes, and Emissions

Considering India's per capita GHG emissions and its historical contribution to the existing emissions stock in the atmosphere, the country's role as a driver of anthropogenic global climate change is still minor. However, with its rapid economic growth, its fast-growing population, and its high rate of urbanisation, GHG emissions are growing at much higher rates than those of the developed countries. India's total GHG emissions have more than doubled between 1990 and 2008 (IEA 2010), and the World Energy Outlook 2009 Reference Scenario projects that CO2 emissions in India will increase by more than 2.5 times by 2030 from 2008 (IEA 2010). With an average per capita value of only 2.44 tonnes per year in 2012 (WRI 2015), India still ranges far below the world average of about 4 tonnes. This together with the prevalent poverty in India has long served as a telling argument to legitimise India's strict rejection of any binding emission-reduction agreement with the Global North, i.e. Annex I countries, being largely responsible for the climate-change problem. While this argument seems justifiable in terms of global environmental justice, the high growth rate of India's total and – to a lower degree – per capita emissions asks for a deeper analysis.

Average per capita emissions mask the fact that in recent years a growing middle class has produced significantly higher per capita emissions – not to mention upper class members, whose emissions dwarf average emissions in Europe or even the USA. This has raised concern among some scholars and NGOs in India and elsewhere, who ask whether more and more of India today is "hiding behind the poor" – in reference to the apt title of a Greenpeace India publication on this issue.

With its growing size and the increasing purchasing power, the middle class is most relevant in regard to this social-cultural transition and its effect on the environment in general and the climate in particular (Myers and Kent 2003, p. 4966). Myers and Kent argue that it is mainly the consumption sectors meat consumption, individual motorised transport (IMT) and electricity use that have the greatest potential to be affected by rising incomes (Myers and Kent 2003, p. 4964ff). As one example, the release of the Tata Nano car – the most affordable car ever produced – has reminded the world of the potential risks of the Western-oriented development and planning paradigm being followed in most of the developing world. The rapidly increasing shares of personal motorised transport users especially in cities, the sharp decline in walking and bicycle trips, and the fact that public transport and non-motorised transport users are mostly captive users illustrate the problem (Tiwari 2011, p. 9). Even if the trends in car ownership still range on a rather low level at less than 2% on national level and at 10–15% in richer cities, aspirations for owning a personal motorised mode of transport are immense. It is also the growing demand for long-distance travel and the rising market in air travel that will have a major impact on the development of future emissions in India, and much of this relates to the dynamics of upward social mobility.

Therefore, the new consumption patterns on the one hand and the high overall carbon intensity of the Indian economy on the other will inevitably lead to growing emissions – and to a growing environmental responsibility on the part of the Indian middle class. The majority of this social segment is located in cities. It may thus be said that India's *future* urban middle class holds a key to both national and, to a lower but still significant degree, global GHG emissions in its hands.

3.4 The Case Study: The Megacity Hyderabad

3.4.1 Location and Early History

Hyderabad, with a population nearly nine million people today, is the sixth largest city in India and is expected to reach the threshold of ten million inhabitants by the year 2020 (United Nations 2014a, p. 332). It is expected to grow to 12.8 million people by 2030 (United Nations 2014a, p. 316). With its rapid population growth and its population still ranging between five and ten million people, Hyderabad can be classified as a city, which has recently emerged as megacity. This feature was one of the reasons to consider the city as case study for a project under the BMBF-funded Future Megacities programme with the project title *"Hyderabad as a Megacity of Tomorrow: Climate and Energy in a Complex Transition Process towards Sustainable Hyderabad"*.

Hyderabad is situated on a hilly terrain of grey and pink granite at an average altitude of 542 m. Apart from the very typical landscape of dotted hills with characteristic granite rock formations, there are quite a number of artificial lakes created by dams on the Musi River. While it is situated along the banks of the Musi River, Hyderabad is located within the crossroads of the two larger rivers, Krishna and Godavari, in the peneplain Telengana (MCH 2005, p. 1). Hyderabad is well connected with other metropolitan areas through a well-developed national and regional railroad network as well as some national and state highways converging in the city. The city's newly established and state-of-the-art international airport well connects the city with destinations outside India and contributes to the attractiveness of the city in regard to foreign direct investments.

The city of Hyderabad was established in 1591 by Muhammad Quli Qutb Shah, the fifth and most celebrated ruler of the Qutb Shahi dynasty of the sultanate of Golconda. The original city plan of Hyderabad was inspired by and incorporated many features of the mythical Islamic heaven as laid out in the Quran (Luther 2008, p. 1). Literally planned as replica of the Qur'anic "Gardens of Eternity", the city was laid out around a monumental central building, the Char Minar, with the crossing of two axial streets oriented along the cardinal directions. This area, which also contained other new buildings, civic spaces, and shopping areas, covered only a quarter of a square kilometre (Das 2015, p. 49). This feature in part accords with the ancient Indian architectural theory of Hindu Vastu Shastra, but there are other features which clearly go back to Islamic and Persian ideas of an ideal configuration of streets and buildings (see Pieper 1984, p. 47). Already during the period of the Qutb Shahi dynasty, Hyderabad developed as an important economic, trade, and cultural centre for the larger region of Golconda.

In the late seventeenth century (1687), Hyderabad was taken over by the Mughal emperor Aurangzeb and was since then ruled by several appointed Nizams (governors), until the Nizam Asaf Jah I declared his independence from the Mughal imperium (Das 2015, p. 49). During the 200 years of Nizam rule, the region and its central trade and cultural hub of Hyderabad became famous for pomp, flaunted wealth, and noble jewellery (Das 2015, p. 49). However, with the East India Company's increasing dominance over the subcontinent in the second half of the eighteenth century, the Nizams' sovereignty declined considerably. But the Hyderabad princely state remained independent until 1948, when it was forcibly integrated into the Indian union. In 1956, the former Nizam region became part of the newly formed state of Andhra Pradesh (AP) with Hyderabad becoming state capital.

The newly formed state of AP integrated three distinct geographical regions – Telangana, the former Nizam state in the north-west, coastal Andhra in the east, and Rayalaseema in the south and south-west. These newly integrated regions in fact share the same language but have very different geographical, historical, cultural, and social-economic backgrounds. While coastal Andhra is richly endowed with fertile agricultural lands, where early types of irrigation were introduced by the British rulers, Telangana and Rayalaseema rather remained backward with much harsher physical conditions. Especially the former Nizam region of Telangana has been left isolated from the rather technology-oriented approach of the British rulers in terms of education and economic development (Benbabaali 2009, p. 689).

The formation of the federal state in the mid-1950s subsequently nurtured a considerable rise in in-migration to the city, with rising tensions especially between migrants from coastal Andhra and people from Telangana. With their often advanced level of education and development, people from coastal Andhra succeeded filling important and influential positions in business, politics, and administration. This has also led to further neglect and the increasing backwardness of the Telangana region (Das 2015, p. 49f). And while diversity has been a constant feature of Hyderabad, the emergent conflict arose not because of diversity, but more due to rising inequalities (Benbabaali 2009, p. 699). The smouldering Telangana conflict came to the boil again in 2010 with the repeated demand to separate Telangana region from AP. This struggle went on until recently in 2014, when Telangana achieved independent statehood with Hyderabad as its capital.

3.4.2 *Hyderabad as an Engine for Growth*

The post-independence urban growth of the city is characterised by a trend of public-sector-based industrial development during the 1960s and 1970s, with enormous employment opportunities for skilled workers. Also, several scientific research institutions and the headquarters of the South-Central Railway zone settled in Hyderabad through to the late 1980s. The resulting massive influx of migrants from the surrounding districts led to considerable pressure on the housing sector and the existing urban infrastructure. During this time, due to the lack of affordable housing, large shares of the population had to make a living in slums or slum-like conditions (Das 2015, p. 51). With the advent of the New Economic Policy (NEP) and the tendencies of decentralisation and strengthening of the political role of the federal states after 1991 (Das 2015, p. 51; Kennedy 2007, p. 97), the character of urban development of the city changed considerably with a strong focus towards the service sector, particularly information technology (IT), IT-enabled services (ITES), and biotechnology.

This changed path of urban development is specific for the case of Hyderabad. It presents an outstanding example of how globalisation and economic liberalisation can affect the development of a city and its hinterland and how a city can work as engine of economic growth and social change. At the same time, Hyderabad exemplifies to accommodate most of the major social and environmental issues and challenges related with rapid urban growth. These two aspects will be dealt with in the following.

The decentralisation and rescaling of provincial states have opened up a range of so far unknown political options and possibilities but also increased the pressure on state capitals and major urban centres in India to compete for foreign direct investment and attract industries to settle (Kennedy 2007, p. 97; Kennedy and Zérah 2008, p. 115). However, this process of decentralisation only had limited effect on the scope of action for local governments, and this, although the 74th Constitutional Amendment Act was ratified in 1993, suggests extended functions and competences in planning for economic and social development (Kennedy 2007: 98). In the case of development of Hyderabad, it is the federal state of AP playing out the major role in adopting a city-centric growth strategy (see, e.g. Kennedy and Zérah 2008).

In 1995, the AP government under Chief Minister Chandrababu Naidu was confronted with a major economic crisis, with an agriculturally dominated economy, very high levels of subsidies and welfare comprising around 10% of the state GDP, and a stagnating manufacturing sector (Das 2015, p. 51). Bound to take a loan from the World Bank under the structural adjustment programme (SAP), the state government was pressured to reduce expenditures and introduce economic reforms. Naidu cut subsidies and welfare programmes, which especially benefitted the poor, and instead focused on attracting foreign investments in tertiary industries such as information technology, biotechnology, finance, and banking (Das 2015, p. 51; Kennedy and Zérah 2008, p. 113). Interestingly, Naidu also began travelling to search for and learn from experiences elsewhere in the world and was especially impressed by the recent development projects in Singapore and Malaysia, especially the Multimedia Super Corridor (MSC) near Kuala Lumpur (Das 2015, p. 51).

Deeply inspired by these neoliberal programmes, Naidu initiated a plan for the development of a "knowledge enclave" in Hyderabad, known as "HITEC City" (Hyderabad Information Technology and Engineering Consultancy City). The completion of the "Cyber Towers" in 1998 has been seen as a first major landmark in the creation of a "world-class city". With the increasing demand and the given incentives, more such intelligent buildings were constructed. Das argues that the "policy initiatives of creating HITEC City provided a boost to Hyderabad's urbanisation, and spawned massive developments of gated residential apartments, 'intelligent' offices and shopping malls around the HITEC City area" (Das 2015, p. 51; Kennedy and Zérah 2008, p. 113f).

In 1999, endorsed by the success of the reforms and new policy initiatives, a visionary document was created, the "Andhra Pradesh Vision 2020", formulated in close consultation with the consultancy company McKinsey. Largely inspired by the Malaysian Vision 2020, it laid out a state development strategy with Hyderabad being anticipated as an engine of growth. With a vision of leapfrogging towards becoming an information society, the emphasis was laid on the development of the service sector and a further attraction of foreign direct investments in IT and related services, biotechnology, tourism, logistics, healthcare, and educational services. Investments in premium urban infrastructure and the promotion of high-tech knowledge enclaves provided the required incentives and proper conditions for domestic and international companies to invest and establish their services in Hyderabad. Two additional policy initiatives in 2002 and 2005 even further raised the incentives for IT-related and other firms, and the hype on the success of Hyderabad was glaring, similarly as in Bangalore (Das 2015: 51; Iyer et al. 2007: 9ff).

This dominant service sector orientation has not stopped the industrial sector from growing; however it has become more concentrated on the periphery of the city. And today, both the Hyderabad Metropolitan Development Plan 2031 and the first Socio-Economic Outlook 2015 of the new Telangana Government underline the importance of a balance between the service sector and industrial development. Especially the new Telangana government puts an emphasis on restrengthening the role of industrial development too, with Hyderabad as "growth engine" (Government of Telangana 2015, p. 6). The trend towards containment of industries on the urban fringe of the city has made it necessary to improve land-use planning and governance in the peri-urban area. This is one reason also for the spatial restructuring made in 2008 and in smaller steps later on (Das 2015, p. 51) as will be outlined in the following.

3.4.3 Administration and Urban Spatial Restructuring

Hyderabad is the capital of the newly formed state of Telangana, which has separated from the state of Andhra Pradesh in June 2014. As an interim solution, it will function as capital for both states for the following 10 years after separation.[8] There are two major urban administrative bodies in Hyderabad, the planning authority, before 2007 known as Hyderabad Urban Development Authority (HUDA), and the municipality, formerly known as Hyderabad Municipal Corporation (HMC). While the planning authority is in charge of coordination and urban zonal planning (Hyderabad Master Plan), the municipal corporation coordinates and manages all basic urban services. In April 2007, the planning agency Hyderabad Urban Development Authority (HUDA) and the Hyderabad Municipal Corporation (HMC) expanded their sphere of influence to cover a greater area. Through integrating the formerly 12 independent surrounding municipalities, HMC (formerly 175 km^2) became the Greater Hyderabad Municipal Corporation (GHMC) that now covers an

---

[8]The main research of this study has been conducted between 2009 and 2012. All references to Andhra Pradesh in this study connote to the erstwhile undivided state of Andhra Pradesh (AP).

area of 650 km² (about as large as the former HUDA region) with a population of about 6.7 million people in 2007. GHMC is the local urban government of Greater Hyderabad that includes 12 municipalities and 8 gram panchayats (village councils). Its area of 650 km² is larger than the municipalities of Mumbai, Chennai, or Bangalore. It is divided into five zones (south, east, north, west, and central). Each zone is subdivided into circles and wards, with the ward being the smallest administrative unit. A ward usually contains a population of about 37,000 people. At the same time in 2007, HUDA was transformed into the Hyderabad Metropolitan Development Authority (HMDA) and reaches out into areas that are far beyond influence of GHMC – more than ten times the area of GHMC – with about 7,228 km². Hyderabad Metropolitan Area has thereby become the second largest urban region in India after Bangalore (Das 2015, p. 53).

3.4.4 *Hyderabad as Symbolic Representation of "World-Class" Infrastructure Development*

As outlined above, Hyderabad serves as a globally connected hub for economic development and employment for a larger region, which was formerly dominated by agriculture and in quite large parts of the region remains relatively backward in terms of economic development. Politically, the city has been envisioned as an engine of growth for the region, and through an advanced neoliberal economic reform policy of the last two decades, the city has successfully become attractive for domestic as well as international corporations to invest in and to operate their businesses and services from this city. Not only due to these newly gained job opportunities, but also because of the still relatively affordable real estate prices compared to Delhi and Mumbai, Hyderabad offers quite favourable living conditions, especially for rather well off professionals and their families. Also in terms of infrastructure, the city has quite a lot to offer: the recently established and very efficiency-oriented international airport, which is well connected to the city through an expressway to Cyberabad and the outer ring road, underlines the image of a globally well-connected cutting-edge city. A number of well-connected railway stations, a relatively well-functioning city road-infrastructure network, and the envisioned Hyderabad Metro Rail Project (HMP) accentuate this representation. Also in terms of quality of life, the city is perceived as attractive: the local climate in Hyderabad is quite pleasant compared to other cities in the country and a number of parks, beautiful hills with Hyderabad-specific granite rock formations, and several natural and artificial lakes offer favourable places for recreation within the city.

Chapter 4
Conceptualisation and Operationalisation – A Social Geography of Climate Change: Social-Cultural Mentalities, Lifestyle, and Related GHG Emission Effects in Indian Cities

Keywords Lifestyles, Social values · Social structure analysis · 'Tradition' and 'Modernity' · Investive consumption · Social position · Carbon calculator · Ethical consumption · Hedonism · Consumer culture

4.1 Theoretical Framework

In Chap. 2, the most relevant theoretical considerations and conceptual developments in the field of personal-level GHG accounting, general lifestyle research, and environment-related lifestyle research have been presented. Based on these theoretical considerations and implications, a completely new and explorative concept for the analysis of social-culturally based differentials of personal-level GHG emissions is laid out in this following chapter. The building blocks and main components of this concept are depicted in Fig. 4.1. The structure and line of argument in this chapter will follow the main components of the given figure.

4.1.1 Multilevel Perspective and Situative Context of Lifestyle

In this study, individual human behaviour and social practices are seen as embedded within a larger social-economic and social-cultural context. As most other lifestyle conceptualisations, this study does not follow a paradigm of methodological individualism. Rather, lifestyle is understood as a consequence of social mechanisms taking place across multiple levels, from the micro to the macro level of society. Hartmut Lüdtke's (1989, p. 71) account on the structural levels of lifestyle and how these levels interact in dynamic group formation processes (see Sect. 2.2.2.4) is very indicative and contributes to a better understanding of the multilevel interactions from micro to macro level. However, it also is quite a rigid model and the macro level of analysis stops with processes that are still directly associated with social structure and differentiation. Therefore, these processes should not be

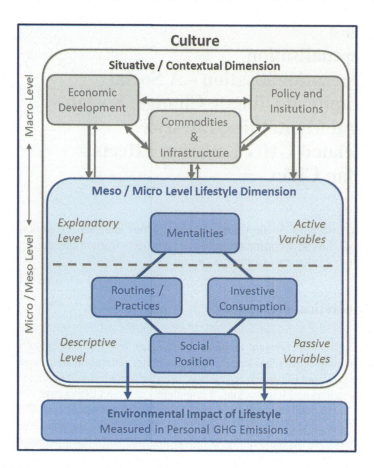

Fig. 4.1 Main components and structure of the concept for the analysis of social-cultural differentials in personal-level GHG emissions

confused conceptually with the broader political-cultural and physical-spatial context. This broader contextualisation is in fact important for an understanding of lifestyle and is rather neglected in most studies. Context includes aspects such as general cultural factors, historical, political, and institutional aspects, level of economic development, and the physical-material dimension – spatial aspects, commodities, and infrastructure. Many studies refer to the embeddedness of lifestyle across levels, as delineated by Lüdtke (1989, p. 71). However, most studies fail to consider the macro level and even more so the contextual and cultural level. This conceptual negligence surprises, as the interpretation of lifestyles can only be based on a deeper understanding of the larger social-cultural embeddedness. This important perspective also includes a consideration of the own role, knowledge, and viewpoint of the researcher(s), e.g. in a post-colonial context (see Sect. 7.2).

In this study, the macro level of social structural differentiation (Fig. 4.1) has largely been incorporated by exploration and qualitative research, mainly during the preliminary phase (see Sects. 4.2.2 and 4.2.3). Albeit difficult to operationalise,

4.1 Theoretical Framework

processes of social group and identity formation, social distinction, segregation, exclusion, and related mechanisms become visible or tangible, e.g. in the settlement structure, in residential patterns, in the use of infrastructure, and in public spaces. Often, these manifestations of social segregation are not so obvious. In the case of Indian cities, the phenomenon of spatial fragmentation is commonly found in most parts of a city. For instance, small and often hidden slum pockets can be found in closest vicinity with very posh areas of the city. Often and especially in more posh areas, vacant land is "occupied" by informal dwellers, who, for instance, rely on an informal agreement with the employer or are based on other arrangements (e.g. servants working in an upper-middle-class area such as Banjara Hills). On the other hand, processes of segregation are in fact taking place in Indian cities apart from these often rather micro-level fragmentation processes. These aspects have to be looked at and become more obvious in broader-based residential patterns.

However, an ascription of purely lifestyle-related factors that drive social-spatial segregation has to be taken with much caution. Such lifestyle-based segregation patterns have not been found empirically so far. Coarse differentials based on social-economic position, such as education, occupation, family background, income, wealth, etc. in fact deliver substantial grounds for selective interaction, distinction, and social closure. And these coarse features operate more effectively than lifestyle, especially concerning social-spatial segregation. The freedom to apply selective demand criteria to processes such as deciding for or against a neighbourhood are primarily based on social-economic possibilities. On the ground of this scope of possibilities, a more subtle difference or distinction ("Ein feiner Unterschied") can be made based on, e.g. aesthetic preferences (see Sect. 2.2.2.4).

For the interpretation, explanation, and contextualisation of the statistically derived lifestyle segments, the contextual and situational dimension (Fig. 4.1) is very important. It extends the macro level of lifestyle as delineated by Lüdtke (1989, p. 71). Macro level *and* context are not only important for the deeper understanding and placement of the identified groups and the involved characteristic behavioural and consumption patterns, they are also crucial in terms of conceptualisation, e.g. for the development of the questionnaire.

4.1.1.1 Culture

The viewpoint on the cultural context (Fig. 4.1) provides an often taken-for-granted analytical basis for any social science study. Of course, Mitchell (1995) has reminded cultural geographers that "there is no such thing as culture" and that "the naming and representation of cultures creates partial, yet globalising, truths" (Mitchell 1995, p. 109). Talking about culture inevitably leads to an abstraction of contentious areas into a partial truth. However, in spite of the reified nature of the concept of culture, social science research focusing on any social-cultural aspect requires consideration of broader contextual factors and conditions. Identifying and defining these conditions is highly subjective and remains the author's responsibility.

For the author of this study, it was seen as crucial to take into account historic and political aspects, aspects of language, religion, and ideas about shared norms and values. Moreover, it was deemed important to take a closer look at gender aspects in the same way as questions on tradition and change. Culture in this study is rather broadly conceptualised; it includes a multiplicity of creative, intellectual, and practical achievements and realisations. It refers to aspects of shared, socially learned knowledge and characteristic patterns of behaviour (Peoples and Bailey 1999, p. 17). The concept comprises "soft" characteristics, such as values, norms, and institutions (understood as rules) as well as material aspects or artefacts such as works of art, tools, buildings, and monuments (cf. Freytag et al. 2016, p. 90). Culture encompasses different lifestyles as well as their contextual components. Lifestyle dynamics as well as associated contextual conditions shape the culture of a group, while cultural changes again re-affect context and lifestyle. The approach taken here highlights how closely culture and lifestyles are intertwined.

Besides the broader concept of culture, there are more specific and more concretely operating contextual aspects operating in the framework of culture. Figure 4.1 highlights these components, namely, policy and institutions, economic development, and infrastructure.

4.1.1.2 Policies and Institutions

Since the early 1990s, India has initiated substantial political and institutional reforms to overcome a major political and economic crisis. The reforms aimed to strengthen market orientation and stimulate private and foreign direct investment through reduction of import tariffs, tax reductions, and market deregulation. Further reforms were implemented over the next two decades, helping to establish a strong and stable economic growth.

The instance of India's liberalisation policies can be seen as a prime example of how political and institutional changes can lead to massive social-economic changes that affect the whole society. Multiple opportunities emerged in terms of new jobs and employment, creative businesses, higher incomes, and interesting future prospects. Steadily, more and more products and services appeared on the market, better and new modes of transport emerged on the scene, and new forms of leisure and recreation arose and became imaginable for many aspiring new consumers. These few lines that attempt to strikingly sum up the effects of political reform in India (for a deeper analysis, see Sect. 3.1.1) are just to convey how important the political and economic boundary conditions are for lifestyle across all levels, from individual to household, via friends and peers up to the overall societal level. And these mentioned reforms only outline economically oriented changes in policies and institutions and that happened on the national level. Also the state and municipality level are important in this regard, as can be exemplified for the case of Hyderabad and, e.g. Bangalore (now Beṅgaḷūru). However, this is only to illustrate the relevance of this contextual component. A more detailed assessment of the different processes and factors will be provided in Sect. 3.4.

4.1.1.3 Economic Development

Moreover, economic development in itself encompasses a multiplicity of much broader implications, especially in the context of an emerging economy like India. The steady economic growth since early 1990s is associated with a whole bundle of dynamic transformation processes that are associated to it. Just to name a few, it affects urbanisation and the dynamics of social mobility and leads to the emergence of a new middle class, new forms of residential patterns, increasing levels of mobility, major changes in consumption and dietary patterns, as well as related impacts on the environment. Along with globalisation, these transformation processes – similarly as in the second half of the twentieth century in Europe – call for new approaches for the analysis of social change and social structure analysis. The most dynamic social-cultural changes concentrate in urban areas and cities in India. Section 3.1 has given an overview of these dynamic processes and has shown how they interact with dynamics of consumption and lifestyle changes.

4.1.1.4 Commodities and Infrastructure: The Symbolic Meanings of Things

There is no doubt about the critical importance of things in our social lives (Shove 2007, p. 4). Things surround us everywhere and things are part of all our daily activities. These material things exist, based on a multiplicity of ascribed symbolic meanings and functions. Analysing consumption involves analysing the functions, character, and meaning of the commodities being consumed as well as the underlying infrastructure that functionally and symbolically serves these consumption patterns.

Many insights in this field come from the sociology of consumption and achievements being made as part of the *cultural turn*. Commodities and most of the material objects that surround us have certain functional purposes and "make sense" to us in various ways. Many of these "things" fulfil de facto basic needs, such as shelter or appliances for cooking. Other things may or may not have direct necessary and crucial purposes to fulfil, and some of their "functions" may have emerged in their own way as being increasingly necessary and taken for granted, such as cosmetics or air-conditioning systems (cf. Shove 2003, p. 3ff). Therefore, the interactions taking place between the individual agent and the material world that surrounds him or her is anything but trivial.

The acquisition of a consumer good may have a variety of ostensible and primary functions, but it may also contribute to an expanded cultural experience for the consumer in terms of "personal self-development and self-expression, and, as with the example of gifts, established and consolidated social relationships" (Warde 2014, p. 281). In consequence, consumer goods and services and other material objects not only carry symbolic meanings and serve as semiotic intermediaries, but they represent valuable "'*resources*' for the construction of individual or collective identities" (Shove 2007, p. 4 emphasis added). From this perspective, the function of consumption is highlighted and the agent takes an individual choice of consumption.

At this point, it makes sense to draw on discussions mainly originating from the field of practice theory. By borrowing a perspective on notions of context, material configurations, and the view on routines from practice theory, fruitful insights for lifestyle analysis may emerge. According to Schatzki et al. (2001, p. 3), social practices can only be understood by taking into account the role of things and their material configurations. Elisabeth Shove et al. (2012) have taken up this perspective and simplify this approach into a model framework. With this quite new approach to practice theory, they bring forward the three key concepts – meanings, competences, and materials. The material dimension – things and their configurations – is seen as having a very important role, following Latour (2000) that artefacts are "in large part the stuff out of which socialness is made" (Latour 2000, p. 113; cited in Shove et al. 2012, p. 9).

With the practice theory's emphasis on routines, it makes sense to examine the interactive linkages between routines, things, and meanings. Many climate-relevant consumption patterns are related to daily routines, such as mobility and commuting, basic dietary patterns (vegetarian vs. carnivore diet), the use of electric appliances, and cooking. Most of the material objects that surround our everyday life – at home as well as in the city, as consumer goods or sociotechnical infrastructures – in the widest sense serve our consumption practices and daily routine. From this angle, commodities can be defined as all those material objects, things, and services that have become part of any (consumption) practice and thereby carry social meaning. Similarly, sociotechnical systems or infrastructures are made and configured out of sets of material structures and things that only gain social meaning through being employed in social practices. In fact, existence, character, and utilisation of infrastructure are closely related to governance, policies, and institutions and can be seen as a factor for and an outcome of economic development. Infrastructure in this specific context includes all those facilities and arrangements, services, and amenities – social and physical – that are privately or publicly established for any group of users or beneficiaries. Infrastructure can be streets, railways, shopping centres, parks, or public toilets in the same way as hospitals, universities, schools, or residential welfare associations (RWA) and so on. Sociotechnical infrastructure in this sense has to be understood as being systematically made of a plethora of components and things. In this view, a "flyover" as well as a gated community can be understood as sociotechnical infrastructural systems, i.e. systems made and configured out of material objects and things.

Lifestyles emerge and operate in selective correspondence with and in selective distancing from the elements and features present in the respective social field or action space. This is why these material objects and the social and physical infrastructure have to be taken into account. This includes getting hold of the multiplicity of symbolic meanings and functions of these things and their configurations.

Altogether, these components, elements, and features can be conceptualised as forming a stage and containing a whole bundle of props or requisites, all of which

4.1 Theoretical Framework

carry a multiplicity of socially constructed and ascribed social-cultural meanings. Differing stocks of capital resources of all involved actors, the market and its development of prices and costs (monetary and transaction), and the resulting power relations cater for differing access rights over the use of requisites as well as specific positions on stage.

The image or abstract concept of a stage and its related requisites allows for an easier explanation of the elements, meanings, functions, and processes at work. The abstract view on having access to infrastructure and other material objects that surround us analytically and purposefully reduces the complexity of class and social-position-related mechanisms to a simple dimension of having access to the material world and being positioned within the material culture of an (urban) society. This level of access to the material world and the character of being positioned within the social space describe the class-related differentials and the resulting power relations. This class-specific positioning is based on a coarse differentiation and confines and limits the action space for each actor. The fine and subtle differentiation ("Feine Unterschiede") can only take place within the confines of the coarsely defined action space.

To stay with the analytically intended simplification and image of the theatre, the play itself allows the actors to refer to and relate themselves with other players and with the material world. Based on their value orientations, preferences, and competences, they observe, perceive, and evaluate the continuously changing stage, its requisites, and the performance of other actors situated on stage. From his or her specific angle, every actor perceives, understands, and interprets the altering character of the material world and the social practices on stage differently. He or she "receives" the variously ascribed meanings of things and practices in correspondence with his value orientation and preferences and based on his social position. This specific reception always happens in reference to others and can be understood as a process of integration and closure, taking place on the meso level of lifestyle.

By taking into account material objects as well as socio-physical systems and infrastructure, symbolic and functional elements and aspects of the urban environment are more directly considered and integrated. Such an approach allows consideration of the relational-spatial aspects of the urban context and broadens the perspective on consumption and lifestyle. On the one hand, it includes physical-spatial features signifying routine spaces, such as the neighbourhood or parks as well as places where a lot of consumption occurs (e.g. supermarkets, corner shops, or takeaways) (Gregory 2009, p. 109, "consumption"). On the other hand, it takes into account more seductive spaces such as leisure parks, cinemas, and shopping malls, where much of the leisured consumption takes place and where mechanisms of spatial social exclusion quite directly become manifest through, e.g. surveillance (Gregory 2009, p. 109, "consumption").

The material configuration of urban areas has substantial impacts on the way of life of the people living in these areas. Not only commodities, such as cars, air-conditioners, mobile phones, and so on, but also places and infrastructure have an exclusive and thereby distinctive character, i.e. access to these material configurations is restricted based on a variety of criteria, formal or informal, subtle or definite.

For instance, gated communities commonly curtail access into the areas quite definitively and formally institutionalised. By contrast, shopping malls, supermarkets, leisure parks, cafes, restaurants, etc. work much more subtle with a diverse set of inclusionary and exclusionary means of framing the public and private spaces of the city.

Malls offer a good example of how commercialised public spaces successfully combine functions of shopping and retail with entertainment, gastronomy, and other "lifestyle-enhancing activities" (Allen 2006, p. 443). This integration of a whole set of social functions into well-designed and subtly well-controlled public spaces are "now more often designed to enable social interaction of a particular kind and to facilitate certain types of reaction to the aesthetic and recreational objects around them" (Allen 2006, p. 443). At the same time, these commercialised public spaces play out a modest form of power that "works through the suggestive pull of the design and layout, offering choices around movement and patterns of interaction" (Allen 2006, p. 445). Interestingly, for many urban dwellers, these commercialised places supersede formerly "real" public places in their functions for social interactions. In this way, an illusionary publicness is staged in a fully privatised space where urban amenities are commercialised and subtle mechanisms function as informal barriers for "those 'who don't belong' or appear 'out of place'" (Allen 2006, p. 442).

Many of these newly evolved commercialised spaces of urban "mall-style living" (Allen 2006, p. 442) and the newly available commodities or "requisites" tend to stand in critical opposition to the remaining traditional, prereform-era features of the urban economy and way of life. In terms of design, ambiance, and lifestyle, shopping in a mall or supermarket tends to stimulate characteristic effects, and it tends to offer a broader set of social-cultural functions as compared to the features related to shopping from a mobile street vendor or at a rather "traditional" corner shop (Kirana). Similarly, different modes of transport or the features related to leisure activities and holidays play out to be quite relevant in terms of distinction and social closure.

Surely, dichotomising such realms bears the risk of oversimplification and remains subjective. However, based on the data, the author has attempted to differentiate between the "new" and the "traditional" material world through principal component analysis (PCA). By means of participant observation and semi-structured qualitative interviews, material aspects of lifestyle and the role of the physical environment have been explored, and building on these results, the author has included these aspects in the quantitative survey. For the concrete operationalisation of lifestyle, the micro and meso levels are directly considered for the analysis. The study thereby attempts to get hold of processes of social categorisations, distinction, and identity formation, as well as material and behavioural aspects that serve as a holistic set of markers for social differentiation (cf. Lüdtke 1989:69). Analytically, these aspects and processes are differentiated according to the dimensions mentalities, conduct of life, and social position.

4.1.2 The Explanatory Level: Mentalities

Mentalities, which rest on subjective determinants such as values, norms, attitudes, and preferences, build the explanatory level of this analysis. An analytical separation of the three constitutive dimensions of lifestyle is not to say that these components do not strongly interdepend. Rather the opposite is the case, as mentalities may be strongly influenced by, e.g. the education and the educational level of parents, moreover by socialisation, caste, employment, etc. The same is true for routines and behaviour, which may strongly depend on attitudes and preferences, i.e. mentalities, but also depend on the social-economic situation, such as income, age, and/or gender. Based on the findings from earlier operationalisation in lifestyle research, it is not suggested to combine different components of lifestyle in the segmentation process. Such an approach thwarts the possibility of getting hold of relational aspects between the analytical elements (see Sect. 2.2.2.4).

Therefore, based on theoretical considerations about the objectives of this study, the author has decided the segmentation or clustering to be based on mentalities. In this study, mentality is operationalised based on a set of variables that measure general attitudes and preferences of respondents. The selected attitudes and preferences touch quite broad and more general issues of everyday life, and they deliberately do not focus on environmental aspects. This set of manifest variables then undergoes PCA in order to inductively reduce the number of variables on mentality into a few dimensions, represented in the form of latent variables. These latent variables that are meant to measure respondents' value orientation or mentality build the basis for the cluster analysis, which means that these variables are handled as an *active* set of variables (see Sect. 4.2.5).

Mentalities can be termed as the "soft" characteristics of lifestyle, i.e. values and general attitudes. Schwartz (1994, p. 21) defines values as "desirable trans-situational goals, varying in importance, that serve as guiding principles in the life of a person or other social entity". (Schwartz 1994, p. 21). Values differ from the notion of attitudes. Values are more abstract and, in fact, attitudes derive from values as life goals as they refer to an "evaluation of a specific object, quality, or behaviour as good or bad, positive or negative"(Leiserowitz et al. 2006, p. 414), i.e. values frame the attitudes of a person.

Surveying human values involves a range of challenges, as one cannot measure values directly. Measuring values requires a set of tangible attitudinal questions that are derived from an assumed targeted value. The set of values and attitudes, however, needs to challenge respondents with statements that touch upon characteristic friction lines and conflicting matters, so that the issues being raised are not evaluated on the basis of largely universally shared general cultural value orientations. Instead, as a matter of fact, the battery items need to deliver enough aspects for a social-cultural differentiation. To figure out enough of these aspects, the author first conducted a qualitative study: he talked to a diverse set of people, e.g. about the city in which they live, about changes they experience in their living environment, about their general expectations, and about their life goals and talked about and encouraged them to evaluate different actions, policies, people, and events (see Sect. 4.2.3). With this data and through participant observation, the author learned about the context, inner logic, motives, and goals of culturally and socially specific attitudes, conflicting lines, and potential cleavages. This study informed and facilitated the exploration and evaluation of social value orientation theories, such as the Schwartz value theory (Schwartz 1994), Rokeach's theory on the nature of human values (Rokeach 1973), or others such as the Kluckhohn-Strodtbeck framework (Kluckhohn and Strodtbeck 1961) or Inglehart's theory of value change in Western societies (Inglehart 1990).

Crucial for developing the survey tool was to keep the questions as simple as possible in order to avoid misinterpretations, both in the translation process of the questionnaire and in order not to overcharge the respondent with complicated questions and phrases. Especially attitudinal questions from known value surveys tend to often contain abstract concepts and complicated issues. In response to the problem of too-abstract question batteries, Schwartz et al. (2001) have developed a tool for measuring ten value constructs of their theory with a new and less abstract tool, the Portrait Values Questionnaire (PVQ). In this theory, the authors identify "a comprehensive set of 10 different types of values recognised across cultures", and they also specify "the conflicts and congruities among these values" (Schwartz et al. 2001, p. 519).

The PVQ consists of short and simple verbal portraits of different people, with each portrait describing "a person's goals, aspirations, or wishes that point implicitly to the importance of a value" (Schwartz et al. 2001, p. 523). For example, "Tradition is important to her. She tries to follow the customs handed down by her family" (Annex I) describes a person for whom traditional values are important, i.e. according to the Schwartz value theory, "respect, commitment and acceptance of the customs and ideas that traditional culture or religion provide the self" (Schwartz et al. 2001, p. 521). To give another example, "She strongly believes that people should care for nature. Looking after the environment is important to her" describes a person who appreciates universalism, i.e. according to the theory, "understanding, appreciation, tolerance and protection for the welfare of all people and for nature" (Schwartz et al. 2001, p. 521).

The attitudinal items are measured on a six-point Likert scale, on which respondents are asked to rate "How much like you is this person?" They can choose from the following ratings: "very much like me", "like me", "somewhat like me", "a little like me", "not like me", and "not like me at all" (Annex I). Respondents' values are inferred from self-reported rating on their similarity "to people that are implicitly described in terms of particular values" (Schwartz et al. 2001, p. 523). Due to the long-term and international experience in its application and due to its simplicity, conciseness, and comprehensiveness, PVQ was assumed to be highly suitable and effective for a segmentation based on social value orientation.[1]

---

[1] In combination with other items, the explanatory value of many of the PVQ items in PCA was quite weak compared to other items (see below). With communality values often below 0.4, the

4.1 Theoretical Framework

Complementary to the PVQ battery, a more culturally specific set of attitudinal questions was constructed in order to cover a measurement of attitudes towards changes in consumption and processes of social transformation. These aspects are particularly important with respect to the framing of lifestyle and sustainability. The Likert scale for these items was similarly created in six levels: "strongly agree", "agree", "somewhat agree", "somewhat disagree", "disagree", and "strongly disagree". Altogether 53 questionnaire items have been included in the final questionnaire. This quite large number of items provided a maximum of flexibility in respect to selecting a suitable set of items, broad enough to touch a variety of differentiating issues. After a number of explorative PCA trials, a set of 21 items have found their way into the final analysis.[2]

With the author's focus on consumption as a determinant of personal GHG emissions, many of the selected items draw on issues related to consumption with assumed targeted value orientations that tend to involve hedonism, materialism, enthusiasm towards consumerism, and stimulation. In presumed opposition to these hedonistic attitudes, stand items that describe orientations based on frugality and thriftiness. These Gandhian ideals of simplicity stand very close to attitudes that are rather driven or forced out of poverty and termed here "culture of necessity".

Moreover, the author was particularly interested in how these consumption-related attitudes correspond with or oppose issues touching upon the fields of tension between change and continuity and between globalisation and Indianness (cf. Brosius 2010, pp. 5, 12). While "modernity" is conceptualised by largely drawing on recent ethnographic literature on the (urban) middle classes in India (Brosius 2010; Donner 2008; Fernandes 2006, 2009; Lange et al. 2009; Säävälä 2010; Upadhya 2009; van Wessel 2001, 2004; Varma 1998), it is the concept of tradition that builds – in addition to the mentioned literature – on a Hindu mythological background. Most of the aspects, however, have also been raised as part of the qualitative study. Moreover, the author was interested to include attitudes that indicate an orientation towards *self-transcendence*, which is highly relevant in the context of consumption, related aspects of sustainability, and environmental and social engagement. To address this field and measure the value orientation of benevolence and universalism, some of these more generally oriented items from the PVQ have been supplemented by self-conceptualised items on social ecology, conscious consumption, and sense of community. Figure 4.2 depicts all targeted value orientations, and based on their relative positions, it roughly indicates how the concepts relate to each other in terms of content. This above-given short overview of the basic components of value orientations used in the survey will be further detailed in the following sections.

---

proportion of the item's variance explained by quite some of the explorative factor models was relatively low. In consequence, many of the PVQ items were left out from the final analysis.

[2] Annex provides an overview of all surveyed items, sources, assumed target value, and reasons for or against inclusion into the analysis.

Fig. 4.2 Concept cloud of all targeted value orientations included in the survey. The sizes of the words *coarsely* point to the number of items that aim to measure the respective targeted value. The relative positions of the concepts *roughly* indicate proximity or distance in respect to content

4.1.2.1 Attitudes Towards Consumption

As delineated in Sect. 3.1, political reforms based on New Economic Policy (NEP) as well as the closely related transformation processes of urbanisation and globalisation have created a context of rapid social-cultural change. Based on other research, especially on the new middle classes and the role of consumption, the author was aware of the lines of conflict between tradition and change that in particular is manifested in the realm of consumption. To measure these impacts and to get hold of respondents' attitudes and values towards a "modern" materialist culture, in which shopping, buying, and consuming are perceived as pleasure, enjoyment, and entertainment, a large set of items were created in order to cover the quite different aspects relevant in this field. Apart from the more direct and explicit hedonist pursuits of consumption, the notion of exclusiveness of shops, locales, and products bears more subtle and in some cases subconscious motives, related to emulation, identity management, self-realisation, and self-expression (cf. van Wessel 2004, p. 99). Also aspects of a virtue of simplicity and thrift, the role of financially forced frugality and last but not least aspects of a social-ecologically motivated consciousness of consumption have been included in the analysis (Table 4.1).

Hedonism and Materialism vs. Simplicity and Thrift

India exhibits a unique case for the analysis of consumption orientation, with its post-colonial accounts of the Gandhian ideal of simplicity, the Nehruvian era of socialism, and the post-socialist phase of economic reform and liberalisation in the early 1990s. With the latter period of economic reforms and rapid social transformations, the notion of consumption has changed substantially with regard to its meaning and its implications for other social issues. And indeed, it can be clearly stated that the occurrence of a materialist consumer culture in India is a rather new phenomenon.

Generally, and in the Indian context in particular, present-day consumer culture has been described as a "condition in which people seek self-realisation or self-expression through goods rather than through spiritual or social pursuits" (van

4.1 Theoretical Framework

Table 4.1 Overview of questionnaire items measuring targeted values towards consumption

Source	Item	Targeted value orientation
OC	It sometimes bothers me quite a lot that I can't afford to buy all the things other people have	Hedonism/ materialism
OC	I quite frequently shop in more expensive and exclusive stores	
OC	I like to buy things just for the fun of it	
OC	Going to shopping malls is one of the activities that I really like, even if I don't buy anything	
PVQ	Seeks for chances to have fun; do things that give her/him pleasure	
PVQ	Wants to enjoy life; having a good time	
OC	Often buying new things, such as clothes, electronics, etc., is very important for me in order to take part in social life of my friends and colleagues	
OC	I dislike a luxury brand when it is used by everyone	
OC	Things have changed a lot. I can now afford almost everything I wish to have and I go for it	
OC	The place where you live almost says everything about a person	
OC	Consumption as such is given too much importance in our society and doesn't make you happier in the end	Simplicity and thrift
OC	It is not good to accumulate wealth and material goods for the satisfaction of personal needs. It is enough to fulfil the needs of one's family	
OC	Seeking truth, harmony, and unity should be given priority over achieving wealth and prosperity	
OC	I believe that everybody should follow the virtues of austerity and modesty. This would make the world a much better place	

OC own conception, *PVQ* taken from Portrait Value Survey based on Schwartz et al. (2001)

Wessel 2004, p. 95). And indeed, with economic liberalisation, availability of consumer goods increased substantially, and with rapid economic development, income levels rose at an unprecedented pace. At the same time, all kinds of foreign products poured onto the so far protected Indian market (cf. Mathur 2010, p. 213), and the Indian middle class became a focus of broad media attention as a great new market for consumer goods (van Wessel 2004, p. 94). Moreover, the ubiquitously visible and sometimes exaggerated political and medial construction (Fernandes 2009, p. 219f; Mathur 2010, p. 212f) of this social-cultural transformation created and still creates a hopeful and aspirational outlook into the future (Aufbruchstimmung).

However, at the same time, this political and social-cultural context nurtures serious doubts and resistance against rapid social-cultural change especially among members of the middle classes, leading to religious revivalism (Section below) and an ambivalence or even denial of mass consumption and its legitimacy (van Wessel 2004, p. 95). Margit van Wessel, in her ethnographic study on the urban middle classes in Baroda, a provincial town in the Indian state of Gujarat, has remarkably shown these moral issues on modern consumption and the related ambivalence among members of the urban middle classes (van Wessel 2004, p. 95).

According to her research, this ambivalence becomes manifest between the observed and the self-experienced practices of modern consumption. Also Säävälä (2010, p. 122ff) with her research on middle-class moralities in Hyderabad illustrates this critical position and the moral attitude towards spending, largely among lower middle-class people in Hyderabad: in the context of a rapidly transforming material and media world, many respondents in her interviews "tended to create an image of the average middle-class person as greedy and excessive, while defending themselves against the same charge" (Säävälä 2010, p. 122). The informative examples of an ethically framed "self-presentation as frugal rather than extravagantly consumeristic" (Säävälä 2010, p. 122) underline the subtle linkages between frugality as a necessity due to financial constraints and a similar stance motivated out of virtue and moral reasons (see further below in this subchapter). These linkages are crucial to consider here, as they point to the importance of morality with respect to money and spending.

Apparently there is a moral ambivalence between the "consumerist urge" (Säävälä 2010, p. 121) being expressed by a "perpetual need to spend money" on the one hand and the construction of money as "an anti-social, destructive and morally dangerous element" (Säävälä 2010:124) on the other hand. While most of the mass and advertising media celebrate consumerism quite unchallenged, there seems to be a common notion among the middle classes which sets the frame more negatively by regarding wealthier and more educated urban people as having lost their moral integrity and being blinded by greed (Säävälä 2010, p. 123). Säävälä (2010, p. 123) points to the remarkable fact "that the idealising image of rural people's moral value" is so evident even "in the discourse of people who do not themselves have direct roots in villages" (2010, p. 123).

Also highly relevant in this realm is the prevalent idealism of simplicity and thrift (Mathur 2010, p. 226; van Wessel 2004, p. 95). While the above-delineated consumer culture orientation represents a rather new phenomenon in the Indian context, it is the latter cultural theme of austerity and simplicity which is largely based in Hindu mythological traits (see subchapter below) and revisited by Mahatma Gandhi. Mahatma Gandhi is "often admired as the sublime practitioner of the ideal of simplicity" (van Wessel 2004, p. 99), and his far-reaching influence cannot be overestimated in the Indian context, both culturally and also politically. However, even if the ideal seems to widely persist today also among the new Indian middle classes, the actual practices exhibit a different approach to consumption. Culturally, money and wealth do play an important role with regard to identity and self-expression. According to van Wessel, a practical realisation of simplicity and austerity is even "seen as standing in the way of survival in the dirty world" (van Wessel 2004, p. 100).

These two contradictory strands and the resulting ambivalence have cautioned the author to select the items that measure attitudes towards consumption carefully and also to get hold of the internal conflicts and ambivalences. Therefore, items have been created addressing hedonism and consumer culture orientation rather

4.1 Theoretical Framework

indirectly, being aware of the fact that respondents tend to downplay their own consumption behaviour. Säävälä in her book on middle-class moralities (2010, p. 122) refers to this tendency of her interviewees to create an image of the average middle-class person as greedy and excessive (Säävälä 2010, p. 122). Similarly, in quite a number of the qualitative semi-structured interviews conducted by the author, respondents rather discuss and expound emulative consumption behaviour of others, while professing to be immune against the enticements of modern consumption themselves. For instance, in the following quotation, a young engineering professional well depicts the impacts of advertising and fashion discourses on the young generation. He makes clear at the same time that he is not being affected himself:

> *When the changes started happening [refers to economic liberalisation], I have already passed my school days and got into professional career. So, I was matured to handle the change. Now, the people who are in the 15s and 16s, they start looking at all these gazettes, which are coming up, so the impact of all this is getting more on them. They can't handle it well and they become addicted, virtually killing their friend to get these things from the gazette; these are the negative aspects of rapid change. (Interview No. 022_2010_5_6: 10)*

Another quote from a quite well-off bank employee thematises the emerging practice of buying branded and more expensive fashion items. Similarly, he refers to IT professionals and interestingly members of the middle class who – according to him – seem to have fallen under the spell of large foreign companies and their expensive products. While he himself has been immune to this cultural change, he observes, understands, and supports his son's positive attitude towards the expensive exclusiveness of his preferred products:

> *Many people, those who are in software industry and middle class people, they prefer only branded clothes, and branded shoes. Many, I mean, these software engineers and all, they prefer branded items, and that is why there are branded shops, which are also very crowded. Of course, we purchase from outside only like local Bata or Action Shoes. But my son, he wears only Reebok or Adidas, quite opposite. [...] They are very light and comfortable. And he says they are durable also. To some extent, he is correct also. [...] To some extent, [...] comfort wise, they are good, when compared to some other brands and they are very light. (Interview No. 018_2010_5_2: 11)*

In order to avoid or reduce any social expectancy bias, which may evolve from this ambivalence towards consumption, the items on consumption have been selected very cautiously. The author has attempted to frame hedonism and the theme of conspicuous consumption under the notion of entertainment, enjoyment, and pleasure as basic feature of shopping and modern consumption.

The items have been formulated to draw on preferences instead of directly measuring the underlying values and motives in order to minimise the role of moral conflicts. One crucially important feature of "modern" consumption is associated with the combined and interconnected realms of shopping and consumption on the one hand and leisure and entertainment on the other hand. Commercialised semi-public spaces such as malls and leisure parks are a quite recent phenomenon in India, having emerged with the political economic reforms of the early 1990s. Apart from the simple act of shopping, these commercialised spaces have a lot to offer concerning individual requirements of enculturation, identity management, and conspicuous consumption. According to Mathur (2010, p. 213) "particular manners, ways of living and conspicuous consumption of goods which are simply not available to all classes of people serve as a means of gaining social status and repute. Conspicuous consumption, in a nutshell, becomes a status symbol in itself".

Subtly exclusive (see Sect. 5.2.3.4) and culturally novel, these places can be seen as mirrored images of the modern Indian consumer culture. The reason is that malls are able to holistically incorporate and exhibit nearly everything that is related to the images of modern lifestyle and consumption. Malls not only function as localities for buying things such as apparel, accessories, gifts, electronic goods, and lifestyle commodities, but they also offer a whole variety of options for recreation, wellness, entertainment, and leisure in form of beauty parlours, coffee shops, pubs, restaurants, and even amphitheatres and cinema halls (Mathur 2010, p. 221). Moreover, it is a typical feature of multiplex shopping malls in India that they are very well kept and extraordinarily clean. They are usually also observed and controlled by security staff and the entrances mostly have security checks.

In short, malls are systematically designed spaces with a very characteristic infrastructure created for customer fulfilment. Very similar to gated communities, malls are privately run, self-contained, secured, and somewhat exclusive spaces. They seemingly work under an invisible force that takes control of nearly everything, starting with air, temperature, and odours via sounds, colours, and light effects to the directions, movements, and even moods of their visitors. These meticulously organised super-worlds have become a physical, social, and cultural opposition to the existing outside world, an outside world that tends to be perceived in contrast as undisciplined, polluted, hot, noisy, crowded, and unsafe. Just as the material infrastructure is expected to function properly, people or visitors of malls are socially expected to function adequately as knowledgeable and well-informed participants of a newly emerging consumer class.

Gaining a sense of recognition as a member of this class therefore entails not only the financial resources but also adequate knowledge and skills to be able to participate. Nita Mathur (2010) in this context refers to the notion of conspicuous consumption. According to her research, it involves a "competition for status based on an individual's socio-economic competence" (Mathur 2010, p. 226). Malls provide free access to this "source of knowledge", and at the same time, they become testing grounds (cf. Brosius 2010, p. 2) to practise and experiment with the newly acquired skills. It is not surprising, therefore, that Christiane Brosius (2010, p. 30) associates this rather new practice of mall shopping in urban India with the concept of flânerie:

> A flâneur is a cultured person, a pleasure- and event-seeking expert of emerging capitalism and urbanism. [...] For her/him, viewing and buying came to be consumption as a stage play of pleasure; the performance of floating through urban arenas of capitalism is more important than the actual act of buying.

Indeed, as Brosius argues, flânerie requires an adequate setting or a stage to perform on, such as shopping malls, tourist and ritual sites, or leisure parks (Brosius 2010,

p. 30). These spaces function in different ways as a learning ground to acquire elementary knowledge on new forms of consumption and lifestyle.

This learning process is part of a much broader process of *enculturation*. This term is not meant to exaggerate. It is based on the assumption of rapid economic growth combined with an unprecedented prevalence of images that are created and exhibited in order to convey new forms of culture and living. Arjun Appadurai, in his path-breaking work on globalisation (1996, p. 53), stresses that "there is a peculiar new force to the imagination in social life today. More persons in more parts of the world consider a wider set of possible lives than they ever did before". In line with advertising and mass media, the above-mentioned arenas of consumption and especially shopping malls continuously create and exhibit a multiplicity of "fashionable" and ephemeral images that are made to be received, consumed, processed, and contextualised.

In addition to the theme of "seeing and being seen" (Brosius 2010, p. 23), there is another important dimension to the newly emerging consumer culture in urban India, namely, the notion of *Erlebnis*. Schulze claims this aspect to be foundational for modern society. It can be at best identified as an "eventful sensuous, deep and immediate experience" (Brosius 2010, p. 23). As shown above, shopping malls are testing grounds for modern lifestyle, but they may also work as stages, on which satisfaction and happiness become visible and one's own "good taste" and a "beautiful life" (Schulze 1992, p. 34ff) can be exhibited to the world (cf. Brosius 2010, p. 22).

Hedonist and conspicuous consumption plays a crucial role in managing one's own identity with regard to some members of the middle class. Taking part and being recognised in practices of modern consumption conveys a sense of belonging to the middle class or even "world class" (Brosius 2010, p. 22). The hedonistic dimension to it can be termed as a "feel-good factor" (Brosius 2010, p. 21), which translates into the behavioural aspects of everyday leisure activities. In this way, pleasure is generated through commercially reproduced recreational activities (Brosius 2010, p. 179). As also shown, however, this culturally rather new phenomenon does not come without due ambivalences and moral conflicts. The resulting factor solution of the PCA has interestingly merged variables delineating the paradigm of morally driven "frugality and thrift" with variables representing the paradigm "culture of necessity" (see below).

Culture of Necessity

Very important for the understanding of urban consumption patterns in a social-cultural context of prevalent poverty is the aspect of financially driven or forced frugality. Under conditions of financial constraints or even severe poverty, people have a very different perspective on consumption. For those who are poor, thrift denotes a determining standard of living and a basic attitude without any alternative:

> *The thing is, however hard I try, I only earn this much; but if I try to cross my limits it would mean a loss for us. Why should we make a loss for us? We should be satisfied with whatever we have. If for example we eat only how much we require it is ok, but if we try to overfill our stomachs, then it is a problem. Why should we try to show off? If I have more desire it is a loss for us. If you ask anybody in Hyderabad they say the same thing. (Interview No. 009_2010_3_23: 2)*

This statement drawn from an interview given by an illiterate middle-aged woman who lives in an informal settlement in Bagh Lingampally is quite illustrative for the assumed paradigm of frugality. In this analogy of food intake symbolising general consumption, the financial situation forbids a person to consume more than their own stomach can take. Fulfilment of desires that are beyond the very basic needs level is perceived to be harmful and seen as loss. And subtly, there is an implicit moral component also that says one should be happy with whatever one has. This financially driven *and* morally framed position on frugality and thrift draws on the Gandhian ideals of simplicity and asceticism (cf. van Wessel 2004, pp. 99ff; Mathur 2010, p. 226). Besides this normative connotation, the quotation implies a somewhat fatalistic view, which corresponds with the logic of religious traditionalism, as laid out in the following section. It conveys an attitude of unconditional satisfaction with the God-given social position. The following interview sequence indicates a similar strand, but with the difference of the respondent having achieved a slightly higher social status:

> *We are poor, definitely, but at the same time we have a little, we can afford. I am not able to explain. Sometimes it is okay, sometimes it is hard. It is like what only we work we will eat. Otherwise there is nothing to eat. It is okay, I have a number of responsibilities and my sisters go to school; like there are lot of bills and other things I have to take care, power bills, and water bills. My father has no income as he is blind. (Interview No. 010_2010_3_24: 4)*

This young man, who works as a car driver and lives in an informal settlement in Old Safilguda, evaluates his and his family's social situation quite positively and humbly. With his job, he himself can feed his family and take care of his sisters' education. Moreover, he was able to take a loan for getting his own house built for his marriage. His humble assessment is based on the assumption that hard work, thrift, modesty, and foresightful investments into basic future projects (education of sisters, having an own house) will most probably lead to an eventual success story for the whole family (see also cluster 1 in Sect. 6.4.1). However, his hand-to-mouth metaphor ("what we work we will eat") does not match his actual situation of having consolidated a quite reasonable status quo and a rather optimistic outlook for his whole family. Table 4.2 exhibits the items created for the measurement of the targeted value paradigm of "culture of necessity".

Social-Ecologically Motivated Consciousness Towards Consumption

Largely drawn from research experience from Europe and Germany (Peters et al. 2013, p. 230ff), the author has included a couple of questionnaire items that address consumption from a perspective of social-ecologically motivated consciousness and

4.1 Theoretical Framework

Table 4.2 Overview of questionnaire items measuring the targeted value paradigm of "culture of necessity"

Source	Item	Targeted value orientation
OC	It doesn't matter how much I earn; when I have a choice between two equal products, I usually go for the lowest priced product	Culture of necessity
OC	My life situation leaves no other option than carefully watching out how much I spent	Culture of necessity
OC	I carefully watch out how much I spend in order to save my money for harder times	Culture of necessity
OC	Whatever I earn goes into maintenance of basic needs	Culture of necessity

OC own conception, *PVQ* taken from Portrait Value Survey based on Schwartz et al. (2001)

responsibility. Conscious and responsible consumption involves an attitude that builds on reason and motivation for individual agency. It comes with a positive cognition about the meaningfulness and efficacy of one's own actions. This is especially the case with regard to a consciousness towards sustainable consumption. A social-ecologically conscious consumer perceives that he or she has an impact on production processes and the market. She or he is convinced of the relevance of individual consumption decisions with regard to the eventual effect on the market. Such a motivation is driven by the assumption that through imitation by others, and the cumulative influence on the demand for sustainable products, a general gradual change in awareness is effectuated. Hence, ethical consumption usually involves a great level of self-efficacy, which is an attitude that assumes one's own actions to be powerful enough to convey a message or even change something.

The earliest accounts of politically motivated boycott of consumer goods in India go back to the "Swadeshi" (indigenous goods) movement. First in Bengal in the early twentieth century and later driven forward by Mahatma Gandhi, the Swadeshi movement aimed to bring forward the Indian economy and make it self-sufficient, particularly against the British colonial textile industry. With the aim of "Swaraj" (home rule), Gandhi promoted the exclusive consumption of hand-spun, hand-woven cloth called "Khadi" (Trivedi 2003, p. 11; cf. also Vedwan 2007, p. 675).

Another more recent account of environment-related consumer concern has emerged over the issue of pesticides being found in products from Coca-Cola and PepsiCo. A report issued in 2003 by the Delhi-based environmental NGO Centre for Science and Environment (CSE) indicated the presence of pesticides in a number of popular beverages greatly exceeding European standards (Vedwan 2007, p. 659). The notion of an often taken-for-granted high quality standard of a well-known "Western" product was used to convey that the nearly unregulated pesticide industry in India has very serious environmental and public health impacts reaching as far as into a product like this (Vedwan 2007, p. 660). In reaction, a broad-based moral boycott of Coca-Cola and PepsiCo products began, with several states issuing a ban on these products and a drive to publicly destroy Coca-Cola and Pepsi bottles, initiated by several cultural nationalist organisations. Moreover, alternative and mainly "indigenous" products such as buttermilk and homemade lemonade gained importance, with some of these soft drinks being made of the same contaminated water as that used by Coca-Cola and PepsiCo (Vedwan 2007, p. 674).

Vedwan (2007, p. 663) contextualises this movement in a new strand of environmentalism in India, which has emerged since the onset of political reforms and economic liberalisation as part of the NEP. Ideologically, this new environmentalism appears to be based on the needs and perceptions of the urban middle classes against the background of deepening environmental problems, especially in urban areas due to rapid urbanisation, changing lifestyles, and industrialisation (Vedwan 2007, p. 660). It is contrary to the historical pattern of environmental movements in India, which essentially focused on the livelihood struggles of the poor and marginalised and the "issue of equity in relation to access and use of natural resources" (Swain 2014, p. 210). Much of this assessment draws on colonial and post-colonial accounts of largely rural-based material and discursive conflicts that have also struck other broader social and political struggles (cf. Mawdsley 2004, p. 79). The new environmentalism in India is rather urban-based and globally oriented "in tracing the origins of and the possible solutions to the environmental problems", aiming at issues such as climate change and related discourses on global climate equity (see Sect. 3.3) (2007, p. 663).

The case of pesticides being found in beverages well illustrates the new political role of the Indian middle class in a context of rapid change. Vedwan (2007, p. 660) argues that "in a postcolonial context characterised by rapid and uneven economic change, largely unrestrained by environmental and social safeguards, the question of how to reduce the often-destructive effects of such runaway growth, in the face of state apathy and even complicity, has never been more important". And the middle class is increasingly successful in setting the agenda, especially in the urban context. Middle-class environmental organisations have grown in number and impact, and in case of being unable to mobilise and sustain broader coalitions, they increasingly resort to involving the courts and forcing the state into action (Vedwan 2007, p. 664).

This newly gained middle-class identity and agency suits well to the notion of ethical consumption and the "participatory rhetoric" associated with it (Vedwan 2007, p. 675). A more general tendency towards taking into account societal and ecological consequences in purchasing decisions has been found among segments of the urban consumers (see also Singh 2009). This has been addressed by the first two items listed within the category of ethical consumption (Table 4.3). In order to also cover the trend of patriotic consumption, the author has included two questionnaire items that address aspects of product origin. *Make in India* is a campaign quite recently initiated by the Narendra Modi government in 2014, aiming to promote brands and products produced in India (see also Sect. 3.1; Table 4.3).

The above-outlined critical dimensions of consumption have indicated at the complexity, multiple dimensions, and ambivalences of modern consumption in urban India. This study attempts to address these dimensions quite comprehensively in order to cover a broad spectrum of attitudes towards consumption.

Table 4.3 Overview of questionnaire items measuring the targeted value paradigm of "social-ecological conscious consumption"

Source	Item	Targeted value orientation
OC	I try not to buy products made by companies which are socially and ecologically irresponsible	Social-ecological conscious consumption
OC	When shopping I regularly pay attention to the environmental friendliness of the products	
OC	I think that all this lifestyle and overconsumption is un-Indian. Foreign companies are just trying to make their profit with us	
OC	Indian people should always buy products made in India instead of buying imports from other countries	

OC own conception, *PVQ* taken from Portrait Value Survey based on Schwartz et al. (2001)

4.1.2.2 Tradition and Change

In a context of rapid and fundamental social-cultural change and in the midst of modernisation and rapid transformation processes, discourses on tradition and continuity tend to gain importance. Tradition, understood as commonly known and accustomed patterns of behaviour and a largely shared knowledge base about a characteristic cultural identity, is perceived to be challenged by processes of rapid change. More people are increasingly apprehensive of this perceived threat, and they become concerned about preserving their traditional values. Also among the Indian urban middle classes, it has been shown, for instance, that there is a tendency of religious revivalism (van Wessel 2004, p. 297; Varma 1998, p. 143). In such a context, measuring this concern and getting hold of the underlying oppositions and fields of tensions is fundamentally important for a segmentation based on social values. In following, the author will give an overview of the major lines of conflict and how these have been operationalised in form of items measuring these oppositions.

Religious Tradition

Very formative in early sociology is Max Weber's account on tradition, where he defines traditional action (traditionales Handeln) as one of four types of social action – one that is determined by settled habits and routines ("durch eingelebte Gewohnheit"). According to Weber, traditional action is quite far from meaningfully oriented action ("'sinnhaft' orientiertes Handeln"), rather being a reaction to a familiar stimulus based on customised attitudes (Weber 1922, p. 11). Thereby, Weber demarcates traditional action from instrumental-rational (zweckrational) and value-rational (wertrational) action (Weber 1922, p. 11). Deeply influenced by the processes of radical modernisation in Europe in the late nineteenth century, Weber conceptualises tradition as non-rational in opposition to rationally oriented modernity (Rosa et al. 2007, p. 23).

With this perspective on early European modernity, contemporary India offers some interesting parallels and features that are similarly challenged by recent processes of modernisation and globalisation. In particular, religious values that originate from a long history of written and orally traded Hindu mythology have a profound basis in the larger value system of the Indian culture. In an initial stage of this research, the author was quite moved by the writing of Max Weber on Hinduism in India (Weber 1986). Aiming to get hold of the question of why features of European enlightenment and modern culture emerged in the occident and not in China or India, Weber made a very thoroughgoing examination of "world" religions. The observations that Weber made on Hinduism start out at very foundational aspects of religion and culture, so that his findings after the passing of a whole century are still relevant, especially for an understanding of mythologically and religiously based traditions.

India's cultural historic background is characterised by a very broad spectrum of religious thought with polytheistic, animistic, pantheistic, monotheistic, and atheistic religious traits and a variety of religions that have emerged as a combination of these traits. Quite a few very important religions have their origin in India, and some of these have had a substantial impact on a global level. The constitution of India subsumes all religions that have their origin in India under Hinduism or Hindu religion and such a broad definition indeed makes sense, due to the historical background of the concept (von Stietencron 1995:143f). Hinduism in fact comprises a conglomeration of different religious thoughts and traditions, which in part also have different origins and which are based on different holy scripts (von Stietencron 1995:144). The author of this study therefore refers to this wider context and notion, when talking about Hinduism.

Max Weber's analysis of Hinduism mirrors a quite bleak picture of the foundations of Indian culture, where he argues that in contrast to, e.g. China, for instance, Indian religiosity gave birth to an ethic which is theoretically and practically the most world-denying of religious ethics that has ever existed:

> *Das Gebiet der indischen Religiosität [...] ist im stärksten Kontrast gegen China die Wiege der theoretisch und praktisch weltverneinendsten Formen von religiöser Ethik, welche die Erde hervorgebracht hat.* (Weber 1986, p. 536)

Based on this general observation, Weber argues that asceticism and contemplation have their earliest roots in Hindu religion. He states that asceticism is religiously rooted and that thereby wealth and worldly pleasures are negatively evaluated based on religious and moral arguments (Weber 1986, p. 536). This aspect is also relevant for a deeper understanding of the values of simplicity and thrift, which have been discussed in the section above. According to Weber, this cultural fact has its roots in the realm of religious tradition. Ghandi revived these ideals through his practised and widely promoted simple living.

Based on the idea of Samsara (transmigration) and the doctrine of Karman (retribution), two closely related aspects of Hindu rationalisation are found to build the foundation of a Hindu-specific theodicy. The answer to the question of how evil and suffering can exist under the eyes of an almighty creator, and the question of why one deserves a certain social situation, is according to Weber very well and very logically laid out in the Karman doctrine (Weber 1986, p. 167). Weber calls this logic of a strictly rational and ethically determined cosmos a result of the most consequent theodicy which has ever emerged in history (Weber 1986, p. 168). Based on the Karman logic, any ethically relevant action inevitably affects the destiny of the actor, and this is linked to the social fate of the individual and her or his position in society and within the organisation of the caste system.

In this logic, illness, affliction, and poverty are result of self-inflicted misconduct in previous lives; through self-determined action, anyone is able to influence his or her destiny after rebirth. The Karman doctrine corresponds with the logic of an eternal world and with the rationality of the caste system. Despite the existence of cyclically recurring eons (yugs; see below), in which chaos, disaster, and destruction stands at the end of each cycle, there is no such thing as the last judgement in most Hindu doctrines. After each closed sequence, there is a restart or resurrection with a new, millions-of-years-long cycle. As part of this eternal system, there remains an important individual agency, which allows everyone to work on and improve her or his social situation in forthcoming lives, well-illustrated in the following quote from Weber:

> *Wenn das kommunistische Manifest mit den Sätzen schließt: »Sie« (die Proletarier) »haben nichts zu verlieren als ihre Ketten, sie haben eine Welt zu gewinnen« – so galt das gleiche für den frommen Hindu niederer Kaste. Auch er konnte »die Welt«, sogar die Himmelswelt gewinnen, Kschatriya, Brahmane, des Himmels teilhaftig und selbst ein Gott werden, – nur nicht in diesem seinem jetzigen Leben, sondern in dem künftigen Dasein nach der Wiedergeburt, innerhalb der gleichen Ordnungen dieser Welt. Die Ordnung und der Rang der Kasten waren ewig (der Idee nach), wie der Gang der Gestirne und der Unterschied zwischen den Tiergattungen und den Menschenrassen. Sinnlos wäre der Versuch sie umstürzen zu wollen. Die Wiedergeburt konnte ihn zwar hinab in das Leben eines »Wurms im Darm eines Hundes« führen, – aber je nach seinem Verhalten auch hinauf in den Schoß einer Königin und Brahmanentochter. Absolute Vorbedingung aber war in seinem dermaligen Leben die strenge Erfüllung seiner jetzigen Kastenpflichten, die Vermeidung des rituell schwer sündhaften Versuchs, aus seiner Kaste treten zu wollen.* (Weber 1986, p. 170)

Weber further argues that especially for members of lower casts, there is no reason for upheaval or for striving for societal progress. The only way of escaping from the eternal cycle of rebirth and re-death is seen in salvation through merits in this world based on the logic of caste ritualism and the doctrine of Karman (Weber 1986, p. 171). In consequence, agency for determining the promising outcomes of rebirth does not involve any possibility of improving the situation for others, for society, or the world as a whole. The scope for improvements therefore does not involve any possibility to change the predetermined character of the current life. Effects remain restricted to subsequent lives and to the personal level. To better understand this rather complex issue, it is instructive to explain some of the related concepts and show how they are connected.

The Karman (also Karma) doctrine builds on a Hindu mythologically informed worldview, which paradigmatically conveys a divine and otherworldly tone, with the concept of Brahman at its fore. Brahman signifies a unitary life force or Supreme Being that "connects all existence, […] [it] has no form nor shape, is timeless and eternal, and is believed to pervade everything (animate and inanimate), and everything is it" (Deshpande et al. 2005, p. 132; cf. von Stietencron 1995, p. 150). This ubiquitous consciousness and witness of all existence is immanent in all humans as conscious Self, called Atman (von Stietencron 1995, p. 150). For humans to realise this unity between Brahman and Atman, i.e. the ultimate enlightenment, means to realise "salvation by release from karma, the wheel of rebirth" (Morris 1967, p. 591). Karman conceptualises the belief "that actions in one life determine fortune and status in the next" (Morris 1967, p. 590), and getting released from this eternal cycle of Samsara means to attain Moksha, the highest goal in Hinduism. How can this state of enlightenment be achieved, i.e. how can one remove "the layers of ignorance preventing one from being aware of the Atman" (Deshpande et al. 2005, p. 132)?

Mythologically, the answer can be found in the concept of Dharma. Dharma is understood as the rule of nature, and it provides an individual framework for righteousness, morality, and ethics. Dharma is the normative foundation of any human action and basis for a higher cosmic and moral order (von Stietencron 1995, p. 145). It provides every individual with his/her place in this hierarchical society, and his/her duties are prescribed by the caste of his/her birth (Sovani 1978, p. 651). Therefore, Dharma can be understood as the guiding principle for the way of life of every person to attain Moksha (Jain 2011, p. 110).

While the role of Dharma is very important on the individual level, the context of society plays a determining role for the individual, too. The Indian epic Mahabharata says: "Dharma is so called because it protects 'dharnat' (everything); Dharma maintains everything that has been created; Dharma is thus that very principle which can maintain the universe" (Lingat 1973, cited in Madan 1989, p. 117). That means, given that everyone follows his or her Dharma and seeks to be righteous, "Dharnat" is collectively protected from disintegration.

However, the Mahabharata and other Sanskrit texts known as Puranas also show that periodic destruction is inevitable as it is predetermined in the law of nature with the earth "being created and destroyed in cycles" (Narayanan 1999, p. 1). Calculations of time in Indian mythology are based on the concepts of Kalpa, Manavantara, and Maha Yuga. As concisely examined by Narayanan (1999, p. 2), 1 day in the life of Brahma represents one Kalpa, which again consists of a thousand so-called great eons, or Maha Yuga. The concept of the recurrent Maha Yuga contains four characteristic periods, (1) Krta or Satya Yuga, (2) Treta Yuga, (3) Dvapara Yuga, and (4) Kali Yuga. Kali Yuga is the age in which we live now. These periods are characterised in the Sanskrit texts, based on research done by Narayanan (1999, p. 2) as follows:

> The golden age (krta yuga) lasts [...] 1,728,000 human or earthly years. During this time, dharma or righteousness is on firm footing. Righteousness is on all four legs, if one uses a quadruped as the analogy. The treta age is shorter, it lasts [...] 1,296,000 earthly years; dharma is then on three legs. The dvapara age lasts half as long as the golden or krta age; it is 864,000 earthly years [...] and dharma is now hopping on two legs. During the kali yuga, the worst of all possible ages, dharma is on one leg and things get progressively worse. There is a steady decline through the yugas in morality, righteousness, lifespan, and

4.1 Theoretical Framework

in human satisfaction. This age lasts for [...] 432,000 earthly years [...] and this present cycle according to traditional Hindu reckoning is said to have begun around 3102 BC. (Narayanan 1999, p. 2)

According to this mythologically drawn law of nature, with cycles of emergence and destruction, of harmony and sorrow, of beauty and hatred, the character of our time is unavoidable and cannot be changed through individual action. Hence, there are both a notion of genuine responsibility for societal and moral decay and a fatalistic view on human existence in relation to nature.

As part of the qualitative research, quite a number of informants drew on this mythologically informed image of a bleak future of humanity that interestingly links the loss of tradition and basic values with a geophysically driven catastrophic termination of human society, as is well described in the following statement:

> *There are many sins, and in the near future the world is going to end up; we are going to face a lot of problems, loss, earthquakes; what I feel, we are having a better life today, but in the future this might change, and also lot of changes within the people who will try to commit many sins, like they don't take care of the parents, they don't respect them, they don't take care of the sisters, killing wives; all such things take place and then there is the end for the world. [...] As we are respecting my in-laws, maybe my children don't respect their in-laws [...] and so on; it carries on and there is no end until the world ends up; like something bad will happen, Tsunami or earth quake. (Interview No. 009_2010_3_23: 11)*

Interestingly, the respondent highlights the role and importance of family values and kin-related moral issues (e.g. respect to elders). But who is seen responsible? It is not humanity or the intended actions of certain groups of people that are responsible for an assumed loss of values and traditions. It is much more understood as a God-given destiny and cyclical termination and re-emergence as described with the mythological concept of Kali Yuga. Another respondent was able to describe the effects of Kali Yuga quite illustratively as follows:

> *As I know from the mythology [...] for every 7 villages there remains back only 1 village, the rest will be destroyed, like people die due to floods, earthquakes, and heat. [...] Kaliyuga is nearing; all these signs are the signs for the world to end, like there are so many bombs, over rains, lot of heat. The time comes, when rocks or huge stones fly off in the air due to the wind. [...] Today, people don't even give water to drink. [...] In the worst days [of Kaliyuga] people will die without food. Right now, [...] it is difficult to eat one meal every day. [People] are trying to kill each other by bomb blast; [...] people die; there is no respect to elders; all these are sins. (Interview 011_2010_3_24: 11)*

Floods, heat, over-rains, storms – weather-related events take an important role in his explanation of Kali Yuga, as the climate is a strong natural player in affecting humans in their everyday life. It seems that the respondent perceives it as nature's strongest weapon to fight humans' mistakes. Moreover, he appears well informed about what actually leads to Kali Yuga and what the impacts would actually look like. Explanations on Kali Yuga from the respondents similarly attribute all kinds of moral misbehaviour (not worshiping God, not respecting the elderly, forgetting about God and moral duties once people become richer, etc.) to the mechanism of purification and eradication of the sinful due to God's revenge.

Kali Yuga is not a marginal issue just known by a few very religious people. It seems to have a space in daily interactions. For instance, one respondent explained that her servant keeps on saying that the world is ending in 2012, and her priest states "God is angry with the people. What sins we have done so we have to reap" (cf. case 23, female, 81 years old, high household income, graduate). When asked directly if this would be Kali Yuga, she answered: "May it be Kali Yuga, I don't believe it, but everyone keeps on talking about it".

Table 4.4 portrays the six questionnaire items that have been selected for surveying the most fundamental orientation of tradition – religious tradition. In the following, the author will give reference to a rather moderate orientation of Indian-specific traditionalism, which is crucially important as it seems to mediate between more religiously rooted traditional values and modernisation. The author assumes that these moderate traditional values – in the following subsumed under the notion of "family tradition" – form a compromise that allows a person to embrace "modern" lifestyle practices and at the same time not lose sight of passed-on core values and basic features of their parents' identity.

Family Tradition

As the above explanations show, the paradigm of religious tradition is deeply rooted in Hindu mythology and conveys a quite radical and fatalistic worldview. In consideration of the findings in respect to "modern" consumption and the associated ambivalences, one gets an idea of the complexity of this realm of value orientations. Based on these observations and the findings in the qualitative research, the author of this study argues that tradition and modernity cannot be understood as two opposites on a linear continuum. Tradition and modernity are rather complex notions with multiple facets that cannot be measured on a linear unidimensional scale. With

Table 4.4 Overview of questionnaire items measuring the targeted value paradigm of 'religious tradition'

Source	Item	Targeted value orientation
OC	I believe that human nature is bad, evil, and wicked. To maintain social order, the only means are coercion and punishment	Religious tradition
OC	We cannot do anything about our future, as the decline of the society is predetermined	
OC	People are like they are due to their actions and ways of living in their previous lives	
OC	We are living in the age of Kali/[for Muslims read "We are running towards the End of Time"]. Evil and immorality dominate and we cannot do anything about it	
OC	Whenever there is a difficult problem, it is better to leave everything to God	
PVQ	Religious belief; tries hard to do what her/his religion requires	

OC own conception, *PVQ* taken from Portrait Value Survey based on Schwartz et al. (2001)

4.1 Theoretical Framework

the above-drawn focus on urban consumer culture, one extremely important realm of "modern" lifestyle has been delineated within the broader paradigm of hedonism and materialism. However, religious tradition is not necessarily an opposition to this hedonistic consumer culture, but there are clear lines of conflict between these two paradigms. Similarly, more moderate notions of Indian-specific traditional values, such as the often-referred-to ideal of the joint family and the norm of respecting elder people, is just an additional dimension of a differentiated view on the Indian social value system. This paradigm sets itself apart from the religious and mythological perspective with a rather secular framing. It emphasises the importance of naturally given in-group affiliations and related ideals of family and community. Thereby, it draws on related social-structural issues such as notions of extended family, related kinship ties, and roles and duties based on family and kin-related hierarchies, such as respect and obedience towards elders.

The dividing line between religion and secularism with respect to Indian traditional values was not intuitive at first. The author assumed that all items – also the more secular-oriented ones – would fall into the same dimension under PCA. In the explorative factorisations, it was therefore surprising to find statistical evidence for respondents making a difference along this line. The secularly framed position on tradition does not comprise otherworldly and fatalist propositions and rather builds on worldly and more rational statements. The differentiation between the two dimensions is very favourable for the analysis of social values with respect to tradition and modernity. This more differentiated framework is better able to depict the "different shades of modernity" (Brosius 2010, p. 31) and reveal more nuanced and socially specific orientations towards modernisation. It also contributes to better expose the ambiguities that arise in the midst of rapid social and cultural change.

For instance, a renunciation of religion does not necessarily include rejection of traditional values, such as those foundational for the family tradition paradigm. The following quotation illustrates this very well. It is taken from an interview with an upper-middle-class social worker, who regards herself to be very open-minded, modern, rational, and secular. She approved straightforwardly following all those norms and values which have been drawn as part of the family tradition paradigm:

> *I think there are some traditions that are really ridiculous, for example, if there is a solar eclipse and I am pregnant I have to lie down under the bed, I am sorry I am not going to follow that, and it means I am a bad Indian or bad Hindu, or bad Muslim? Sorry, than I am a bad Hindu, a bad Muslim, so what. But when I talk about traditions, the way we do Namaste or the way we are caring for our old people, [...] or respect the old, regardless what they have done for me or how they have treated me for example, the typical mother-in-law and daughter-in-law kind of syndrome, if I have problems with my mother-in-law, still, she knows and I know that I will look after her when she is old, regardless of whether I am wearing a pair of jeans or a saree. Regardless of whether I have a tattoo, I will look after her, when she is old. I will not put her into an old people's home. I will sacrifice everything to ensure that this woman is looked after. So, those traditions, yes, I will uphold, but if you are talking about religious traditions, don't ask me, I am not religious, I don't believe in God! (Interview No. 014_2010_3_25: 18)*

In response to a more general question on tradition, she makes an upfront statement against religion and blind faith in God, and she very clearly differentiates between religious and secular traditional values. Quite strikingly, without making any reservation, she emphasises the importance of giving unconditional respect and service to elderly members of the family, in this particular case, her own mother-in-law. She appears to quite strictly argue this traditional norm as being based on an inevitable law that has been inscribed everlastingly into the Indian culture and its basic value system. Consequently – according to her view – the same norm operates persistently and independently of any social-cultural change, meaning that a modern way of life (e.g. jeans, tattoos) does not counteract or oppose ideals of community and the joint family. And nor does a renunciation of religion – in her view – lead to a loss of general Indian traditional values that are not based on religion. This view is emblematic for all those members of the Indian urban middle classes who tend to have a rather global and cosmopolitan orientation and usually higher levels of education and income.

With some evidence, it can be stated that much of the social reality of the upper middle classes in urban India takes place in rather exclusive, closed-up, and air-conditioned spaces, which are elementary and determining for the overall socially fragmented urban landscape. This spatial context of modern urban lifestyle in India is very characteristic, and it structurally interacts with common features of consumption practices and ways of living: young urban professionals in Hyderabad are often forced into a nuclear family setup, they tend to bring up their children in internationally oriented private schools, they are usually bound to work 6 days a week, and usually they spend this time in suitable and air-conditioned office buildings and often tend to live in highly safeguarded spaces such as apartment complexes or gated communities. Quite a lot of their leisure time is spent within gated community areas or in similarly exclusive shopping malls, cinema halls, or secured parks.

Many of these controlled and well-organised spatial fragments of the city are created to convey a sense of distinction and exclusiveness that evidently differentiates from the so-far "unmodernised" India. And just in the way these exclusive spaces interact with "modern" lifestyle, so does the still ubiquitously visible traditional India oppose and challenge those who have distanced themselves through their social practices from this old way of life. While the rapid transformation processes drive many urban middle-class people away from passed down practices and their traditional way of life, they will still uphold well-known and very basic traditional values. And probably this fear of losing one's own cultural identity makes people even more likely to uncritically embrace family traditional values, maybe as an ethical compensation to the substantial changes in the practical ways of living and consuming.

The following quote from a middle-class woman, who lives with her husband, her mother in law, and her 3-year-old son, illustrates a critique that is often raised and which can be seen as representative for a more general discourse that turns around common features of modern urban lifestyle in Hyderabad:

> There are a lot of changes, people are not respecting their elders and this is a tremendous change. All this is because of the higher studies and going abroad and coming over from there people are not giving respect to their parents, also elders. Overall, the way of talking to the elders, the way they speak without any respect. They forget the culture where they were

4.1 Theoretical Framework

born and brought up; they try to follow the culture, which they see abroad and try to adopt. This is not good; we should follow our own culture. (Interview No. 008_2010_3_23: 6)

The respondent directly refers to an increased interaction with and influence from the Western world through migration and higher education. In her view, globalisation and the advent of modernity have started to transform basic cultural norms that are foundational for the Indian culture. Her bleak analysis addresses an allegedly unprecedented intergenerational conflict based on a loss of respect towards elders, which is – according to her observations – driven forward by people who try to emulate a Western lifestyle (cf. van Wessel 2001, p. 133f). As in some other instances of such moral ambiguities (see above; cf. van Wessel 2001, p. 243), she excludes herself from this diagnose and takes a strong position of resistance against this charge, being aware of the continuity of this norm that resides within herself. This continuity remains not only because she herself lives the ideal of the joint family. It is even more so because the underlying values persist through derived images of an ideal situation that is continuously contrasted with the present reality.

Kin-related hierarchies ("respect to elders") and the commonly upheld ideal of the joint family are quite striking in this context. In culture and personality studies, the sense of affiliation to a naturally given in-group such as the family is conceptualised as "collectivism" and in some cases even more specifically as "vertical collectivism" (Triandis 2002, p. 139). According to Triandis (2002, p. 139), "vertical cultures are traditionalist and emphasise in-group cohesion, respect for in-group norms, and the directives of authorities". Moreover, vertical collectivism is regarded to "correlate with right wing authoritarianism [...], the tendency to be submissive to authority and to endorse conventionalism. Both vertical collectivism and right wing authoritarianism correlate positively with age and religiosity, and negatively with education and exposure to diverse persons" (Triandis 2002, p. 139). The author rejects the statement that the Indian culture is a vertical culture. However, it is indicative that some of the mentioned studies have figured out cultural traits in the Indian context that are very closely related to the structure of the family tradition paradigm in this study.

Table 4.5 depicts an overview of the included questionnaire items that were created in order to measure the value orientation paradigm of family tradition. A more item-specific discussion of this and all other paradigms and their associated variables is given in Sect. 5.2.3.

4.1.2.3 Other More General Values: Schwartz Values and Values Towards Community and Sharing

Values are abstract concepts or beliefs that refer to motivational goals or guiding principles in the life of a person, i.e. values express aspects of foundational meaning and identity of a person. Apart from the very critical value dimensions outlined above, i.e. values informing attitudes towards tradition and modernity and towards

Table 4.5 Overview of questionnaire items measuring the targeted value paradigm of "family tradition"

Source	Item	Targeted value orientation
OC	I believe in our society marriages should take place within one's own caste/community	Family tradition
PVQ	Tradition is important; tries to follow the customs handed down by her/his family	Family tradition
OC	Obedience and respect for elders is a very important value and should be maintained	Family tradition
OC	In matters of marriage, boys and girls may be consulted, but the final decision should be taken by the parents	Family tradition
PVQ	Always behave properly; avoid doing anything people would say wrong	Family tradition
PVQ	Be obedient; believes s/he should always show respect to her/his parents and to older people	Family tradition
OC	I believe in the joint family system. One should subordinate one's needs, wants, desires, and aspirations to those of the family	Family tradition
OC	I think it is not possible to maintain the Indian tradition of the joint family system. Old age homes are a good alternative	Family tradition

OC own conception, *PVQ* taken from Portrait Value Survey based on Schwartz et al. (2001)

consumption and thrift, the author has included more general values from the Schwartz Portrait Value Questionnaire (PVQ) (Schwartz et al. 2001, p. 520).

The value model suggested by Shalom Schwartz (1994) builds on ten general social value orientations. The values are conceptualised in regard to their final goals in opposition to each other. According to the theory, these oppositions between competing values become clearer in Fig. 4.3. It organises the value orientations along two bipolar dimensions with two respective higher-order value types. In this way, "self-transcendence" contrasts with "self-enhancement"; "openness to change" opposes to the higher-order value type of "conservation". Schwartz' theory has been tested in several countries, and the model has proven to be transferable, albeit different cultures differ in their specific structure of values (Schwartz et al. 2001).

The first bipolar dimension contains self-transcendence in opposition to self-enhancement. Values of self-transcendence (universalism and benevolence; see Table 4.6) are critically important in regard to sustainable lifestyles as they define the extent to which values motivate people to transcend their own interests and promote the welfare of others.[3] Social-ecological concern is closely related with the general social value of universalism. Schwartz (1994, p. 22) describes the value with notions of "understanding, appreciation, tolerance, and protection for the welfare of all people and for nature". It conveys a rather extroverted attitude and implies

---

[3] Apart from the dimensions outlined in this chapter, four additional items have been included in the survey based on the Portrait Value Questionnaire (PVQ). Two items were selected covering the dimension of "security", which range closely with the dimension of tradition. Another two items addressing "self-direction" fall in the dimension of "openness to change" and neighbours with stimulation and hedonism according to Schwartz et al. (2001, p. 521f).

4.1 Theoretical Framework

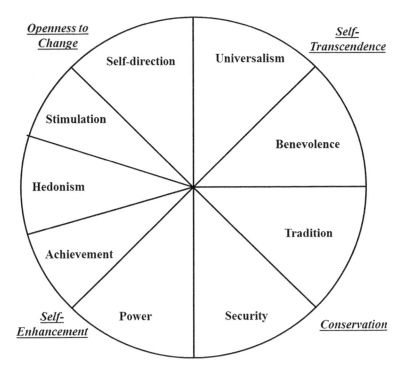

Fig. 4.3 Adapted model of Schwartz' theory on universal aspects in the structure and contents of human values, their relations, higher-order value types, and bipolar value dimensions. (Figure adapted from Schwartz 1994, p. 24)

interaction with society and environment. The locus of control over the outcomes of events and situations is therefore rather internally situated, and respondents supporting this paradigm tend to also have much higher levels of self-efficacy compared to, e.g. supporters of the rather fatalistic paradigm of religious tradition (see above). With its quite explicit social and ecological perspective, it also counters the hedonist conspicuous consumption orientation (see above). According to Schwartz' theory, universalism is highly oppositional to hedonism, as depicted in Fig. 4.3.

Benevolence, according to the Schwartz value theory, builds the second component of self-transcendence. It embraces "preservation and enhancement of the welfare of people with whom one is in frequent personal contact" (Schwartz et al. 2001, p. 521). The values benevolence and universalism are conceptualised in opposition to the value dimension of self-enhancement (achievement and power, see Table 4.6). The orientation towards self-enhancement tends to motivate people to improve their own situation also against the interests and benefits of others (Schwartz 1992, p. 42f). This is "power" on the one hand, which delineates "social status and prestige, control or dominance over people and resources". On the other hand, it is achievement, which involves "personal success through demonstrating competence according to social standards" (Schwartz et al. 2001, p. 521).

Table 4.6 Overview of questionnaire items measuring the targeted more general values based on Portrait Value Questionnaire (PVQ) and others

Source	Item	Targeted value orientation	Higher-order value dimension
PVQ	Believes that people should care for nature; looking after the environment	Universalism	Self-transcendence
PVQ	Every person in the world should be treated equally; everyone should have equal opportunities in life		
PVQ	Everyone should be treated justly; the weakest in the society should be protected		
PVQ	Help the people around her/him; wants to care for their well-being	Benevolence	
PVQ	Respond to the needs of others; tries to support those s/he knows		
PVQ	Be in charge and tell others what to do; wants people to do what s/he says	Power	Self-enhancement
PVQ	Wants to be the one who makes the decisions; likes to be the leader		
PVQ	Being very successful; hopes people will recognise her/his achievement	Achievement	
PVQ	Wealth and prosperity; believes that it is a sign or doing it better than others		
PVQ	New ideas and being creative; doing things in original way	Self-direction	Openness to change
PVQ	Make her/his own decisions about what he does; likes to be free and not depend on others		
PVQ	Looks for adventures; wants to have an exciting life	Stimulation	
PVQ	Likes surprises and is always looking for new things to do; do a lot of different things in life		
PVQ	Government ensures her/his safety against terrorism; government should be strong to protect its citizens	Security	Conservation
PVQ	Live in secure surroundings; avoids anything that might endanger her/his safety		
OC	I would be willing to work together with others to improve my neighbourhood	Community and sharing	n.a.
OC	I really appreciate sharing and exchanging things with friends and neighbours instead of buying so much new stuff		

OC own conception, *PVQ* taken from Portrait Value Survey based on Schwartz et al. (2001)

4.1 Theoretical Framework

Interesting also for the analysis of attitudes towards change and modernisation are values within the dimension "openness to change". These are "hedonism", which has been already addressed in the chapter above (values towards consumption; see Sect. 4.1.2.1 above), "stimulation", and "self-direction". "Stimulation" is an orientational pattern, which strives for excitement and novelty. A person who follows this paradigm tends to seek for challenges in life (Schwartz et al. 2001, p. 521). Similarly, self-direction is a value that creates an attitude of independent thought and action-choosing, creating, and exploring (Schwartz et al. 2001, p. 521). Situated in direct opposition to this dimension is "conservation", with the values of "tradition" and "security". Tradition has been treated already above as part of Sect. 4.1.2.2. Security as conceptualised in the Schwartz theory is situated very close to traditional values, with strong preferences for "safety, harmony and a stable society. People following this paradigm seek for stability of relationships, and of the self" (Schwartz et al. 2001, p. 521).

In sum, Schwartz' values form a comprehensive set of general social values, and the model allows measuring and comparing the structure of individual value orientations. In addition to these rather general and field-tested attitudinal items, the author has included two more self-conceptualised items that build a value dimension, which increasingly gains importance in the light of global challenges, such as climate change, environmental degradation, and economic crisis. This value dimension is an attempt to measure motivational goals that seek answers and try to directly respond to the limits-to-growth dilemma (Meadows et al. 1974) in search of alternative approaches to the future of humanity. The trend of sharing and collaborative consumption, which is more prevalent in Europe, is based on the principle of sufficiency and aims to avoid unnecessary consumption of resources (Belk 2014, p. 1596f). Also in urban India, a few initiatives refer to the principles of sharing and collaborative consumption. However, this trend still operates very subtly and becomes visible only on neighbourhood level. In Bandra West, Mumbai, for instance, a small neighbourhood-based "free market" (give-away shop) has been established by a group of citizens who got to know about and experienced the idea themselves during a visit to Berlin. Other initiatives and ideas have been communicated as part of informal talks by the author to students and young professionals in Hyderabad. The idea of voluntarily working together with and for the community was an interesting aspect also in reference to the ideal of sharing and collaboration.

The conceptualised items measure first, the willingness to work together with others to improve one's own neighbourhood. The second item addresses the issue of sharing by asking the respondent to indicate the level of appreciation of sharing and exchanging things with friends and neighbours instead of buying new things. Table 4.6 gives an overview of all included general value items as delineated in this chapter.

4.1.3 The Descriptive Level: Routines and Behaviour, Investive Consumption, and Social Position

Measuring values and attitudes is a challenge as results highly depend on the selection of items, which in itself is subjective. The range of general issues and everyday-life problems that can be drawn upon in order to form a broad and comprehensive measurement tool is nearly infinite. In consequence, the author had to be very careful in selecting the dimensions of measurement. The above-given theoretical and contextual considerations illustrate the complexity of this endeavour. As the measurement of values and attitudes builds the substance for the segmentation of value orientation groups, the author has put due emphasis and consideration on this conceptualisation. Much of this research process was deeply informed by the wider context of the research as it was outlined in Sect. 4.1.1 above.

The conceptualisation of the descriptive level was more straightforward and involved mainly methodological challenges. In particular, the carbon calculator and the measurement of investive consumption posed some methodological problems, as it was important to arrive at a solution with a clearly laid-out number of dimensions. This clarity was not trivial, because a too-large number of dimensions for the description of the value segments would involve problems of comparability. In the following, the conceptual aspects of the descriptive level will be laid out.

4.1.3.1 Social-Economic Position (SEP)

The social-economic position (SEP) is a crucial aspect for analysis of lifestyle as well as for any other social structure analysis. It is also widely used in more applied research domains, especially in epidemiological research (Howe et al. 2008, p. 2). It aims to differentiate the population based on life chances and living situation. SEP subsumes a multiplicity of advantages and disadvantages of a person, and unlike the concept of social stratum (Schicht), it allows the whole population to be categorised, including those, e.g. who are retired or those who have no income, such as housewives or students (Hradil 2001a, p. 371). SEP delineates all those aspects of living that can directly be experienced, quite similarly to with the German sociological concept of *Lebenslage*:

> Der Begriff der Lebenslage [richtet sich] auf die unmittelbar erfahrbaren Lebensbedingungen eines Menschen (auf die jeweilige Kombination seines Einkommens, seines Bildungsabschlusses, seiner Wohnbedingungen usw.). (Hradil 2001a, p. 374)

Lebenslage is as comprehensive as SEP, but it is usually conceptualised as an alternative to class or social stratum. In this study, relevant social-economic factors are consulted as individual descriptive measures in order to describe and compare the value orientation clusters.

In this study, SEP is conceptualised quite broadly with variables such as education, employment, caste, age, gender, religion, marriage, household size, income, as well as household and personal assets.[4] Education is measured on personal respondents' level, but in addition, the author has included questionnaire items that aim to get hold of the parents' educational level to measure cross-generation educational mobility. The classification of employment is based on the Goldthorpe classification (Evans 1992).

Caste is another highly relevant marker of vertical social-cultural differentiation. Based on communications with experienced social science researchers in Hyderabad and respondents in the qualitative interviews, the author was aware that surveying caste bears huge risks of social desirability bias. Nevertheless, the author had to include this aspect in the survey and in the evaluation of SEP. Responses on caste and tribal background were therefore kept open at first, without providing the respondent with a fixed choice set of answers. From the answers, the author has classified respondents based on Indian most-basic census categories of scheduled caste or scheduled tribe and general caste.

Respondents were also asked whether the household housed daughters who were not yet married. This aspect was surveyed, although it was not assumed to have such an effect on consumption and behaviour as was discovered in the final analysis (see Sect. 6.4.1). An item was also included to ask respondents whether they had own children still living in their household.

In evaluating SEP in empirical research, monetary measures such as income or consumption expenditure are often used as an exclusive measure. This is despite the general recognition that these measures usually fail to capture the diversity of well-being (Howe et al. 2008, p. 2). Moreover, income and consumer expenditure require extensive resources for household surveys (Vyas and Kumaranayake 2006), and especially household income is very difficult to measure, as explained in detail, e.g. by Shea Rutstein and Kiersten Johnson (2004): *First*, most people do not know their income or only know it in broad ranges; this is especially true for aggregate household income. *Second*, people are likely to hide their income for fear of government intervention. *Third*, income is not always pooled together by all household members. *Fourth*, household income can have various sources and is often variable daily, weekly, or seasonally. *Fifth*, it is also difficult to value home production and unpaid production of goods and services in a household. And *sixth*, there is the problem of unearned income, such as that gained through interest on loans, property rents, or gambling winnings (Rutstein and Johnson 2004, p. 2f). However, income is and remains a very crucial factor for lifestyle, because income determines the scope of choices concerning consumption. Income, for instance, allows one to understand the phenomenon of respondents with high levels of income and even so low levels of consumption. Neither consumer expenditure nor asset indices allow for a revelation of this highly relevant phenomenon.

---

[4] The author has also included other questionnaire items, some of which have not been mentioned here. All those variables were included, which have proved to have a significant differential effect from being member of the value orientation clusters. The results of these variables in regard to effects on cluster membership have been depicted in Annex VII.

Therefore, income has been included as one important factor of SEP. In order to allow for an income evaluation of respondents without personal income (e.g. housewives, students, etc.), household income has been surveyed and equivalised according to the household size (see Sects. 5.1.4 and 6.2). However, due to the mentioned drawbacks, an alternative measure has been searched for to also assess SEP in non-monetary terms. One such approach is the asset-based approach that has arisen from demographic studies such as the Demographic and Health Survey (DHS) (Rutstein and Johnson 2004). The asset-based approach builds on collecting information on ownership of a range of durable assets (e.g. car, refrigerator, television), housing characteristics (e.g. material of dwelling, floor and roof, toilet facilities), and access to basic services (e.g. electricity supply, source of drinking water) (Howe et al. 2008, p. 2). As it allows for flexibly integrating a whole set of multiple variables, this measure allows for social-economic differentiation across all social segments. It equally includes very basic amenities such as a mattress or a biomass cooking stove (chullaha) in the same way as air-conditioners and cars. However, these assets are statistically weighted through the well-established methodological approach of PCA based on intercorrelations. Many surveys use this index as a proxy for income or as a reliable indicator for consumer expenditure (see, e.g. Filmer and Pritchett 2001; Howe et al. 2008; Rutstein and Johnson 2004; Vyas and Kumaranayake 2006). In this study, it is used as a subsidiary measure of income in order to identify reporting inconsistencies in respect to income. It also functions as an indicative variable among others for the evaluation of SEP.

In addition to this subsidiary function of the index as part of SEP, the author has conceptualised the dimension of asset ownership in a different and more lifestyle-specific way. The ownership of assets – be it on personal or household level – indicates to the level of a quite specific realm of consumption, which the author defines as "investive consumption". Characteristic for this realm of consumption is the rather long-term use and the long-term lifestyle-defining character of these goods and services in this category. The specific features of "investive consumption" will be delineated in the following.

4.1.3.2 Investive Consumption

The concept of "investive consumption" builds on the idea of measuring and weighting ownership of durable consumer goods that symbolise quite long-term financial investments of a person or household into wealth and status. Asset-based indices have been widely applied for evaluating social-economic position (SEP) and for providing an alternative proxy for income. In this study, the asset-based index is additionally conceptualised as a tool to reflect and evaluate the level of long-term investive expenditures on assets and amenities, be it on personal or the household level. For this purpose, the asset-based index is even more accurate and much closer to evaluating a person's or household's capability of gaining and exhibiting social-economic status based on ownership of durable assets (e.g. air-conditioner, car), infrastructure, and housing characteristics (e.g. employment of servants, source of water). For instance, Filmer and Pritchett (2001, p. 116) argue that the asset index is not a measure to account for current consumption expenditures; it is much more viewed as a "proxy for something unobserved: a household's long-run economic status". Hence, the asset-based index is a tool to measure a highly relevant dimension of consumption. It captures social-economic status independent of short-term fluctuations in income and expenditures.

Conceptually most critical in this context is the intermediating role of "investive consumption" for the foundational elements of the lifestyle concept: a household's stock of material assets (measured with the asset-based index) can be viewed as a function of financial-economic capability (often evaluated on the basis of income or consumer expenditure) on the one hand and values and attitudes towards consumption and lifestyle on the other hand (see Fig. 4.4).

High levels of income do not necessarily translate into high levels of consumption. Overall, household consumption levels highly depend upon decisions being made with regard to saving and investments. Existing household infrastructure, market conditions, the cultural context, other contextual issues and boundary conditions, and in particular values and attitudes of (influential) household members play a significant role in consumer decision-making (see Fig. 4.4 for a very simplified model). For instance, quite a few respondents in the qualitative survey have raised concern against modern consumption, referring to notions of simplicity and thrift (see Sect. 4.1.2). However, as will be shown in the following (see also Sect. 4.1.3), behaviour and consumption is not necessarily based on choice and decision-making. Routines and patterns of behaviour, in particular, are not repeatedly reflected and consciously decided upon by the individual, and hence, they are not so much based on choices being made. On the contrary, larger investments are usually based on conscious decisions. For instance, whether an air-conditioner is being switched on when it is hot is not so much a matter of choice. In many cases it is rather based on repeated, reactive, and known patterns of behaviour. However, buying an air-conditioner is based on a usually well-thought-out consumption decision. What is most relevant in this illustrative example is the fact that the individual has access to an air-conditioner (AC), i.e. a routine of using the AC has evolved largely because an AC has been purchased for the household. Contrary to the routine-based use of

Fig. 4.4 Simplified model of consumer decision-making with regard to investive expenditures (investive consumption). (Source: own draft)

an AC, which happens to be rather subconscious and unreflected, it is this purchase decision of buying an AC that is based on a consciously made choice.

This differentiation is one of the foundational assumptions being made in this study. It is the attempt to address a fundamental critique being raised especially from the viewpoint of social practice theory, that much environment-related social science research is based on the ABC paradigm (see Sects. 2.2.2.5 and 4.1.3.3). Investive consumption is based on rather consciously made choices to invest in or to purchase certain long-term consumer goods (e.g. AC, car, and motorbike), contract-based long-term service arrangements (e.g. employment of servants), or household infrastructure (e.g. a private water connection). These investive consumption goods tend to be relatively more costly, and they are usually used within longer time horizons.

Most critical in this regard, the ownership of many of these durable assets tends to create path dependencies for conduct of life, daily routines, and everyday consumption (as shown for the case of the AC). Many durable consumer goods are able to influence and in some cases even determine certain aspects of living. The above-given example of the purchase decision to buy an AC has illustrated the potential of path dependencies associated with the investment in long-term consumer goods. And such a decision has therefore several implications for the analysis of lifestyle. Owning an air-conditioner in the Indian context is not so common, and therefore having one involves a high potential for a gain in status. As an air-conditioner is highly visible and as it provides a means to please the guest with comfort, it is an item which exhibits a certain social-economic status.

Moreover, also concerning personal-level carbon footprint analysis, a differentiation between everyday and investive consumption is of due relevance. As the ownership of certain durable consumer goods tends to translate into particular long-term patterns of consumption, it is in fact these path dependencies that allow for an evaluation of the long-term environmental or climate-related impact of investive expenditure decisions. The author has therefore developed an approach to estimate the personal-level average emission effect of selected investive expenditure decisions, such as the purchase of a car, motorbike, air-conditioner, washing machine, etc. With this estimation, the study provides an overview of key points of intervention (cf. Bilharz and Cerny 2012). It indicates the relevance of path dependencies that investive consumption decisions pose, and it allows for assessing the reduction potential that avoidance of certain investive consumer goods may have. The latter point may be of interest in particular for consumers, who thereby are provided with relevant information in regard to the long-term effects of investment decisions (see Sect. 4.1.4).

4.1.3.3 Routines and Patterns of Behaviour

Environmental research that focuses on the impacts of human behaviour on the environment and climate change has been criticised to emphasising consumer choices too much and neglecting the relevance of routines, habits, and patterns of

4.1 Theoretical Framework

behaviour. Following Giddens, Spaargaren (1997, p. 30), for instance, argues that consumers are deeply involved in producing and reproducing structural constraints and opportunities and that as a result, domestic practices are both "actor driven and system imposed". Similarly, Elisabeth Shove states that environment-related social science research commonly follows the "dominant paradigm of 'ABC' – attitude, behaviour, and choice" and challenges the assumption that behaviours are largely being motivated by beliefs, values, attitudes, and preferences (Shove 2010, p. 1273). Shove, Watson, and Ingram contend, "consumption is embedded in relatively inconspicuous routines occasioned by the characteristically mundane socio-technical systems of everyday life".

This argument is duly relevant for an understanding of the cultural dimensions of ordinary consumption. Large shares of personal and household GHG emissions can be traced back to those areas of consumption that are – as Shove (2003, p. 9) rightly states – customary and based on everyday practices. These everyday practices are "undertaken in a world of things and sociotechnical systems that have stabilising effects on routines and habits" (2003, p. 9). The theory on social practices, however, focuses on these aspects of everyday life, while the individual level, values, attitudes, and preferences are less important or even neglected. The lifestyle concept, on the other hand, allows for an integration of both, taking into account the more conscious consumption choices (in this study framed under the concept of investive consumption; see above) as well as behaviour, which functions more implicitly, subtly, and based on routines and habits.

Routines and patterns of behaviour are conceptualised in this study as human activities that are conducted by the same person repeatedly in very similar ways and patterns. The conduct of routines and behaviour patterns tend to take place subconsciously and largely without reflection. Reflection upon these habits and routines can be triggered, for instance, when the actor is confronted with the electricity bill (e.g. in the case of using an air-conditioner) or when any other irregularity disturbs the routinised act of doing something (cf. Spaargaren 1997, p. 28).

Lüdtke (1989, p. 40) states that lifestyles primarily evolve on the basis of private investments and consumption decisions. In fact, larger investments are not only most visible for the social environment and instrumental for social distinction, but larger investments to a great extent also provide the long-term and path-dependent infrastructure for lifestyle (e.g. car, washing machine, air-conditioner, see Sect. 4.1.4). Accordingly, operationalising the dimension of performance with a strong focus on consumption and investments is in line with Lüdtke's observations and allows for a more targeted analysis along the lines of the most constitutive elements of lifestyle-specific practices.

These everyday patterns of behaviour and daily routines are among the most relevant aspects of lifestyle for environment-related lifestyle research. In particular, it is all those areas of everyday routines where consumption of resources and the release of GHG emissions play a role. Any of these routines usually involves a multiple set of complex direct and indirect impacts on the environment. By means of the qualitative study, the author has identified a number of dimensions in the everyday life of people, which are, *first*, relevant for people in Hyderabad across all social segments. This aspect is important for the purpose of comparison among and between different social-cultural segments of the urban population. *Second*, it was maintained that the selected dimensions of routinised behaviour involve sufficient variance for the purpose of differentiation. And *third*, attention was paid to the fact that the selected areas of consumption and behaviour involve aspects relevant to the subject of climate change, most importantly in regard to the effects (GHG emission).

The identified areas are *shopping, leisure,* and *holidays. Media use* has also been included, as it represents an additional indicator interesting for environment and climate-related lifestyle research, namely, education, knowledge, and awareness with regard to environment and climate change. Moreover, *expenditure on internet* as an indicator for the use of the internet, the level of *meat consumption*, and the *dominant mode of transport* were included in the analysis. All areas of consumption behaviour and everyday routines have been selected based on the insights gained from the qualitative study. Participant observation has also been highly relevant in this regard. Especially concerning shopping, leisure, and holidays, the author had to cover a broad spectrum of possibilities in order to allow for differentiation across all social-cultural segments.

4.1.3.4 Carbon Calculator

Besides the segmentation of value orientations, it is the carbon calculator which represents the centrepiece of the overall analysis in this study. The carbon calculator was developed based on experiences from the GILDED Project (Peters et al. 2013, p. 226ff) and based on the findings of the preliminary qualitative study. Apart from the conceptualisation, the most fundamental input for the calculator was based on the consultation of a database that contains a comprehensive set of Indian-specific and in part regional-specific emission factors. This database has been computed and compiled by no2co2 (Gilani 2010, 2012), a core project partner of the Sustainable Hyderabad Project. These emission factors were compiled using a number of different approaches, including primary research of industry data and technical literature review. Some of the emission factors were also suggested by the IPCC Tier1 Emission Factor Databases. The database of factors does not claim covering the entire product life cycle. For some resource uses, only direct emissions from fuel combustion and electricity consumption have been considered (Gilani 2012, p. 4). The database has been verified and validated by the Indian Institute for Management (IIM) and by the School of Public Affairs, University of Colorado Denver. According to our research, the provided factors are the most accurate factors available for India, even though they should be used for indicative purposes only, have a finite degree of uncertainty, and are expected to vary with time (Gilani 2010, 2012).

4.1 Theoretical Framework

The carbon footprint of a person or household is an approximated reflection of all direct and indirect GHG emissions released as an effect of a person's or household's overall consumption. It is measured in tonnes of CO2equivalents (CO2e) for a time horizon of generally 1 year. The amount of GHG emissions released as an effect of an activity (e.g. consumption of a particular good or service) is expressed in sum through an emission factor. Hence, an emission factor is the rate of GHG emissions of an activity measured per unit with an inclusion of output as well as the input of all resources and by-products.

A direct measurement of personal-level GHG emissions by means of a household survey requires a well-balanced approach between maximising of accuracy and practicality. It is not feasible to cover the whole life cycle of all goods and services being consumed by each person of a household. Nor should the calculation be too vague and cover only direct emissions. At the time, when the survey was conceptualised, there were only a limited number of emission factors available, and these related mainly to those activities and goods that are most relevant for carbon footprinting, such as motorised transport and electricity consumption. Besides restrictions in regard to the availability of emission factors, the author had to limit the number of questionnaire items, partly because surveying of consumption data demands quite some time and space.

In order to justify all the above-mentioned requirements within the given scope of possibilities, the author selected the following realms of consumption for the carbon calculator: (1) private motorised transport, (2) public transport (intracity transportation), (3) long-distance travel (train, bus, and air travel), (4) food consumption, (5) electricity, and (6) cooking. The set of selected aspects is on par with existing carbon calculators in Germany (e.g. Umweltbundesamt Calculator, see www.klimaktiv.co2-rechner.de/) or in India (e.g. no2co2 calculator, see www.no2co2.in/). Overall, this calculator offers a simple measurement of the most relevant domains of consumption in regard to household-level and personal-level GHG emissions. Ideally – based on the surveyed data of this study – it would have been feasible to also estimate the amounts of emissions associated with the incorporated emissions of consumer durables, such as electronic items, white goods, etc. However, reliable emission data on such incorporated emissions of major household equipment was – at the time of this analysis – not available. However, from this research and from the emissions calculations, the author has developed a new approach to estimating average emissions associated with the use of certain technologies. This methodological approach is outlined in the following.

4.1.4 Consumption-Practice-Oriented Carbon Footprinting to Measuring GHG Emission Effects of Consumer Decisions

In Sect. 4.1.3.2, the author has highlighted the relevance of investive consumption for the analysis of lifestyle. Investments in household and personal, more durable assets and amenities tend to determine associated consumption practices for longer time horizons. Such investive consumption decisions often create path dependencies that substantially structure consumption patterns of households and individuals. The decision to buy a car or the fact that a car is accessible for an individual is likely to considerably affect the person's mobility pattern and is likely to determine the person's dominant mode of transport. The same applies, e.g. to the purchase of an air-conditioner that is usually bought for using it for cooling the house or apartment. As an effect, other lower-carbon room conditioning practices tend to lose relevance. The author has therefore considered the possibility of estimating the effects of such investive consumption decisions in terms of average increases in the overall carbon footprint of individual consumers. Similarly, it is possible with such an approach to get hold of emission effects of lifestyle-related consumption decisions or routines, such as the regular consumption of meat or dairy products.

With the focus on GHG emissions, the spectrum of consumption practices relevant for the analysis is straightforward. Moreover, almost all of these relevant practices can be traced back to a certain social-technical system, the material and technical basis of a consumption practice. Only with regard to food emissions is the material dimension nontechnical. For all other sectors, the practices are coupled to a material dimension that is based on technology, and this is in particular true for all those practices that are most relevant in regard to GHG emissions: almost 50% of all individual carbon footprints are from the household-based use of electricity and therefore can be traced back to the usage of electric household appliances. Another 16% per average goes back to individual motorised transport and the use of motor vehicles which are owned in common by members of the household. Moreover, there are emissions from the use of cooking fuels, making up 8% of all surveyed personal GHG emissions. These emissions are also related to and require durable cooking appliances such as LPG or kerosene stoves.

Other than the above-mentioned sectors of consumption, taking up a specific practice in the realm of public transport and long-distance travel does usually not require an initial personal investment in any technical equipment or any long-term contract (except, for instance, monthly bus tickets). That people often remain true to established practices in this sector is rather related to routines and the familiarity and knowledge that people gain from the recurrent character of the practice.

The approach does not require other data than those surveyed for the carbon calculator and the wealth index. With such a perspective, it would also be possible to analyse the practice-related symbolic meanings and individual motivations of people who 'carry' specific social practices. But that would require a more practice-specific approach concerning the survey. With the focus on lifestyle as in this study, an analysis of such scope is not possible, but the author has made the attempt to measure practice-specific average GHG emissions. The author exemplifies this attempt in the form of an "excursus" that is assumed to facilitate an understanding of lifestyle-specific differences in consumption patterns and personal GHG emissions. It also aims to explore the relevance and scope of this simple methodological approach.

4.2 Methodological Approach

4.2.1 Research Design and Methodological Approach

The research process of this study builds on three important methodological components: first, an explorative phase, which informed the author's understanding of the research context in Hyderabad, of its major stakeholders engaged in the field of sustainable urban development and climate change mitigation (cf. Reusswig et al. 2012), and of the role and quality of climate change perception among the public as well as among relevant experts and stakeholders (Reusswig et al. 2009; cf. Reusswig and Meyer-Ohlendorf 2010, 2012). Second, a qualitative household survey with semi-structured interviews was conducted in order to assess the public awareness, understanding, and perception of climate change (cf. Reusswig and Meyer-Ohlendorf 2010). The qualitative survey also provided the basis for developing and testing the proposed carbon calculator that represents a core element of the following quantitative analysis. Most important for this study, however, was to explore, assess, and test relevant indicators for measuring cultural and context-specific attitudes and values as well as consumption and behaviour patterns, relevant to anthropogenic climate change.

The third and most critical component of this study builds upon the quantitative survey. It comprises assessment of social-demographic data, aspects of general values and attitudes, routines, and behaviour patterns, data on investive consumption, such as personal and household amenities, and last but not least data as basis for the carbon calculator, i.e. everyday consumption. The data from this survey provides the data source for the proposed segmentation of value orientation patterns, for the carbon calculator, and the analysis of lifestyle-related consumption and behaviour patterns. Figure 4.5 outlines the research design with all the involved steps and components of the overall research process.

The following chapter will present the methodological and analytical steps of the above-mentioned research design in more detail. It will first summarise the aspects of the explorative phase. It will give an overview of the methodical and analytical approach with respect to the qualitative study, and it will then give a comprehensive description of the quantitative survey, its sampling process, data collection, processing, and analysis.

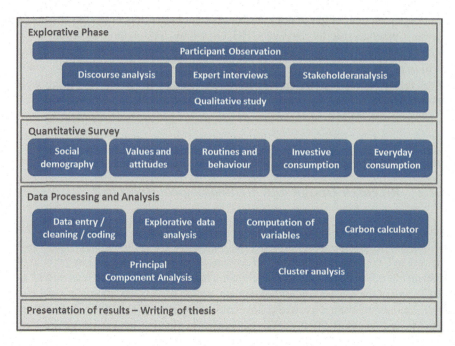

Fig. 4.5 Research design: outline of the research process. (Source: Own draft)

4.2.2 Preliminary Research and Gaining Access to the Field and Topic

A helpful source of foundational information and knowledge was gained through the project-related, rather explorative studies as part of the initial project phase between February 2009 and January 2010. To better understand the relevance of climate-related issues and climate change in the public discourse in India in general and in Hyderabad in particular, the team of researchers conducted expert interviews with administrative and technical staff, NGO officials, university teachers and professors, urban planners, and journalists (Kimmich et al. 2012; Reusswig et al. 2012). In order to better understand some of the root causes and also the cultural specificity of the public understanding and perception of climate change and sustainable development, a comprehensive media analysis was conducted on the topics of climate change, sustainability, and the impacts of extreme weather events, both in national English newspapers (1-year coverage of *The Hindu* and *The Times of India* (Reusswig et al. 2009)) and local Telugu-language newspapers (3-month coverage of the *Eenadu, Vaartha, Andhra Jyothy*[5] (Reusswig and Meyer-Ohlendorf 2012)).

---

[5] Because of the language barriers, the research, selection, and translation of articles were conducted by our partner Centre for Media Studies (CMS). I would like to thank CMS for their support.

4.2 Methodological Approach   125

Moreover, a methodologically informed stakeholder analysis with a focus on urban development and planning, sustainability, and climate change mitigation and adaptation gave the author useful insights into the urban governance structure of the city and helped him to gain expedient knowledge for some of the major project objectives: to initiate an effective stakeholder process and develop a participative perspective action plan for the future development of the city (Kimmich et al. 2012; Reusswig et al. 2012).

These research steps were accompanied by extensive field visits in Hyderabad, which allowed the author to realise a methodologically structured participant observation process. Photographic documentation and informal talks accompanied by taking field notes were the techniques applied in order to deepen understanding of and familiarisation with the research context.

4.2.3 Qualitative Study

Intertwined with the preliminary research outlined above, the most critical element of the preparations for the lifestyle survey was the qualitative study that was conducted between March and May 2010. It was meant *firstly* as an exploratory study that provides insights into the group-specific differentials of individual carbon footprints, their perception and knowledge of the local climate conditions and the concept of global climatic change, and their relation to strong climate signals (such as heat waves and strong rain events) (published in Reusswig and Meyer-Ohlendorf 2010). It *secondly* aimed as the foundational data source to build the standardised questionnaire for the representative survey that was conducted between September 2011 and April 2012. And *thirdly*, the qualitative study facilitated the interpretation of the results of the quantitative survey, especially with respect to the analysis of values and attitudes and the understanding of the cluster solution (see Sect. 4.2.5.2). The key thematic foci, objectives, tools, and aspects of this qualitative study are outlined in Table 4.7 below.

In total, 27 semi-structured interviews were conducted in this way, also containing a quantitative part addressing personal and household energy consumption and consumables, relevant in respect to GHG emissions. Additionally, social-demographic information such as household income, personal assets and consumer durables, migratory background, caste, religion, etc. was surveyed.

The respondents were selected through theoretical sampling (Flick 2006, pp. 125, 128; Strauss 1987, p. 39) based on the following aspects: respondents gender, age, income, education, and area's distance to the core city. The location aspects were included in the sampling in order to cover various localities and to understand location preferences of different social groups. The location analysis was done through GPS logging of the household and GPS-tagged photographic documentation of the

Table 4.7 Key aspects and foci, objectives, tools of the qualitative survey

Thematic focus	Key objectives	Additional supportive tools	Key aspects
1. Planned lifestyle segmentation	Understand the relevance of the lifestyle concept and the character of distinctive behaviour in the Indian urban context based on social position, value orientation, and conduct of life	Household location analysis (GPS logging, mapping: photographic documentation, observation protocol)	Value orientation (attitudes, world view, aims in life)
		Informal talks	Conduct of life (social practice, behavioural and consumption patterns, endowment)
		Participant observation	Social position
	Identify segmentation indicators for planned quantitative survey	Newspaper screening	
2. Social representation of climate change	Understand group specific perception of climate change	Informal talks participant observation newspaper screening	Perception of environmental pollution
			Energy-saving behaviour
			Perception of "weather changes" (cf. Reusswig and Meyer-Ohlendorf 2010)
			Reasoning of "weather changes"
			Knowledge and ideas of the concept of climate change, reasoning, emotional responses, perceived need for change
			Solutions with respect to climate change mitigation and adaptation
3. Climate affectedness	Understand differing levels of affectedness from climate related impacts		Affectedness from and coping with heat waves, strong rain events and flooding
4. Carbon footprinting	First assessment of group-specific carbon footprint		Household electricity use
			Inner-city mobility
			Long-distance mobility
			Cooking fuel
			Meat consumption
			Investive consumption/consumer durables

area around the interviewed household.[6] Care was taken that a good coverage of the different locality types and different social-economic groups was achieved.

A translator was present during most of the interviews. The interviews were audio recorded and additionally recorded by hand in an interview protocol. The audio data was processed through verbal transcription by the researcher with the support of a professional transcriber. For the purpose of this study, the transcribed data was initially examined completely. In a second step, only the relevant material for this study was extracted for further analysis. The actual analysis was then carried out based on qualitative content analysis (Mayring 2002).

In sum, this data and information laid the groundwork for the development of the lifestyle survey that integrates the assessment of general values and attitudes, environment and climate perceptions, behavioural and consumption patterns, a carbon calculator, and a wide array of demographic characteristics. This comprehensiveness required a complete standardisation of the survey in order to reduce the length of the questionnaire and therefore not to overstress the commitment of respondents.

4.2.4 Quantitative Survey

The quantitative survey represents the core element of this study. The following sections will outline the methodological steps of data collection with the development of the survey instrument, training of enumerators, translation of the questionnaire, the sampling process, and the actual survey.

4.2.4.1 Data Collection: Questionnaire, Translation, Training of Field Assistants, and Pretest

As stated above, the construction of the questionnaire was mainly informed by previous explorative research, especially the qualitative study. To facilitate greater consistency between the involved research assistants, the questionnaire was prepared in a way that the interaction between interviewee and interviewer was mainly based on a prescribed structure. The main structure of the survey instrument was fourfold (Annex I): the *first* part of the survey contained a contact sheet, providing the interviewer with an introduction to the subject. This part also involved a short part for metadata collection about the respondent and a standardised observation protocol (to be conducted by the enumerator after the completed interview). The observation included locational characteristics with respect to the immediate neighbourhood, the house itself, and its interior appearance.

---

[6] In some cases it was not feasible, e.g. when the respondent was not willing to invite the interviewer to his home and the interview took place in the office or in a café.

The *second* part of the questionnaire addressed general value orientations (items based on Schwartz' Portrait Value Survey, PVQ) as well as more specific attitudes with regard to city development, religion, tradition, environment, climate change, lifestyle, and consumption. The *third* segment involved questions on energy use, food consumption, and transport for the carbon calculator as well as other behavioural and consumption characteristics such as mobility patterns, shopping, leisure, holidays, personal assets, and household characteristics. Last but not least, the survey closed with a fourth and closing segment addressing social-demographic information (e.g. education, employment, migratory background, expenditures, religion, caste, etc.).

One of the greatest challenges was the scope of the envisioned survey, covering the whole social spectrum of the city, i.e. all social classes. Therefore, the questionnaire was conceptualised for face-to-face interviews, as there are still high rates of analphabetism and as face-to-face interviews allow higher response rates. Moreover, due to the comprehensiveness of the instrument, filter questions were applied in order to keep a limit on the length of the questionnaire. After finalisation, the questionnaire was professionally translated to Telugu by a professional translator from Hyderabad.

4.2.4.2 Sampling

Access to social-demographic data disaggregated to the city level was very limited. This limitation also narrowed down the options for a precise sampling method. The most reliable source of data was available from the Greater Hyderabad Municipal Corporation (GHMC) 2009 assembly elections with numbers of electors for each ward, i.e. all citizens above 18 years old residing in the respective election ward. From this data, the number of voters for zones, circles, and wards was available. All 5 zones were considered in the overall sample selection, from which a random selection of 12 out of 18 circles was made (first-stage sampling) with overall 127 wards. In a second stage, 60 wards (number determined by researcher) were randomly selected, proportionate to the size of the population in each selected circle. From a total number of 605 interviews, the number of interviews per ward was determined proportionate to the ward population size. In a Google Maps-based GIS, start points for random route household selection were determined within each selected ward (for every three interviews one start point). The start point was selected by the author based on an easily distinguishable feature on the map such as a healthcare unit, a Kirana store, a pharmacy, or a school. The random route procedure instructs the field investigator to start a random route from the start point. The random route instruction says, e.g. "as you stand in front of the start point building, walk left, and take the first right, then take the second left and on this lane or street, approach the fifth house on the right". After finishing an interview or after taking an appointment for an interview, the field investigator starts with a new random route from the house that she/he has interviewed and carries on with this procedure until three interviews have been completed and then approaches the next start point in the ward. To secure random selection on the household level, the birthday method was applied, i.e. selecting that person for an interview who is above 18 years old, who lives regularly in this household, and who had her/his birthday most recently.

This three-stage proportionate geographical cluster sampling approach allowed the researchers to cover the Greater Hyderabad Municipal Corporation area of approx. 650 km² by still concentrating only on a selected number of wards (60 out of 150) without compromising considerably on representativeness. The approach reduces survey costs and permits a spatial analysis with regard to, e.g. differences between urban and peri-urban areas.

4.2.5 Data Analysis

Based on the research design and the conception of lifestyle, the data from the survey was analysed as outlined in Fig. 4.6. This flowchart depicts the major components and methodological steps in a sequential order. It differentiates between manifest and latent variables. Manifest variables are those that were measured directly by means of the survey and that directly correspond with questionnaire items. Latent variables are those variables that are composed and built from a set of manifest variables through, e.g. factor analysis or cluster analysis.

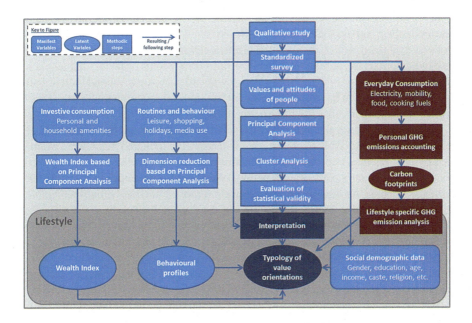

Fig. 4.6 Methodological and analytical steps for the lifestyle segmentation and lifestyle-specific GHG emission accounting. (Source: Own draft)

The analysis builds on four main strands: (1) the typology of value orientations; (2) a measurement of investive consumption based on personal and household amenities (wealth index); (3) profiles of behaviour and routines building on dimensions such as leisure, shopping, holidays, and media use; and (4) the personal carbon footprinting originating from everyday consumption. In addition to these four strands of analysis, that each condense a set of manifest variables into a single dimension, there are social-demographic data that remain in their original form as passive and manifest variables.

4.2.5.1 Building a Typology of Value Orientation Patterns: Cluster Analysis with Principal Component Analysis (PCA)

Principal Component Analysis (PCA)

A larger set of attitudinal items (51 statements measured with a 6-point Likert scale) were used to get hold of respondents' values and attitudes towards broader dimensions of everyday life. To structure this large set of variables in a content-specific way and to reduce them to a manageable number of dimensions, PCA was used, first as an explorative operation and secondly confirmatory run, controlling for the number of dimensions and included items. As suggested by Backhaus (2011, p. 378), the components were rotated with an orthogonal Varimax Rotation, i.e. the orthogonal factor axes do not correlate with each other.

In the initial explorative analysis, all 51 items were included. To decide upon and determine the number of dimensions to be extracted in the final solution, the scree plot was examined, which displays the eigenvalues associated with a component in descending order. The scree plot visualises those components or factors that explain most of the variability in the data. A steep downward curve up to a bend includes all those components that have considerably higher eigenvalues compared to those following the bend in a rather flat or horizontal line.

Moreover, based on the communality values, which indicate the proportion of variability of each variable explained by the factors, the included items were examined by how far they contribute to explaining variance within the overall solution. Items with low communality values below 0.4 and items that do not fit well into the extracted dimensions in terms of content were removed iteratively, until a suitable solution was arrived at. For the evaluation of the final factor solution, the measure of sampling adequacy (MSA) was applied, showing the Kaiser-Meyer-Olkin (KMO) measure and the Bartlett test of sphericity (Fromm and Baur 2008, p. 325).

After completing the formal tests, the most important step of analysis is the interpretation and labelling of each component or factor. This is done based on reading and understanding the coherence between high-loading items in each of the factors. In the case that the factor solution does not make sense in terms of content, the result needs to be dismissed. For the purpose of interpretation of and understanding the evolved patterns of orientation, the explorative phase of the project and the

4.2 Methodological Approach

qualitative survey were very critical. All dimensions, together with the underlying manifest variables, had to be put into a broader context.

The PCA was conducted to arrive at a manageable set of variables as basis for the cluster analysis. As preceding step to the cluster analysis, PCA allows the most important dimensions to be distinguished that are represented by the applied set of attitudinal items. A preceding PCA has great advantages compared to non-factorised clustering. It reduces a large number of variables to a minimum and provides for a centralised set of few uncorrelated, independent variables. For this analysis, five dimensions were identified, which were then used for the next step, the cluster analysis.

Cluster Analysis

Based on previous exploration of different algorithms, such as two-step cluster analysis, the author decided for a k-means clustering. In order to obtain start values for this centroid-based algorithm, a hierarchical cluster analysis (ward) was previously conducted. Without definite start values, which the k-means algorithm requires, random start values are computed. Therefore, the combination of both procedures makes sense here, because the k-means procedure builds on the ward method and improves and refines the results of the hierarchical algorithm. For both methods, the k-means and the ward method, the group memberships of ten clusters were computed. In order to determine a formally valid number of clusters, different statistical tests were conducted, the Eta^2, the PRE-value, and F-max (cf. Schendera 2010, p. 119). Based on these statistics, a final cluster solution was decided.

Based on Fromm (2010, p. 214) further criteria were considered: (a) examination of differentials of means of the underlying variables between the clusters and (b) an evaluation of the standard deviations of each variable within the cluster. Moreover, some authors suggest the computation of F-values, which measure the proportion between the cluster-specific variance and the variance of the overall sample for each of the underlying variables (Fromm 2010; Schendera 2010; Backhaus 2011). According to the statistical tests, a six-cluster solution was most suitable and with regard to its content well interpretable. The first level of interpretation was done through reading and comparing the cluster means of the underlying components.

In order to evaluate and compare the clusters with regard to the underlying distribution of factors, an analysis of variance is required. As this cluster solution produced more than two independent groups or samples, in which the factor values were not normally distributed, a one-way analysis of variance (ANOVA) is problematic. In this case, it was suggested to test the group-specific distribution of variables and validate the significance of difference through the non-parametric Kruskal-Wallis test (Bortz and Schuster 2010, p. 214; Field 2011, p. 559). The Kruskal-Wallis test is a non-parametric counterpart to the one-way independent ANOVA. It tests differences between several independent groups (Field 2011, p. 559). The test confirms significance in the case that the underlying variable differentiates at least between two of the involved clusters. The test proving significant therefore gives confidence that the significant effect is genuine, but, just like a one-way ANOVA, it only tells that a difference exists; it does not tell exactly where the differences lie (Field 2011, p. 564). However, comparison of medians for each variable in each cluster allows for a quite reliable differentiation between the clusters (Annex II, see also in Andy Field (2011, p. 565), who suggests looking at boxplots for a more specified differentiation).

Besides formal criteria for the evaluation of the cluster quality, it is most important to find the clusters interpretable. This includes generating theoretically meaningful names for the value orientation patterns. Once the interpretation of the clusters based on the active variables is completed, the so-called passive variables are taken into account. The methodology behind this step of interpretation and description of clusters is explained in Sect. 4.2.5.6.

4.2.5.2 Carbon Calculator

The questionnaire items have been created based on availability of reliable emission factors (EF). The selection of EFs was carried out in close consultation with a direct project partner no2co2, a research institute which has developed the first India-specific web-based carbon footprint calculator. Many of the selected EFs have also been provided by no2co2 (Table 4.8). According to the author's research, the provided factors are the most accurate factors available for India, even though they should be used for indicative purposes only, have a finite degree of uncertainty, and are expected to vary with time (Gilani 2010, 2012).

4.2.5.3 Calculating Average Long-Term Emission Effects of Specific Consumption Practices

To arrive at an average value of GHG emissions associated with a specific consumption practice, there are different approaches available: first, with regard to the use of electronic items such as air-conditioning, washing machine, and television, it is useful to draw on available data of consumption averages of Indian-specific appliances. no2co2 has computed these consumption averages for a large number of appliances based on Wattage and estimations of average annual usage times. This data have been provided by no2co2 in personal correspondence. They can be shared on request.

The resulting value was then extrapolated to the individual annual mean based on the average number of appliances per household and adjusted to the mean household size. In this way, the author arrived at a per person emission value, that indicates the per capita average amounts of emissions associated with the use of a specific appliance.

Second, for all other domains of consumption, more specific data is available from this study. For the use of any technology or the adoption of a consumption practice, the author has summed the practice-specific emissions from all respondents

4.2 Methodological Approach

Table 4.8 Underlying emission factors (EF) and their sources

Activity	EF common name	Measurement	Weight EF	Units EF	Source EF/ comments
Private transportation	Two-wheeler	Duration of usage	0,65	kg CO2e/h	Direct correspondence with Vivek Gilani
	Four-wheeler	Duration of usage	4,20	Kg CO2e/h	Direct correspondence with Vivek Gilani
Public transportation	Tan AC	Time spent in taxi	4,66	Kg CO2e/h	Direct correspondence with Vivek Gilani
	Taxi-non-AC	Time spent in taxi	3,30	Kg CO2e/h	Gilani (2010)
	Auto-rickshaw	Time spent in auto-rickshaw	1,80	Kg CO2e/h	Gilani (2010)
	Local bus- AC	Time spent in local bus (AC)	0,40	Kg CO2e/pass./h	Gilani (2010)
	Local bus-non-AC	Time spent in local bus (non-AC)	0,20	Kg CO2e/pass./h	Gilani (2010)
	Shared auto-rickshaw	Time spent in shared auto-rickshaw	0,49	Kg CO2e/pass./h	Direct correspondence with Vivek Gilani; three-wheel auto-rickshaw two-stroke engine, divided by three passengers
	MMTS/local train	Time spent in local train	0,60	Kg CO2e/pass./h	Gilani (2010)
	Car sharing	Time spent in shared car	1,77	Kg CO2e/pass./h	Direct correspondence with Vivek Gilani; we assumed AC cars, petrol car, shared by three people
	Chartered office/school bus	Time spent in office school bus	0,22	Kg CO2e/pass./h	Direct correspondence with Vivek Gilani

(continued)

Table 4.8 (continued)

Activity	EF common name	Measurement	Weight EF	Units EF	Source EF/ comments
Lang distance travel	Train	Time spent in long distance train	0,70	Kg CO2e/pass./h	Gilani (2010); for long distance, i.e. longer than eight hours. 24 h are assumed
	AC bus	Tim; spent in AC Bus	0,60	Kg CO2e/pass./h	Gilani (2010)
	Int. air travel <4 h flight	Number of flights	304,00	Kg CO2e/pass./flight	Gilani (2010)
	Int. air travel – 4 to 8 h flight	Number of flights	625,00	Kg CO2e/pass./flight	Gilani (2010)
	Int. air travel > h flight	Number of flights	1070,00	Kg CO2e/pass./flight	Gilani (2010)
	Dom. air travel <45 min flight	Number of flights	71,00	Kg CO2e/pass./flight	Gilani (2010)
	Dom. air travel – 45 min to 1 h 15 min flight	Number of flights	100,00	Kg CO2e/pass./flight	Gilani (2010)
	Dom. air travel >1 h 15 min flight	Number of flights	128,00	Kg CO2e/pass./flight	Gilani (2010)
Food	Milk	Amount of milk consumed	0,83	Kg CO2e/l	Gilani (2012)
	Mutton	Amount of Mutton consumed	12,69	Kg CO2e/kg mutton	Gilani (2012)
	Chicken	Amount of chicken consumed	4,48	Kg CO2e/kg chicken	GEMIS (2010); frozen chicken, average international
	Beef	Amount of beef consumed	8,61	Kg CO2e/kg beef	Gilani (2012)
	Pork	Amount of pork consumed	5,53	Kg CO2e/kg pork	Gilani (2012)
	Rice	Amount of rice consumed	0,92	Kg CO2e/kg rice	Gilani (2012)

(continued)

Table 4.8 (continued)

Activity	EF common name	Measurement	Weight EF	Units EF	Source EF/comments
Electricity and cooking	Electricity	Amount of electricity used	1,33	Kg CO2e/kWh	Brander et al. (2011)
	LPG – domestic	Number of cylinders used	44,50	Kg CO2e/cylinder	Gilani (2010)
	Kerosene	Litres used	2,46	Kg CO2e/l	Direct correspondence with Vivek Gilani
	Wood	Amount of firewood used	1,78	Kg CO2e/kg	Gilani (2012)

that follow the particular practice through their consumption. The mean of this total amount for each of the examined practices represents the annual personal mean of emissions associated with the respective consumption practice.

This calculation does not directly support the lifestyle analysis. It is a measure that seeks to highlight specific consumption effects of certain climate-relevant consumption practices based on average personal emissions. It represents an alternative explorative approach to estimate GHG emission effects of practices. In particular, it indicates the long-term impact of certain investive consumption decisions, such as the purchase of a car, motorbike, fridge, or air-conditioner.

4.2.5.4 Investive Consumption: The Wealth Index

The conceptualisation of investive consumption as one of the important pillars of the lifestyle concept is new and unique. However, the method to create the index is based on experiences from many studies, where it is used as an indicator for social-economic position (SEP) and in some cases as a proxy for income. A number of studies build the index as a basis of a set of subjectively selected wealth indicators (e.g. household amenities, personal assets, access to services) that are aggregated into a sum score with the aim to reflect household "wealth". Most studies, however, use more sophisticated approaches that weight the included variables. The simplest way is to limit the aggregation to a linear index by assigning equal weights which are summed up for each owned asset (for this method see, e.g. Razzaque et al. 1990) or by assigning weights based on a subjective relative rating of each owned asset (e.g. see Butsch 2011, p. 97). Both these approaches are appealing due to their simplicity and apparent objectivity. However, they have been criticised for the fact that the imposition of numeric equality is too arbitrary, since different assets are unlikely to have an equal effect on households' wealth (Filmer and Pritchett 2001, p. 116). This same critique is to be considered if assets are classified into pre-determined socio-economic categories and weighted accordingly, as suggested by Butsch (2011, p. 97).

Particularly in epidemiological research, a rather new approach is gaining broader acceptance as a measure to weight wealth indicator variables and aggregate these into a one-dimensional measure of SEP by means of principal component analysis (PCA). This method has now become an increasingly routine application (Vyas and Kumaranayake 2006, p. 460) that has been tested in various contexts, especially in low-income countries. The approach to build an asset-based index is rather simple and assessing the underlying amenities and assets through a household survey is straightforward (Annex I). The wealth index is a flexible tool, able to differentiate inequalities in social-economic status within a broad social spectrum based on assessing significant inequalities in durable consumption (Filmer and Pritchett 2001, p. 128; Vyas and Kumaranayake 2006, p. 467).

The procedure of building the asset-based index is mainly based on suggestions given in Filmer and Pritchett (2001), Vyas and Kumaranayake (2006), and Ruthstein and Johnson (2004). Table 4.9 lists all included indicator variables. The wealth index in this study includes binary variables (e.g. ownership of a flat screen TV; yes/no) as well as counted variables (e.g. number of air-conditions equivalised per number of household members). Counted variables have been included in cases with expected higher variance.[7] The framework of this study requires a specific differentiation both in the lower as well as in the higher social-economic segments. Therefore, the author did not aim for a simplified assessment of wealth, but was rather interested in a tool that is able to cover the whole social spectrum of the city and the inherent complexity of its overall social differentiation.

The author examined and explored in a few runs of the procedure different measurement levels. The best results in terms of coherence were achieved if the metric variables were not dichotomised and the involved additional information was thereby not removed. The inclusion of counted, metric variables is also done in other studies, such as in Filmer and Pritchett (2001, p. 117) as well as in Vyas and Kumaranayake (2006, p. 463).[8] Using metric variables is statistically not problematic, if all variables are z-standardised (Vyas and Kumaranayake 2006, p. 463). All included counted (metric) variables (Table 4.9) that measure the number of underlying assets were equivalised according to household size, i.e. the number of items in a household equivalised (adults = 1; children below 13 years = 0.5) to a per person level. The procedure of building the index is as follows:

- For the selection of relevant items, potential indicator variables were explored concerning mean, frequencies, and standard deviation (SD).
- The variables with low levels of SD were excluded.
- The missing values have been exchanged by the variables' overall mean.
- All raw data was then z-standardised in SPSS.
- The z-standardised variables then underwent PCA to compute the indicator weights to all included items (Filmer and Pritchett 2001, p. 116).

---

[7] Only selected indicators have been surveyed as counted items.
[8] In most cases, the variable number of rooms is included as metric variable together with dichotomous variables.

4.2 Methodological Approach

Table 4.9 Overview of included amenities building the wealth index and respective measurement levels

	Variable description	Dichotomous vs. metric measurement
Kitchen	Electric cooking range	Dichotomous
	LPG Stove	Dichotomous
	Kerosene stove	Dichotomous
	Chullaha	Dichotomous
	Fridges one-door	Metric
	Fridges two-door	Metric
Cooling/heating	Fan	Metric
	Air-cooler	Metric
	Air-condition	Metric
	Warm water geyser	Dichotomous
Washing machine	Semi-automatic washing machine	Dichotomous
	Full-automatic washing machine	Dichotomous
Mobility	Car	Metric
	Two-wheeler	Metric
	Cycle	Dichotomous
Media/telecommunication	Black-and-white TV	Dichotomous
	Colour TV	Dichotomous
	Flatscreen TV	Dichotomous
	Landline	Dichotomous
	Computer	Dichotomous
	Internet	Dichotomous
	Mobile phone	Dichotomous
Various assets	Mattress/cot	Dichotomous
	Pressure cooker	Dichotomous
	Mixer grinder	Dichotomous
	Credit card	Dichotomous
Servants	Part-time maid	Dichotomous
	Full-time maid	Dichotomous
	Cook	Dichotomous
Housing	Number of rooms	Metric
	Semi-pucca house type	Dichotomous
	Pucca house type	Dichotomous
Access to water	Own tap	Dichotomous
	Shared tap	Dichotomous

- The factor coefficient scores, i.e. the factor loadings of the first component, were then multiplied by the variables' z-standardised values for each respondent.
 - The sum of these indicators' values represents the value of the wealth index. A standardised score, which differentiates each respondent based on their underlying asset structure.

The index therefore estimates the relative level of difference in terms of personal and household investments in durable goods and services. It indicates therefore the relative level of investive consumption.

4.2.5.5 Patterns of Consumption and Behaviour: Mobility, Shopping, Leisure, Holidays, and Media Usage

For the operationalisation of everyday conduct of life in the mentioned areas, mobility, shopping, leisure, holidays, and media use, a broader set of questionnaire items was developed based on the results of the qualitative survey. For the selection of variables, it was important to consider collinearity with the measurement of carbon emissions. This rule came without any problems concerning shopping, leisure, holidays, and media use, but it was an issue with regard to mobility. Mobility-related emissions are directly measured and calculated based on the time spent on transportation of a particular mode, e.g. on a person's own four-wheeler. The same approach for measuring, e.g. the preferred mode of transport was most evident concerning the adoption of everyday practices of mobility. In spite of this problem of collinearity, the author was not aware of any other choice than measuring respondents' most dominant mode of transport. And in fact, by looking at the dominant mode of transport, it is not the summed carbon emissions from all used modes, but it is the most preferred mode of transport that is taken into account. It depends both on the financial means and availability (e.g. motorised vehicles) as well as on lifestyle-related preferences and behaviour patterns.

For this purpose, mobility data was analysed concerning the time spent using different modes. The results were then classified through analysis of the frequencies distributed across different classifications. The author then decided on a classification with four broader categories, (a) walking and cycling, (b) public transport, (c) two-wheeler, and (d) four-wheeler. The summed values of time spent on different modes of transport within each of these categories were taken as the basis for determining the dominant mode of transport for each of the respondents.

For the other realms of conduct of life, the author decided to conduct a PCA under consideration of the methodological aspects, explained above in Sect. 4.2.5.2. First, an explorative PCA allowed the author to decide on the number of components extracted (scree plot) and the items to be included (communalities and content-wise interpretational criteria). In the second, confirmatory round of PCA, two components were extracted, respectively, for the four areas of conduct of life: *first*, the use of media for information gathering; *second*, preferences and patterns of usage of shopping facilities; *third*, practices and preferences of leisure activities;

4.2 Methodological Approach 139

and last but not least, practices of vacation and holiday. All PCA have been examined in terms of the Bartlett test of sphericity, the KMO measure, and total explained variance in order to check quality and validity of the results.

4.2.5.6 Description and Analysis of Clusters with Passive Variables

The above-delineated procedures give an overview of the methodological steps associated with constructing the foundational aspects of lifestyle and of the personal carbon footprints, which are to be explained through the concept of lifestyle. These four blocks of analysis led to the output of latent variables that represent the building blocks of this study. Additionally, there are subsidiary variables within the theme of social demography. These variables were prepared and in some instances recoded but represent manifest variables, such as caste, income, religion, and employment. All variables that have not been included in the construction of the value orientation clusters (Sect. 4.2.5.2) are treated as passive variables. Passive variables are taken into account for the further analysis and interpretation of the value orientation clusters and for the construction of the lifestyle typology. For this purpose, the single clusters were analysed concerning the distribution of all passive variables across clusters.

Chapter 5
Results Part I: Descriptive Analysis of Manifest Variables and Preparation of Latent Components for the Lifestyle Analysis

Keywords Income · Investive consumption · Consumption · Frugality · Principal component analysis · Kaliyuga, Caste

Building typologies for the analysis of environment-related social research is a challenging task. The analysis of lifestyle and related GHG emissions involves a number of dimensions, which altogether reproduce and demarcate a quite complex but still tangible and therefore simplified and rough representation of a typical social-cultural group. It is a balancing act between the problem of lacking tangibility due to over-complexity and the compromise on probably important parameters that need to remain unconsidered. In addition to the core approach of the value- and attitude-oriented segmentation process (Sect. 4.2.5.2), passive variables are consulted to describe and delineate the lifestyle clusters in respect to social position and behavioural and consumption patterns. To facilitate this second step of analysis, multiple ways have been taken into account in order to reduce the number of dimensions and simplify the underlying variables but still get the most out of these additional variables. The following sections will introduce the most relevant components, related data preparations and dimension reductions, as well as a first overview of the distribution in the overall sample.

5.1 Manifest Variables

5.1.1 Gender, Age, Religion, and Caste

As mentioned in Sect. 4.2.4.2 in respect to the sampling method at the respondents' household (birthday method), a sampling bias has occurred leading to a skewed dataset concerning gender balance. It is assumed that male household members have in some cases pushed to the front to give the interview, ignoring the random sampling selection. Consequently, there are relatively more male respondents in the overall sample, with female respondents making up 32% against 68% of male interviewees. The Census 2011 reports a balance of 51% males against 49% females

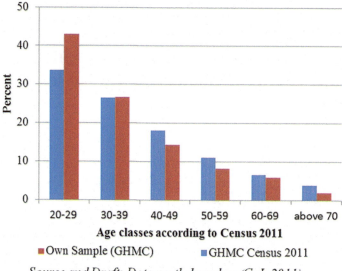

Fig. 5.1 Break-up of age groups of overall sample compared with data from Census 2011 for the Greater Hyderabad Municipal Corporation (GHMC) (The Census data have been categorised into bins of 5 years. While the author's study has only respondents starting from 18 years, the respective Census data bin contains people from 15 to 20 years of age. For this reason, respondents below 20 years of age (11 respondents) have been removed from this figure). (Source and Draft: Data partly based on (GoI 2011))

within the Greater Hyderabad Municipal Corporation (GoI 2011). This imbalance is remarkable and should be kept in mind for the further analysis and especially for the lifestyle segmentation.

In respect to age, the sample distribution is less skewed, but still biased, as delineated in Fig. 5.1.

The youngest age group in this comparison is over-represented in this study, having almost 10% points more (43%) than the Census data (34%). This is against an under-represented distribution among the higher age classes above 40 years of age. The median for this study's sample is 31 years, while the mean is about 35 years.

Hyderabad, with its history of the Nizam dynasty, historically has a relatively large share of Muslims compared to most other cities in India. Data from the Census 2011 disaggregated to the GHMC area indicate a share of above 30% of Muslims compared to almost 65% of Hindus. The author's study has a somewhat biased sample distribution with around 75% of Hindus against only 17% of Muslims (Fig. 5.2). This imbalance can be traced back to the fact that the questionnaire was issued only for Telugu- and English-speaking respondents. Many of the Hyderabad-based Muslims, however, are not firm in these languages as they quite commonly speak Hindi and Urdu. In all those cases in which the respondent was unable to respond to one of these two issued languages, the interview had to be cancelled.

5.1 Manifest Variables

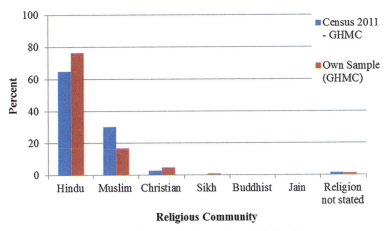

Source and Draft: Data partly based on (GoI, 2011)

Fig. 5.2 Break-up of religious groups of overall sample compared with data from Census 2011 for the Greater Hyderabad Municipal Corporation (GHMC). (Source and Draft: Data partly based on (GoI 2011))

Furthermore, the study's sample depicts a slightly larger share of Christians, who make up about 4.6% compared to the 2.8% in the Census data. All other religious groups represent a negligible share in Hyderabad with altogether less than 1%.

Respondents were also asked to report their affiliation to a community or caste. Depending on different factors, as, for instance, the caste background, education, or dialect of the interviewee, caste is difficult to survey and bears high risk of biased measurement. The overall sample of this study has a share of around 16.5% of respondents reporting to be scheduled caste or scheduled tribe. All others have either refused to answer (3.1%) or have reported to belong to a general caste (80.3%).

5.1.2 Marriage, Family Structure, and Household

The overall sample of this study (only adults above 18) has a share of 44% of people in the age group 18–29 years.[1] Hence, it is not surprising that the share of singles is quite large. But compared to an approximation from the Census 2011 data, which indicates a proportion of about 23% of adults never having married, the share of 29% in this study is quite remarkable. The author assumes the reason to be the relatively higher share of the youngest age segment.

---

[1] From Census 2011 data, the author has approximated that 40% of all adults are between 18 and 29 years old.

From all those respondents, who are or who have been married (divorced and widowed have been also counted in this category), 85% have children. This is more than 60% of the overall sample. The gender difference is particularly high in this sample, with an overall sex ratio – males per 100 females – of 117.34 males. The Census 2011 indicates a ratio of males per 100 females of only 104.68. This result should be kept in mind, as the subject of female discrimination and the role of dowry will play a role as part of the lifestyle segmentation. With the cultural practice of dowry, it is common for families with female children to have early concerns on the issue of arranging a proper match for their own daughter(s), and – as will be shown in this study – this concern leads to changes in the overall consumption patterns.

In respect to household size, this sample has a median of 5.0 in terms of the total number of people living in a household. The equivalised household size, which is an adjusted measure with adults counting as 1.0 and children below 13 counting as 0.5, has a median size of 4.5 people. Compared to the results from the Census 2011 that gives a mean for urban Andhra Pradesh (AP) of 4.1 (un-equivalised total number), these figures are quite high. It may relate to the higher costs of living and other cultural factors that the density in a metropolis like Hyderabad is higher than in other urban areas in AP.

5.1.3 Education and Employment

With regard to education, this study's results show a substantial degree of educational mobility. Figure 5.3 depicts the highest educational level achieved by the respondent and his or her father. It is remarkable that more than 40% of respondents' fathers were reported to have never attended any school. This figure is

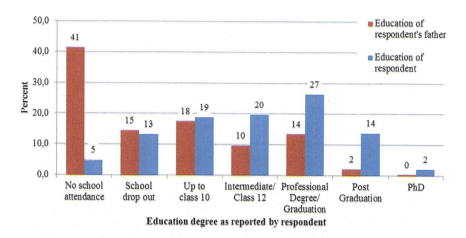

Fig. 5.3 Bar chart of educational degrees of respondent and her/his father in percent

striking when seen against the tremendous difference in the share of respondents without school attendance, which was just 5%.

In respect to school dropouts without any final school qualification, the levels are quite balanced between fathers and their offspring (15 vs. 13%, respectively). Concerning all other qualifications, the subsequent generation gains substantially against the educational status of their parental house. Especially from the level of intermediate/12th class upwards, the share has doubled across almost all education qualifications up to the level of PhD. The highest gains are found in PhD and in graduation.

Concerning employment, respondents were asked to report in which profession they do most of their work. If they were not currently employed, they were asked to characterise their major work in the past. Table 5.1 depicts the employment categories according to the Goldthorpe classification (Evans 1992).

The figure demarcates quite a large share of women reporting to be housewives, without being otherwise employed (18.5%). Overall, 55% of all female respondents work as housewives. The other 45% of women are split into quite a large share of professionals (e.g. IT professionals; 14.5%) and students (12.3%). Quite some are also self-employed as small proprietors, with (2.8%) and without employees (4.5%). Some women are also employed as workers, unskilled (2.8%) but also skilled workers (3.4%). Interestingly, 38.5% of all semi- and unskilled workers are women, representing quite a large share, considering the relatively small proportion of women against men in this sample.

Overall, students make out to be 13.1% as part of the total sample. Also the number of professional employees is quite remarkable (20.5%); however, for the Hyderabad context, it is not surprising (see Sect. 3.4). All workers taken together (skilled and unskilled workers as well as lower-grade technicians) make out a share of almost 16%. The class of proprietors with or without employees also has quite a substantial portion of together almost 22%.

Table 5.1 Break-up of major employment category, past and present

Employment	Frequency	Percent	Valid percent
Housewife and not otherwise employed	100	16,5	18,5
Retired pensioned	25	4,1	4,6
Unemployed	25	4,1	4,6
Student	71	11.7	13,1
Semi- and unskilled worker	13	2,1	2,4
Skilled manual worker	52	8,6	9,6
Lower-grade technicians: supervisors of manual workers	21	3,5	3,9
Small proprietors, artisans, etc., without employees	76	12,6	14
Small proprietors, artisans, etc., with employees	41	6,8	7,6
Routine non-manual employees	7	1,2	1,3
Professionals	111	18,3	20,5
Nor reported (missing)	*63*	*10,4*	*n.a.*

5.1.4 Income

A presumably important category is income. For reasons of comparison and illustration, the author has decided to classify the variable of income in this study according to the income brackets defined by McKinsey Global Institute in their study on India's future consumer markets "The Bird of Gold" (2007). They built their classification on the NCAER publication "The Great Indian Middle Class" (2004) and recalibrated it to the examined time period. Problematic in this case is that the study bases all its analysis and projections on the so-created classification of annual household income, but without considering the size of households. This adjustment, however, is essential, against the background of a substantial level of variance in household sizes in India. For instance, the author's study shows a standard deviation of 3.5 for the total size of households. Even with the adjusted measure, which counts adults as full members and children below 13 by a factor of 0.5, the standard deviation is 2.5. With this variance in sizes of households, a comparison based on household income without an equivalisation distorts the analysis considerably.

In order to reduce this bias, the author has equivalised the variable of annual disposable household income by the adjusted size of the household (adults = 1; children below 13 years = 0.5). For categorisation in analogy to the McKinsey classes, the income brackets themselves have been adjusted to a factor of 4.1, which is the average household size in urban Andhra Pradesh according to the Census of India (2011). Table 5.2 gives an overview of the underlying classes and their ranges of total and equivalised annual household income applied in this survey.

The study from McKinsey offers disaggregated data of urban and rural populations in India. Data for urban India coarsely indicate the actual income distribution in Hyderabad. A comparison with the distribution of this study offers a possibility to validate the reliability of the sampling process. Due to the lack of socio-economic data in Hyderabad, this is one of the few possibilities to do this validation. For this purpose, the McKinsey projections were extrapolated to the year 2011 and then compared with the distribution of combined annual household income (here not adjusted to household size) of the author's survey. Figure 5.4 depicts both distributions and shows well that the two samples are quite similarly distributed, with two highest-income groups being slightly over-represented against a slightly smaller

Table 5.2 Original and equivalised income brackets based on McKinsey Global Institute

Income class	Original income brackets defined by McKinsey (household/year)	Equivalised income brackets (based on household income/year)
Deprived	<90,000	<21,951
Aspirers	90,000–200,000	21,951–48,780
Seekers	200,000–500,000	48,780–121,951
Strivers	500,000–1,000,000	121,951–243,902
Globals	>1,000,000	>243,902

5.1 Manifest Variables

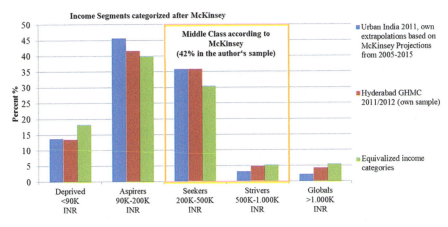

Fig. 5.4 Comparison of income distribution in Hyderabad (author's survey, combined household and equivalised categories) and urban India. The groups in the yellow box (seekers and strivers) represent the Indian middle class according to McKinsey Global Institute. (Source: own draft, in part based on MGI projections, McKinsey Global Institute 2007)

Table 5.3 Descriptive statistical overview of combined household and equivalised net income

Annual income			
(Based on combined household income)			
		Combined household income	Equivalised income per capita
N	Valid	541	527
	Missing	64	78
Mean		354,262	95,540
Median		180,000	40,000
Std. deviation		775,122	240,450
Range		7,188,000	2,997,333
Minimum		12,000	2667
Maximum		7,200,000	3,000,000

share of the second poorest income group (aspirers). It also shows the share of the middle class according to McKinsey Global Institute (MGI).

The median of annual combined household income with a value of 180.000 INR is located within the upper third of the second poorest income segment (aspirers; see Table 5.3). This value is still situated somewhat below the threshold to becoming middle class, according to MGI.

Figure 5.4 also gives a picture of the distribution of equivalised net income per capita. As poorer households tend to have larger household sizes (Pearson correlation coefficient $r = 0.16$ is significant at the 0.01 level), the income equivalisation also has an obvious effect on the ratio: the two highest-income brackets grow slightly by a few percentage points, and the lowest-income group grows quite substantially. The middle-income groups thereby lose some of their share. This comparison shows that un-equivalised income distribution data tend to mask income inequalities. And, in particular, they are able to make the gap between the very rich and the very poor appear smaller, as relatively rich households are more likely to share the combined household income with fewer people.

As discussed in Sect. 4.1.3.1, it is a serious challenge to measure income adequately, and the reliability income values provided are often questioned. This study also faces the problem of missing data due to respondents being unable or unwilling to report their household income (Table 5.3).[2] The wealth index, which will be introduced in the following, is often used in supplement to income for the measurement of social position (see Sect. 4.1.3.1).

5.2 Construction and Descriptive Analysis of Latent Variables and Components for Lifestyle Analysis

5.2.1 Investive Consumption: The Wealth Index

In this study, the wealth index serves as an indicator for investive expenditure and consumption, because it measures the scope and the level of personal and household amenities. The method to develop the index from quite a substantial number of household and personal amenities is outlined in Sect. 4.2.5.5. Figure 5.5 gives an overview of the basic statistics of the index. Some variables, such as the number of fans or air-conditioners or as the number of rooms, have been equivalised by the number of household members. In all those cases in which the size of the household was missing, an average value for household size was applied (based on census data, disaggregated for urban Andhra Pradesh; 4.1 persons per household, see also subchapter above; Census 2011). The index was then scaled, with the lowest wealth index at 0 and the highest wealth index at 100. The median value is situated at about 20, while the mean is located slightly above at 23 points. The distribution is slightly clustered around the mid-lower end of the scale (Fig. 5.5). A differentiation between respondents in these lower strata is therefore difficult. However, among those households with higher amenity levels, differences are more significant.

Moreover, Table 5.4 gives an overview of the distribution of ownership shares of durable assets and household characteristics specified for each wealth class. The results are found to be internally coherent across quintiles in most cases. It should be kept in mind that a few of the durable assets are personal assets. For instance, a mobile phone may be owned by one of the household members, but it is possible that the respondent has reported to have no personal mobile phone. It is quite remarkable, though, that there are more respondents without a mobile in the richest segment compared to the next lower two segments (third and fourth).

---

[2] Missing data of household size increases the number of missing values in terms of equivalised income (Table 5.3).

5.2 Construction and Descriptive Analysis of Latent Variables and Components...

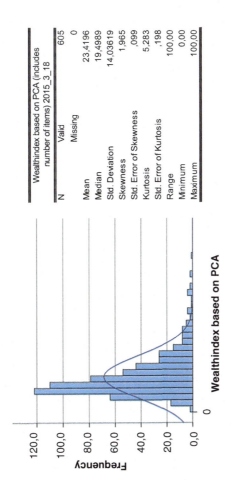

Fig. 5.5 Histogram and descriptive statistical overview of wealth index

Table 5.4 Ownership of durable assets and housing characteristics by wealth index quintile

Variable description		Wealth classes					Total
		Poorest	Second	Third	Fourth	Richest	
Kitchen	Electric cooking range	0	0	0,042	0,1	0,182	*0,065*
	LPG stove	0,826	0,983	1	0,992	0,967	*0,954*
	Kerosene stove	0,306	0,116	0,101	0,066	0,05	*0,128*
	Chulha	0,207	0,058	0,042	0,025	0,05	*0,076*
	Fridges (one-door)	0,157	0,554	0,661	0,719	0,62	*0,542*
	Fridges (two-door)	0	0,033	0,058	0,198	0,446	*0,147*
Cooling/heating	Fan	0,884	0,752	0,909	0,95	0,983	*0,896*
	Air cooler	0,041	0,132	0,364	0,521	0,612	*0,334*
	Air-conditioner	0	0	0,009	0,1	0,453	*0,123*
	Warm water geyser	0	0,033	0,094	0,273	0,785	*0,224*
Washing machine	Semi-automatic washing machine	0,017	0,041	0,182	0,372	0,355	*0,193*
	Fully automatic washing machine	0	0	0,033	0,107	0,446	*0,117*
Mobility	Car	0	0,017	0,025	0,223	0,603	*0,174*
	Two-wheeler	0,289	0,628	0,76	0,901	0,851	*0,686*
	Cycle	0,223	0,132	0,149	0,175	0,074	*0,151*
Media/ telecommunication	Black-and-white TV	0,124	0,05	0,017	0,025	0,05	*0,053*
	Coloured TV	0,769	0,926	0,95	0,917	0,612	*0,835*
	Flatscreen TV	0	0	0,025	0,091	0,488	*0,121*
	Landline	0,025	0,05	0,132	0,298	0,463	*0,193*
	Computer	0	0,008	0,174	0,479	0,818	*0,296*
	Internet	0	0,008	0,066	0,273	0,675	*0,204*
	Mobile phone[a]	0,843	0,926	0,983	0,975	0,95	*0,936*
Various assets	Mattress/cot	0,661	0,917	0,959	0,983	0,967	*0,898*
	Pressure cooker	0,603	0,884	0,967	0,967	0,959	*0,876*
	Mixer grinder	0,413	0,793	0,909	0,975	0,959	*0,81*
	Credit card[a]	0,017	0,033	0,132	0,256	0,628	*0,213*
Servants	Part-time maid	0	0,033	0,025	0,231	0,504	*0,159*
	Full-time maid	0	0	0	0,008	0,149	*0,031*
	Cook	0	0	0	0	0,083	*0,017*
Housing	Semi-pucca house type	0,132	0,074	0,025	0,058	0,017	*0,061*
	Pucca house type	0,868	0,917	0,967	0,942	0,959	*0,931*
Access to water	Own tap	0,405	0,826	0,884	0,901	0,959	*0,795*
	Shared tap	0,587	0,207	0,182	0,124	0,066	*0,233*

[a]Personal assets

5.2.2 Latent Dimensions of Preferences, Behaviour, and Consumption Patterns

In order to reduce the number of dimensions that measure and describe preferences, behaviour, and consumption, a number of thematically related variables have been factorised through PCA. Four areas of conduct of life have been taken into account, *first*, the use of media for information gathering; *second*, preferences and patterns of usage of shopping facilities; *third*, practices and preferences of leisure activities; and last but not least, practices of vacation and holiday. All PCAs have come out well, each area with two telling dimensions. The PCAs have been examined in regard to the quality and validity of the outcome: the Bartlett test of sphericity and the KMO measure were applied, and total variance was explained. Table 5.5 depicts the results of the first group of variables, namely, media use.

In respect to the use of media, the first dimension delineates the use of more differentiated media like books, English newspapers, and the Internet. The second dimension condenses the use of red-top media such as local newspapers, radio programmes, and TV. The Bartlett test of sphericity is significant ($p < 0.001$), and the KMO measure is usable with a value of 0.66. The total variance explained is 62%.

The preference and practice patterns in regard to shopping are shown in Table 5.6. The first dimension contains all those shopping preferences that concentrate around more centralised locations such as malls and supermarkets. The second factor comprises rather short distance shopping in kiranas, i.e. mom-and-pop grocery stores and street vendors. The Bartlett test of sphericity for this factor solution is significant at a level of $p < 0.001$, and the KMO measure is usable with a value of 0.503. The total variance explained is 71%.

The next area of conduct of life is leisure activities as depicted in Table 5.7. The factor solution divides the various activities between two dimensions, action- and consumption-oriented leisure activities, which subsume, e.g. shopping in malls, dining, or cinema, and the other dimension with home-oriented leisure activities, such as watching TV and chatting with friends and neighbours on the street. This solution also comes out well with the Bartlett test of sphericity being significant at the level of $p < 0.001$. The KMO measure is usable with a value of 0.73, and the total variance explained is 44%.

Table 5.5 Rotated component matrix of PCA of media usage for information gathering

	Component	
	1	2
Internet, email	0.829	
Daily English newspaper	0,795	
Books	0,762	
News broadcasts on radio or TV		0,854
In-depth reports on radio or TV		0,839
Daily local newspaper		0.505

Table 5.6 Rotated component matrix of PCA of shopping preferences and practices

	Component	
	1	2
Shopping at supermarket	0,877	
Shopping at mall	0,857	
Shopping at kirana		0,822
Shopping at street vendor		0,803

Table 5.7 Rotated component matrix of PCA of leisure activities

	Component	
	1	2
Going to a shopping mall	0,730	
Going out to restaurants for dinner or lunch	0,673	
Shopping	0,651	
Going out for cinema	0,614	
Going out to a cafe	0,587	
Doing sports	0,560	
Visiting parks	0,532	
Chatting with friends and neighbours on the street		0,782
Watching TV at home		0,762
Sleeping		0,477

The fourth and last realm of lifestyle practices comprises activities around holidays and vacation. Table 5.8 shows the PCA results of holiday preferences and practices. Similarly as in the analysis of leisure activities, the practices split up into action- and consumption-oriented activities, such as hiking, trekking, and beach holidays and a more traditional, family, and religion-oriented dimension, with activities such as visiting relatives and religiously oriented travel. The Bartlett test of sphericity is significant ($p < 0.001$), and the KMO measure is usable with a value of 0.674. The total variance explained is 66%.

With these four areas of conduct of life and their respective dimensions, a broad basis for the analysis of clusters in respect to lifestyle practices is laid. These passive variables are crucial, because they serve as the basis for a description and content-wise validation of the cluster analysis. This important fact will become clearer at the end of this chapter.

5.2.3 Latent Dimensions of Value and Attitudinal Orientations

In PCA and cluster analysis, the author has built a typology of individuals based on a number of selected attitudes and values. As a preceding step to the cluster analysis, PCA allows one to recognise the most important dimensions that are

5.2 Construction and Descriptive Analysis of Latent Variables and Components...

Table 5.8 Rotated component matrix of PCA of holiday preferences and practices

	Component	
	1	2
Hiking/trekking/mountaineering	0,811	
Beach holiday	0,782	
Sight seeing	0,727	
Visiting relatives		0,866
Religious-related travel	0,307	0,747

Table 5.9 Parameters for the Measure of Sampling Adequacy (MSA)

KMO and Bartlett test		
Kaiser-Meyer-Olkin measure		0,843
Bartlett test of sphericity	Approximated Chi-square	2925,21
	df	210
	Sig.	0,00

represented by the applied set of attitudinal items. For this analysis, five dimensions were identified through PCA and then utilised as the data basis for the next step, the cluster analysis. A preceding PCA has great advantages compared to non-factorised clustering, as it reduces a larger number of variables to a much smaller and easier to handle, centralised set of few uncorrelated, independent variables.

The factor solution from PCA shows good results, statistically as well as content wise. In an explorative PCA, a five-factor solution is reasonable, with the fifth factor having an eigenvalue of 1.126, while the sixth factor's eigenvalue even falls below 1.0–0.895. The scree plot shows a first bend at the third and a second bend at the sixth factor. Due to reasons of interpretability and the high eigenvalue of the fifth factor, a five-factor solution was chosen. The overall variance explained by this solution is 59%. The communalities, which represent the sum of the squared factor loadings of an item on all factors and which indicate the variance of each item explained by all factors, are rather high, with only two items showing communalities below 0.5 ("Tradition is important; tries to follow the customs handed down by her/his family", 0.497; "I believe that human nature is bad, evil and wicked. To maintain social order the only means are coercion and punishment", 0.442). For the evaluation of the sample regarding its adequacy for factorisation, the measure-of-sampling-adequacy (MSA) was applied (see Table 5.9). In this formal test, the Kaiser-Meyer-Olkin (KMO) measure showed a value of 0.843, which is a "meritorious" result (Kaiser 1974, p. 35). The Bartlett test of sphericity is significant and therefore allows the statement that there are correlations between some variables, which is a precondition for factorisation (Fromm and Baur 2008, p. 325).

With this existing PCA, the author examined the attitudinal and value data in searching for an inherent structure, i.e. a structure based on a number of latent dimensions represented by the extracted components (cf. Fromm and Baur 2008, p. 320). Table 5.10 shows the matrix with the underlying items and their respective

Table 5.10 Matrix of rotated components from principal component analysis (PCA)

Rotated component matrix	Component				
	1	2	3	4	5
Fa_1_a.) We cannot do anything about our future, as the decline of the society is predetermined	**0,833**	0,041	−0,058	0,023	−0,031
Fa_1_b.) People are like they are due to their actions and ways of living in their previous lives	**0,796**	**0,128**	−0,023	0,081	−0,007
Fa_1_c.) We are living in the age of Kali [for Muslims read "We are running towards the End of Time"]. Evil and immorality dominate, and we cannot do anything about it	**0,726**	0,206	−0,035	−0,044	0,065
Fa_1_d.) Whenever there is a difficult problem, it is better to leave everything to God	**0,72**	**0,145**	−0,052	0,053	0,03
Fa_1_e.) I believe that human nature is bad, evil, and wicked. To maintain social order, the only means are coercion and punishment	**0,616**	0,112	0,216	0,005	0,052
Fa_1_f.) I believe, in our society, marriages should take place within one's own caste/community	**0,541**	−0,099	**0,329**	0,043	**0,420**
Fa_2_a.) Whatever I earn goes into maintenance of basic needs	0,211	**0,823**	0,064	−0,018	0,068
Fa_2_b.) I carefully watch out how much I spend in order to save my money for harder times	−0,010	**0,726**	0,252	−0,049	0,278
Fa_2_c.) Consumption as such is given too much importance in our society and doesn't make you happier in the end	0,240	**0,683**	0,22	0,053	0,142
Fa_2_d.) It doesn't matter how much I earn; when I have a choice between two equal products, I usually go for the lowest price	0,209	**0,673**	0,039	−0,092	**0,161**
Fa_2_e.) It sometimes bothers me quite a lot that I can't afford to buy all the things other people have	0,145	**0,623**	**0,377**	0,037	−0,186
Fa_2_f.) My life situation leaves no other option than carefully watching out how much I spent	−0,078	**0,613**	0,232	0,025	**0,316**
Fa_3_a.) I try not to buy products made by companies which are socially and ecologically irresponsible	0,133	0,204	**0,725**	−0,054	0,058
Fa_3_b.) Believes that people should care for nature; looking after the environment	−0,022	0,266	**0,692**	−0,002	0,104
Fa_3_c.) When shopping I regularly pay attention to the environmental friendliness of the products	−0,048	0,248	**0,649**	0,073	0,244
Fa_4_a.) I quite frequently shop in more expensive and exclusive stores	0,04	−0,082	0,032	**0,805**	−0,058
Fa_4_b.) Going to shopping malls is one of the activities that I really like, even if I don't buy anything	0,087	0,019	0,171	**0,732**	−0,048
Fa_4_c.) I like to buy things just for the fun of it	−0,024	0,028	**−0,391**	**0,668**	0,079
Fa_5_a.) Obedience and respect for elders are very important values and should be maintained	−0,030	**0,335**	0,092	−0,052	**0,682**

(continued)

5.2 Construction and Descriptive Analysis of Latent Variables and Components... 155

Table 5.10 (continued)

Rotated component matrix					
	Component				
	1	2	3	4	5
Fa_5_b.) Tradition is important; tries to follow the customs handed down by her/his family	0,067	0,285	0,075	−0,054	**0,635**
Fa_5_c.) In matters of marriage, boys and girls may be consulted, but the final decision should be taken by the parents	**0,470**	−0,063	0,220	0,062	**0,514**

Extraction method: analysis of principal component. Rotation method: varimax with Kaiser normalisation. a. Rotation converged in six iterations

loadings on the components or factors (factor scores). The factor scores (Table 5.10) give evidence of interrelations between items and factors: the closer the score gets to ±1, the higher the correlation to the factor. To interpret a factor and its content, it is required to observe all those items that score high on the respective factor. By means of interpretation of the combinations, it is the task of the researcher to define a common denominator between the high loading items. Interpretation of such content is bound to be subjective. However, for the interpretation of the statistically derived dimensions, the author has made use of extensive ethnographic and sociological work being published during the last decade. This material has been combined with the material from the author's qualitative interviews and both supports and illustrates the interpretation of the five value dimensions. The underlying assumptions and implications have been discussed and contextualised in more detail in Sect. 4.1. Here in this chapter, the content of the factors will be described, analysed, and discussed more specifically, based on the underlying items. In order to increase transparency and traceability of each interpretational step, the author documented the building blocks of the interpretation process in the form of interpretation matrices for each individual factor and its associated items (Annex III). The final results and factor loadings of the PCA can be depicted from Table 5.10.

It has to be stressed that respondents were asked to rate their agreement or disagreement to the given statements on a 6-point Likert scale. This scale allows respondents to coarsely indicate their position, but interpreting the scale is subjective. The applied factor analysis is not supposed to counterfeit an accuracy of measurement that it is not possible to achieve. It can only provide tendencies of answering behaviour.

Finally yet importantly, the author wants to emphasise again that the following five dimensions or factors represent meaningfully associated item batteries that serve as scales to estimate respondents' attitudinal orientations and values. They do not describe types of people, nor do these dimensions exist independent of each other. The factors build the foundation for the following cluster analysis, i.e. every cluster can be described based on a certain structure of agreement (or disagreement) in respect to each of the five factors.

Table 5.11 Rotated component matrix: factor 1 – religious tradition

	1	2	3	4	5
Fa_1_a.) We cannot do anything about our future, as the decline of the society is predetermined	**0,833**	0,041	−0,058	0,023	−0,031
Fa_1_b.) People are like they are due to their actions and ways of living in their previous lives	**0,796**	0,128	−0,023	0,081	−0,007
Fa_1_c.) We are living in the age of Kali [for Muslims read "We are running towards the End of Time"]. Evil and immorality dominate and we cannot do anything about it	**0,726**	0,206	−0,035	−0,044	0,065
Fa_1_d.) Whenever there is a difficult problem, it is better to leave everything to God	**0,72**	0,145	−0,052	0,053	0,03
Fa_1_e.) I believe that human nature is bad, evil, and wicked. To maintain social order, the only means are coercion and punishment	**0,616**	0,112	0,216	0,005	0,052
Fa_1_f.) I believe, in our society, marriages should take place within one's own caste/community	**0,541**	−0,099	**0,329**	0,043	**0,42**

5.2.3.1 Factor 1: Paradigm of Religious Tradition

Factor 1 builds on six items. The last item of this factor also loads on factor 5 and to a minor extent on factor 3 (see Table 5.11). Factor 1 is the factor most specifically related to Indian culture, referring to various aspects of Hindu mythology, however framed within a rather fatalistic paradigm. The interpretation will include the context of the mythological and religious concepts that are foundational for it.

The factor consists of statements that give a religiously framed, passive, and resigned outlook to the world and its future. The first item states that nothing can be done about the future, as the decline of the society is predetermined. This statement takes a view on a changing society, which inevitably leads to the termination of the world as we know it, no matter how hard one may try to intervene. Likewise, the third item points in the same direction by stating that the current period of Kaliyuga is dominated by evil and immorality and that nothing can be done about it.

This pessimistic and fatalistic future outlook is indicative for understanding rejection of pro-environmental behaviour as well as for a deeper and more cultural-specific interpretation of climate change perceptions in India (cf. Reusswig and Meyer-Ohlendorf 2010). Very characteristic and similarly symptomatic for this understanding is the fact that all items of this factor tend to construct the locus of control in and over various situations external to the individual scope of action. Respondents who agree to this paradigm tend to believe in supernatural forces that control the outcomes of situations, events, and actions. For instance, FA_1_d.) underlines this attitude by stating an approach that prefers to leave difficult problems to God instead of wearing oneself out in solving them. In the same spirit, there also seems to be a lack of urgency to act upon certain problems, which indicates towards a rather soft work ethic.

Moreover, FA_1_b.) refers to the rationality of karma. It states that people cannot influence their standing in society very much as they earned this rank due to their actions and ways of living in their previous lives. According to this notion, any action not only has an external impact but also creates an imprint on one's own soul, which then determines consequences in the next life.

In contrast to most of the other items in this factor, FA_1_b.) implies a more active and self-responsible role of humans. However, this responsibility only refers to the personal caste-related and predetermined functionality, which is based on the norms subscribed to one's own dharma. Therefore, the ideology of karma is limited in its scope of agency to the individual or at the utmost family and kin-related level. And adding to the rather fatalistic logic of the religious paradigm, it says that if there is any chance of a change, it can only be achieved through an abiding orientation towards one's own dharma and righteousness. Consequently, societal or environment-related issues are beyond this scope and not worth caring about too much.

The two last and lowest-loading items slightly differ from the other items: FA_1_e.) draws a very dark image of human nature, which relates to distrust and a belief in using punishments and coercion as a solution for social order. With this conservative call for authority, the item takes a rather passive and submissive standpoint that ignores the possibility of righteousness and virtuousness of other people. Beyond the limits of one's own family and kin group, there is no reason for trust in other people, because "human nature is bad, evil, and wicked". Consequently, people who agree to this image also tend to be more egoistic and centred around their own family, as it is well pointed out by Varma (1998, pp. 138–139, cf. Säävälä 2010, p. 155). This law-and-order approach also ignores the potential of social responsibility, and it again constructs the locus of control beyond the individual level.

FA_1_f.) refers to the phenomenon of endogamy and the caste system. The Indian caste system represents a wide-reaching system of social stratification, which structured the Indian society into innumerable hereditary and traditionally endogamous segments. This item alludes to this traditional system by referring to the norm of intra-caste marriage that is based on the rationale of maintaining the "purity" of hereditary lines. An agreement to this item implies the acceptance of a God-given hierarchical order of society (the caste system), and this greater structure then becomes the determining factor for the decision on potential marriage partners. This feature matches well with the inner logic of the paradigm of religious tradition, as the locus of control is shifted beyond the self to the external level, just as conveyed by all other items building factor 1.

Interestingly, the item does not only load on factor 1, but it also has a quite strong connection with factor 5. This is very striking and evidently makes sense with regard to the content of the two factors, which both discriminate discrete paradigms of tradition, namely, religious tradition (factor 1) and the rather secularly oriented paradigm of family tradition (factor 5). Both these dimensions emerged from the PCA with each of them containing a last low loading item that statistically connects the two dimensions with each other: the last loading item (FA_1_f.) of religious tradition (factor 1) also loads on family tradition (factor 5, 0,42). Simultaneously, the last loading item (FA_5_c.) of family tradition (factor 5) also loads on religious tradition (factor 1, 0,47; Table 5.10). Content specifically, the connection of FA_1_f.) to factor 5 is evident with its implicit family-related notion of an intra-caste marriage, which also means that the marriage is usually arranged by the parents. FA_5_c.) is associated with factor 1 through its implicitly stated external locus of control, which is a very typical feature of all items that load on the paradigm of religious tradition.

In conclusion to the above-delineated structure, the first factor represents a latent dimension of general attitudes that follow a traditional, mythological, and fatalistic paradigm, with a rather passive and resigned attitude towards life. In the following, it will be referred to as the paradigm of religious tradition.

5.2.3.2 Factor 2: The Paradigm of Frugality

Factor 2 subsumes six items, with all items load high on the factor (>0.6). The last two items also load slightly (<0.38) on other factors. Frugality and thriftiness build the central theme here, which is implicitly raised in all six items. Most items apparently indicate an attitude that is actuated by a perceived or de facto situation of being left without any choices due to financial constraints and an obligation to save money. The highest loading item in this factor refers to consumption being limited to the maintenance of basic needs, followed by a proposition about saving for harder times (Table 5.12).

This factor, which draws on the value orientation of frugality and thrift, is an important scale for the Indian situation, as poverty and financial constraints represent determining factors for many people, also in the urban context. However, it was assumed by the author that the quantitative data – through PCA – would allow one to discriminate between frugality as a reaction to the financial situation (culture of necessity) and frugality as a morally driven virtue (frugality and thrift). Therefore, an equal number of items were included in the quantitative survey in order to clearly

Table 5.12 Rotated component matrix: factor 2 – frugality

	1	2	3	4	5
Fa_2_a.) Whatever I earn goes into maintenance of basic needs	0,211	**0,823**	0,064	−0,018	0,068
Fa_2_b.) I carefully watch out how much I spend in order to save my money for harder times	−0,01	**0,726**	0,252	−0,049	0,278
Fa_2_c.) Consumption as such is given too much importance in our society and doesn't make you happier in the end	0,24	**0,683**	0,22	0,053	0,142
Fa_2_d.) It doesn't matter how much I earn; when I have a choice between two equal products I usually go for the lowest price	0,209	**0,673**	0,039	−0,092	0,161
Fa_2_e.) It sometimes bothers me quite a lot that I can't afford to buy all the things other people have	0,145	**0,623**	0,377	0,037	−0,186
Fa_2_f.) My life situation leaves no other option than carefully watching out how much I spent	−0,078	**0,613**	0,232	0,025	**0,316**

differentiate between these two paradigms. Following this logic, it was expected that the paradigm of "culture of necessity" would depend to the greatest extent on the actual social facts of income and social status as foundational drivers of such an attitudinal system. However, in the explorative procedure of the PCA, a clear differentiation was not seen in any one of the factor solutions. Rather the two sets of variables tended to get merged within one and the same dimension.

A revisitation of the qualitative interviews after conducting the survey and exploring the data by PCA allowed for a better understanding of this phenomenon. For many lower class and lower middle class families, and especially those who have recently climbed up the social ladder, frugality often remains a determining principle driven from a fear of social decline. This fear may prevent many people from changing their familiar patterns of consumption, and it is often accompanied with an obligation to save money for more essential expenditures such as future school fees for children, healthcare for the family, house construction, and dowry (cf. Säävälä 2010, p. 121).

And interestingly, there were even quite a few respondents from the upper middle classes, who argued in a very similar pattern underlining the importance of thriftiness and saving money for the future. For an illustration, the following line of argument was given by a middle-aged (27–36) upper middle class respondent who works as computer engineer in the so-called HITEC City of Hyderabad:

I think if my income doubles I would put the money that adds into the future of my children. And I think this is more typical for everyone in this country, that's what I find. It is more of a saving society then a spending one, so whatever it is, you try to invest and safe for your children and pass it on to the next generation. So you see, that is not a big lifestyle change again, as we already now put the extra money towards the future. (Interview No. 016_2010_5_1: 7)

And more specifically, with regard to leisure and holidays, there were respondents with substantial income stating that they cannot imagine spending a substantial amount of their hard earned money on fun, recreation, and holidays:

Holidays? Still we don't have the concept of going on a holiday and spending that much money. We prefer going to our native place where all families are. We keep attending all the functions, like marriages, or house-warming ceremonies, and that is how time goes. But we have not been on any international holiday with my family. I have been to so many countries on official tours but nothing personal. We are still not ready for that kind of spending or spend on an international holidays. (Interview No. 021_2010_5_6: 8)

This IT professional with a yearly household income of above 1 million INR argues that his financial situation does not allow him to think of recreational holidays. Moreover, he is bound to visit his native place instead, as his job in the IT sector more or less pushed him into a nuclear family set-up in a megacity such as Hyderabad, and he needs to reconnect with his joint family during vacations.

In most of these above given examples, frugality is rather seen as a virtue of simplicity independently from the specific social position. It is therefore not surprising that there is no significant income effect in terms of disagreement with the paradigm of frugality and thrift. The factor values of this dimension do not correlate with income. That means income does not play much of a role in affecting one's attitude to the paradigm of frugality. This unexpected result will be also discussed as part of the cluster analysis in the following chapter.

Equally relevant here is the fact that the value orientation of frugality does actually translate into low levels of investive consumption. There is a significant negative correlation between frugality and the wealth index (see Sect. 4.2.5.5): high levels of agreement with frugality correlate significantly with low levels of the variable of wealth (Pearson correlation coefficient $r = -0.29$ significant at the level of 0.001), which represents investive consumption. Consequently, frugality as measured in this study and being received as virtue has a much stronger moral connotation than assumed beforehand. It is a paradigm that reflects an attitude of prudence and modesty in regard to (modern) consumption, and it rests – as already stated above – on the Gandhian ideal of simplicity and the principles of Nehruvian socialism (cf. Mathur 2010, p. 225; van Wessel 2004, p. 99). Generally, this notion is highly relevant in the Indian context today.

This result of a rather morally driven paradigm of frugality is also supported by an item that explicitly raises a critique towards mass consumption (FA_2_c.). The item only loads high on the dimension of frugality and thrift and has not made its way into the dimension of social-ecology or into either one of the traditional paradigms. The item argues that consumption has been given too much weight and importance in society and it is misconceived to make people happier in the end.

Interestingly, this moral framing of frugality contradicts an attitude that embraces consumption as means of identity management. Apparently, there is a moral ambivalence between the quite recent medially constructed urge to consume and spend money on the one hand and the more traditional perception on money and consumerism as being lavish and immoral. This ambivalence even becomes more obvious in consideration of item FA_2_e.), which states the unsatisfied desire of keeping up with others and being able to participate in social life through consumption. While money renders it possible to achieve and buy a sheer infinite variety of sensual experience and materiality that is visible everywhere in the urban environment, it simultaneously creates unsatisfiable desires due to its constraints (Säävälä 2010, p. 124).

Keeping this duality of money in mind, it becomes conceivable that the above-mentioned item FA_2_e.) hides an elusive and unspoken jealousy of an allegedly prevalent lavish lifestyle of mass consumption. Stating that consumption is given too much importance in society making no one happier in the end (FA_2_c.) may sound just like modesty and prudence. But as part of factor 2 which quite clearly draws on the value of frugality, it could also be interpreted as driven by social expectancy, i.e. respondents feel morally pushed to pretend an acceptance and relativisation of their rather humble or even precarious situation.

The same contradictions are taken up again as part of the interpretations of factor 4, hedonistic conspicuous consumption. The theme of hedonism clearly opposes the paradigm of frugality, and many of the reference points given here are further discussed in the section below.

5.2.3.3 Factor 3: Social-Ecological Orientation

The next factor comprises three items which quite clearly delineate a social-ecological orientation, mainly drawn from a consumers' perspective. The first item raises the issue of consumer responsibility by refusing products made by companies which are socially and ecologically irresponsible. Another item (FA_3_c.) aims in a similar direction by asserting to pay attention to the environmental friendliness of products. In addition to these two consumption-specific items, the factor also builds on a broader assertion made with item FA_3_b.). This item draws on a more general social norm, namely, caring for nature and being concerned for the environment. It rounds up the factor's paradigm and supports the social-ecological dimension more generally (Table 5.13).

All three items implicitly involve reason and motivation for individual agency and a positive cognition about the meaningfulness and efficacy of one's own actions. This is especially the case with regard to sustainable consumption. A social-ecologically conscious consumer perceives that they have an impact on production processes and the market. She or he considers the relevance of individual consumption decisions with regard to the eventual effect on the market, i.e. through general changes in awareness, imitation by others, and the cumulative influence on the demand for sustainable products. The locus of control over the outcomes of events and situations is therefore rather internally situated, and respondents supporting this paradigm tend to also have much higher levels of self-efficacy compared to, e.g. supporters of the rather fatalistic paradigm of religious tradition (factor 1). With its explicit socially and ecologically concerned consumer perspective, it also tends to counter the hedonist conspicuous consumption orientation of factor 4 (see below). However, this factor is able to measure social and environmental concern with a focus on responsible consumption. It addresses the importance of individual responsibility towards the environment and social issues and will be termed the paradigm of social-ecological consumption and behaviour.

5.2.3.4 Factor 4: Hedonist Conspicuous Consumption

The PCA has delineated another very important factor on consumption. By embracing the materialistic theme of mass consumer culture, it clearly opposes the two other factors that frame the notion of consumption very differently, namely, factor

Table 5.13 Rotated component matrix: factor 3 – social-ecological orientation

	1	2	3	4	5
Fa_3_a.) I try not to buy products made by companies which are socially and ecologically irresponsible	0,133	0,204	**0,725**	−0,054	0,058
Fa_3_b.) Believes that people should care for nature; looking after the environment	−0,022	0,266	**0,692**	−0,002	0,104
Fa_3_c.) When shopping I regularly pay attention to the environmental friendliness of the products	−0,048	0,248	**0,649**	0,073	0,244

2, frugality, and factor 3, social-ecological orientation. The items that build this dimension were selected to measure, by the degree to which respondents tend to agree to a materialist culture, in which shopping, buying, and consuming are perceived as pleasure, enjoyment, and entertainment. Apart from these direct and explicit hedonist pursuits, the notion of exclusiveness of shops, locales, and products bears more subtle and in some cases subconscious motives, related to emulation, identity management, self-realisation, and self-expression (cf. van Wessel 2004, p. 99).

The first of the three loading items refers to the practice of shopping in more expensive and exclusive shops (FA_4a.). It articulates an attitude that plays with the expression of wealth and success. The item conveys the ability to possess and consume expensive, modern goods and fashionable clothes. Agreement to this statement reveals notions of conspicuous consumption. This theme being addressed as in the highest-loading item of this factor indicates that distinction of status through consumption is very central for the interpretation of this dimension. The factor thereby combines a direct statement referring to the issue of conspicuous consumption (FA_4_a.) with a theme of shopping in pursuit of pleasure and enjoyment (Fa_4_b./Fa_4_c.). This combination is highly interesting for the understanding of consumption and behaviour patterns that tend to express quite a level of independence from financial or timely constraints. Such a relaxed and carefree attitude towards consumption is unlikely to unfold in a social context characterised by rather limited economic resources. Table 5.14 depicts the factor-building items.

Moreover, many of the exclusive and often foreign brand-name shops and show rooms are particularly found in multiplex shopping malls. It is the second loading item of this factor that measures the inclination towards visiting malls as means of recreation, pleasure, and entertainment (Con_V_17). As shown in Sect. 5.2.3.4, malls symbolically represent structured and ordered spaces of the modern Indian consumer culture. Malls fulfil various functions for a medially constructed new way of life in India, and they are well able to combine and interconnect the social-cultural realm of leisure and entertainment with the individual requirements of conspicuous consumption, enculturation, and identity management. These privatised consumption-oriented spaces that are well designed and controlled to subtly manipulate their visitors and customers also convey a sensuous experience that is well described by Schulze as "Erlebnis" (Schulze 1992) (see Sect. 4.1.2.1).

Both the theme of exclusive and conspicuous consumption and the latter drawn image of playing sensuous experience (Erlebnis) on the new stages of modern urban

Table 5.14 Rotated component matrix: factor 4 – hedonist conspicuous consumption

	1	2	3	4	5
Fa_4_a.) I quite frequently shop in more expensive and exclusive stores	0,04	−0,082	0,032	**0,805**	−0,058
Fa_4_b.) Going to shopping malls is one of the activities that I really like, even if I don't buy anything	0,087	0,019	0,171	**0,732**	−0,048
Fa_4_c.) I like to buy things just for the fun of it	−0,024	0,028	−0,391	**0,668**	0,079

lifestyle are supported and further taken up by the last loading item of this factor. The item refers to an explicit practice of buying and shopping things just for the fun of it (Con_V_10). Again, consumption is framed under the notion of hedonism and – in this item – implicitly combined with easiness and light-heartedness that conveys a sense of having no financial anxiety towards the future. Moreover, fun and enjoyment are expressed through the practice of shopping and buying. With this PCA, all consumption-related items have been factorised into an in-itself consistent solution. The three resulting factors, frugality, social-ecologically oriented consumption, and hedonistic conspicuous consumption, meaningfully differentiate attitudes and values towards all dimensions of consumption relevant in the Indian context. For an understanding of the dynamics of Indian consumer culture with regard to sustainability, this factor solution is invaluable. With respect to the cluster analysis, the underlying factors may correspond or conflict with each other, they may depict meaningful combinations and patterns of attitudes and values, but they may also reveal inherent ambivalences, especially – as shown above – between different attitudinal dimensions of consumption. These combinative and interactive patterns will be shown and interpreted as part of Sect. 6.4.

5.2.3.5 Factor 5: Family Tradition

The fifth and last factor demarcates a dimension that the author defines as the paradigm of family tradition. This traditional paradigm sets itself apart from the religious and mythological perspective of factor 1. With its rather secular framing, it emphasises the importance of naturally given in-group affiliations and related ideals of family and community. Thereby, it draws on related social-structural issues such as notions of extended family, related kinship ties, and roles and duties based on family and kin-related hierarchies, such as respect and obedience towards elders.

The dividing line between religion and secularism with respect to Indian traditional values was not intuitive at first. The author initially assumed that the items of factor 1 and factor 2 built only a single dimension, namely, the dimension of tradition. It was surprising to find statistical evidence of respondents making a difference along this line. Compared to factor 1, the secularly framed position on tradition does not comprise otherworldly and fatalist propositions and rather builds on worldly and more rational statements. The differentiation between the two factors is very favourable for the analysis of social values with respect to tradition and modernity. Such a factor solution is much better able to depict the "different shades of modernity" (Brosius 2010, p. 31) and reveal more nuanced and socially specific orientations towards modernisation. It may also contribute to better expose the ambiguities that arise in the midst of rapid social and cultural change.

According to Schwartz, an orientation towards traditional values is very closely related with conformity, as both concepts share "a single motivational goal – subordination of self in favour of socially imposed expectations" (Schwartz 1994, p. 24). This subordination derived from tradition and conformity is spelled out in the first item (Fa_5_a.), obedience and respect to elders) as well as in the third item (Fa_5_c.), arranged marriages). The item Fa_5_c.) subsumes both these topics under the general consideration of tradition and family.

The first and thereby highest loading item of this dimension takes up an issue that has been recurrently raised as very critical in qualitative interviews with regard to the topic of social change, modernisation, and tradition. The item is concerned with the importance of maintaining obedience and respect for elders (Fa_5_a.). It refers to the ideal of the joint family and kin-related intra-family hierarchies. In many of the informal talks and the semi-structured interviews, people drew on their fear about how modernity at present affects family life and related aspects based on the ideal of the joint family (cf. van Wessel 2001, p. 30ff).

The second item raises a more general concern in respect to the importance of tradition and customs being handed down by one's own family (Fa_5_b.). With this rather general view on family tradition, the item integrates the other two more specific items of this factor. As in the other items, religious connotations are not raised here.

The third and last item (Fa_5_c.) refers to another important aspect of family tradition by stating that in matters of marriage, boys and girls may be consulted, but the final decision should be taken by the parents. Arranged marriages actually involve two different but interconnected questions of ideology: The first is closely related to the theme also raised in the first item of this factor (Fa_5_a.), namely, unconditional respect of parental authority over children. The second aspect is implicitly inferred and therefore less obvious: marriages that are arranged by the parents usually follow the norm of caste endogamy quite strictly (cf. van Wessel 2001, p. 95). With this reference to caste, the item connects the so-far purely secular dimension of tradition with the religious traditional paradigm of factor 1 (see Sect. 5.2.3.1 above): just as in the loading structure of factor 1, this last item loads high on both factors (factor 5 with 0,51 and factor 1 with 0,47; see Table 5.15).

However, the traditional norm of arranged marriage is still widely shared and accepted in India, and there is quite a tendency that people perceive arranged marriages not contradictory to a so-called "modern" orientation. This was, for instance, well indicated in the following interview sequence by a young female IT engineer who lives in a well-off area of Secunderabad in a gated apartment complex:

> We consider ourselves as modern. We are getting used to modern living, but we still have those traditional values and our core values are so strong that we pray to God; we keep going to temples and we marry as per our elder's wishes. We still prefer arranged marriages. We also train our kids in the same way. (Interview No. 021_2010_5_6: 5)

Table 5.15 Rotated component matrix: factor 5 – family tradition

	1	2	3	4	5
Fa_5_a.) Obedience and respect for elders are very important values and should be maintained	−0,030	**0,335**	0,092	−0,052	**0,682**
Fa_5_b.) Tradition is important; tries to follow the customs handed down by her/his family	0,067	0,285	0,075	−0,054	**0,635**
Fa_5_c.) In matters of marriage, boys and girls may be consulted, but the final decision should be taken by the parents	**0,470**	−0,063	0,220	0,062	**0,514**

There are quite a number of studies indicating that the traditional patterns of arranged marriages are slowly making way for more cooperative patterns, in which the individual has a say in the selection of her or his spouse (Jaiswal 2014, p. 142). This is also indicated in the factor's item that attenuates the traditional norm of arranged marriage to a more cooperative pattern. This means that respondents who disagree to this item rather take quite a strong position against family-oriented tradition.

To summarise, obedience, propriety, and the concept of family and kin are essential cornerstones of this attitudinal dimension. Agreement to this paradigm goes along with an agreement to traditional family values and a recognition of a hierarchical ideal within one's own kin group as well as within society. The individual is subordinate to the group and has its determined functional role in relation to all others. Dharma and class-specific morality provides direction for thought and action. By and large, the factor solution with its two discrete dimensions of tradition offers a good tool to measure in a quite differentiated way respondents' level of agreement to Indian-specific traditional values. Combined with the agreement levels of the other factors, this solution represents a good foundation for the complex analysis of clusters based on differentiated social values and attitudes.

5.2.3.6 Intermediate Summary of Findings

Overall, the distribution of agreement levels on the factor variables shows good results (Table 5.16). "Fatalism" was the factor with the highest level of agreement, followed by "family tradition" and "frugality". The factor "hedonist conspicuous consumption" represents the most balanced variable with respect to the different agreement levels. The factor frugality shows the most polarised distribution with only 22% of all respondents showing an average level of agreement. Family tradition has the smallest number of respondents with high levels of disagreement.

For classifying the factor variables (ratio scale with a mean of 0), levels have been defined by making cut-points at ±0.4 deviation from mean, high agreement $\triangleq$ > 0.4, average level $\triangleq$ −0.4 – +0.4, and high disagreement $\triangleq$ > −0.4.

Overall, the interpretability of the factor solution is highly satisfactory. The five extracted factors are foundational for the cluster analysis, and the specific distribution of factors within each cluster carries a very specific and discriminative value orientation for each segment. This combined pattern covers aspects of tradition and modernisation more generally, and at the same time it provides a more specific analysis of attitudes towards and preferences for consumption. Both are highly relevant for the understanding of milieu-specific consumption patterns and related dynamics.

For the purpose of this differentiation, the cluster analysis was the most promising analytical method on the basis of the extracted attitudinal dimensions. In the following, the author will present the statistical results of the cluster analysis.

Table 5.16 Distribution of agreement levels on the five-factor solution

		Count	Column N%
Fatalism	High disagreement	179	31%
	Average level	111	27%
	High agreement	127	43%
	Total	417	100%
Frugality	High disagreement	161	39%
	Average level	90	22%
	High agreement	166	40%
	Total	417	100%
Social-ecological orientation	High disagreement	138	33%
	Average level	125	30%
	High agreement	154	37%
	Total	417	100%
Hedonist conspicuous consumption	High disagreement	134	32%
	Average level	140	34%
	High agreement	143	34%
	Total	417	100%
Family tradition	High disagreement	120	29%
	Average level	132	32%
	High agreement	165	40%
	Total	417	100%

5.2.4 Cluster Analysis: Factors of Attitudes and Value Orientations as Building Blocks for a Typology of Lifestyles

The aim of the cluster analysis was to merge the overall sample of respondents into homogenous segments to meaningfully structure the dataset based on the selected attitudinal characteristics. The factor analysis provided a good foundation for this analysis, because it reduces the number of variables to a manageable and more interpretable number of latent dimensions based on the response pattern in the dataset. This dimension reduction allows for a much easier interpretation of the clusters.

Based on previous exploration of different algorithms, the author decided for a k-means cluster analysis. In order to obtain start values for this centroid-based algorithm, a hierarchical cluster analysis (Ward) was previously conducted. For both methods, the k-means and the Ward method, the group memberships of ten clusters were computed. In order to determine a formally valid number of clusters, different statistical tests were conducted, the Eta^2, the PRE-value, and the F-max (cf. Schendera 2010, p. 119). Based on these statistics, a final cluster solution was decided upon.

The line diagram of the criterion of explained variance (ETA^2) in Fig. 5.6 shows remarkable bends at the sixth and the eighth cluster solution with an explained

5.2 Construction and Descriptive Analysis of Latent Variables and Components... 167

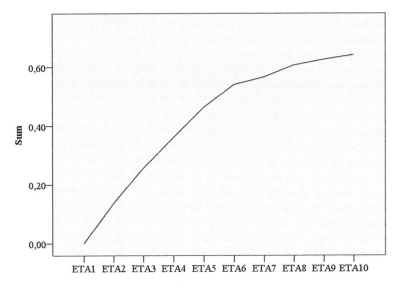

Fig. 5.6 Criterion of explained variance (ETA^2)

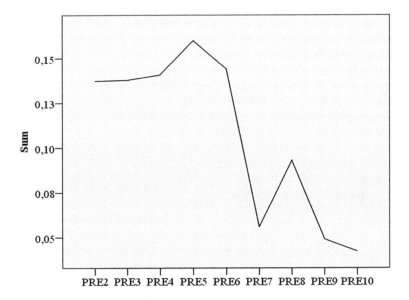

Fig. 5.7 Criterion for the relative improvement of explained variance (PRE)

variance of 54% and 61%, respectively. As ETA also depends on the number of clusters, both solutions can be taken into account (cf. Schendera 2010, p. 130). With PRE5 and PRE6 being the highest PRE-values (Fig. 5.7), this test indicates that the fifth and possibly also the sixth cluster solution are the most sound solu-

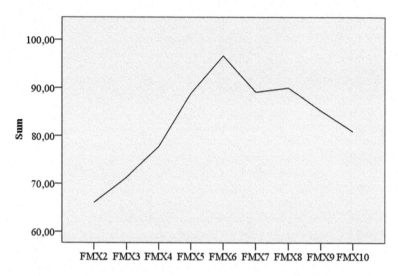

Fig. 5.8 Criterion for the best variance proportion (F-max)

tions formally. The criterion for the best variance proportion (F-max) is independent of the number of clusters. According to the F-max criterion (Fig. 5.8), the sixth cluster solution shows the best results. On the basis of these three tests, the author selected the sixth cluster solution as the formally most reliable one. Even though, ideally, all test statistics should arrive at the same conclusion (Schendera 2010, p. 131), the six-cluster solution shows satisfactory results.

Sabine Fromm (2010) mentions further criteria: a good cluster solution is given if (a) the means of the underlying variables between the clusters differ significantly and if (b) the standard deviations of each variable within the cluster are smaller than the standard deviation of the variable within the overall sample (Fromm 2010, p. 214). All extracted clusters fulfil these criteria (Table 5.19). Moreover, some authors suggest the computation of F-values, which measure the proportion between the cluster-specific variance and the variance of the overall sample for each of the underlying variables (Fromm 2010; Schendera 2010; Backhaus 2011). If all F-values for each variable in each cluster remain at a level below 1, the clusters can be evaluated as being completely homogeneous. This is the case, as shown in Table 5.17.

The final cluster solution involves nearly 70% of the overall sample of 605 respondents. Only 417 respondents have been included in the lifestyle segmentation process. As mentioned above in Sect. 4.2.5.2, the criteria to include respondents into the analysis were strict. The author had to omit all those respondents who had too many missing values on the item battery or who had used only one or a few response categories of the six-level Likert scale. This quite influential decision was taken, because it would have been difficult to assess attitudinal and value priorities accurately enough, if there were too many missing values or the respondent had not really tried to discriminate between attitudes by only using one or two levels on the

5.2 Construction and Descriptive Analysis of Latent Variables and Components... 169

Table 5.17 F-values for each variable of the six-cluster solution

F-value					
	Fatalism	Frugality	Universalism	Hedonist conspicuous consumption	Family tradition
Cluster 1	0,49	0,26	0,27	0,54	0,51
Cluster 2	0,34	0,44	0,53	0,53	0,80
Cluster 3	0,43	0,75	0,74	0,76	0,41
Cluster 4	0,93	0,39	0,29	0,44	0,41
Cluster 5	0,22	0,44	0,36	0,43	0,36
Cluster 6	0,40	0,22	0,43	0,84	0,36

Table 5.18 Frequency distribution of the six-cluster solution

	Frequency	Percent	Valid percent	Cumulative percent
Cluster 1	62	10,2	14,9	14,9
Cluster 2	72	11,9	17,3	32,1
Cluster 3	57	9,4	13,7	45,8
Cluster 4	76	12,6	18,2	64
Cluster 5	105	17,4	25,2	89,2
Cluster 6	45	7,4	10,8	100
Cluster analysis total	417	68,9	100	–
Missing	188	31,1	–	–
Total sample	605	100	–	–

Likert scale. Table 5.18 gives an overview of the size of clusters, which varies between 45 respondents in cluster 6 and 105 respondents in cluster 5.

Besides formal criteria, the selected cluster solution even more needs to withstand criteria in regard to content. Eventually, interpretability of the cluster solution has priority over formal criteria. In Sect. 6.4, the six-cluster solution will be presented with regard to content and interpretation. As the cluster analysis is solely based on attitudes and values, each cluster represents a quite homogenous group of individuals who share common beliefs and attitudes on the selected issues and areas. The interpretation of the clusters' inherent structure, which is based on the above-delineated five factor variables, is a very thoughtful and highly subjective process. To assure the highest degree of transparency possible, the author has documented every step of the interpretation process together with the underlying database. This documentation can be found in Annex IV.

The interpretation is based on three consecutive steps of data examination. The first and foundational step was in fact the interpretation and understanding of the

Table 5.19 Final cluster centres of the six-cluster solution (mean)

	Cluster					
	1	2	3	4	5	6
Fatalism	−1,247	0,197	0,392	−0,800	0,603	0,851
Frugality	1,213	−0,490	−0,590	−1,002	0,351	0,733
Universalism	−0,394	−0,494	−1380	0,909	0,274	0,908
Hedonist conspicuous consumption	0,073	−1509	0,808	0,280	0,368	−0,043
Family tradition	−0,065	0,213	−0,525	−0,195	0,899	−1,354

five dimensions, which were extracted through principal component analysis from the larger set of attitudinal survey items. The results of this examination have been expounded above (Sect. 5.2.3). Building on this understanding of the underlying factor variables, the second step was to examine the levels of agreement to these factor variables in each cluster. This is most essential, because the whole existence of each cluster segment is based on these "active" variables. Only when this interpretation leads to a meaningful outcome and every cluster can be given a characteristic name has the cluster analysis actually succeeded. Eventually, in a third step, the interpreted segments can undergo a broader analysis that looks at other, so-called passive variables that have not gone into the cluster analysis, such as social-structural data, data on behavioural and consumption patterns, and GHG emissions. The result of this procedure is a lifestyle typology that largely builds on values and attitudes that structure behaviour and consumption within a given scope of action.

As already stated, the second step of interpretation is laying out a description of clusters based on the underlying factor variables. Table 5.19 depicts the final cluster centres. The underlying factor scores represent standardised scales with a mean of zero and a standard deviation of one. It is therefore obvious that the means of factor scores within the clusters either have negative or positive signs.

For the interpretation and illustration of the underlying intra-cluster structure of attitudes and values, the distribution of respondents' agreement levels to the five dimensions was visualised through bar charts and boxplots. The given structure of each cluster can be understood as a representation of the underlying attitudes and values of people who belong to this group. Thereby, one needs to keep in mind that this structure only represents an average of each group, with a highest possible degree of homogeneity within the cluster and a highest degree of dissimilarity between the clusters. Table 5.19 visualises the cluster-specific average levels of agreement with the underlying factors based on the mean values within each cluster. Error bars (95% confidence intervals) illustrate the variance for each variable in each segment (Fig. 5.9).

As mentioned, to statistically validate the differences between different values of the clusters, the non-parametric Kruskal-Wallis test was made, and the null hypothesis was tested to examine whether the underlying variables differentiate between at least two of the involved clusters. This was the case with a significance level of 0.05

5.2 Construction and Descriptive Analysis of Latent Variables and Components...   171

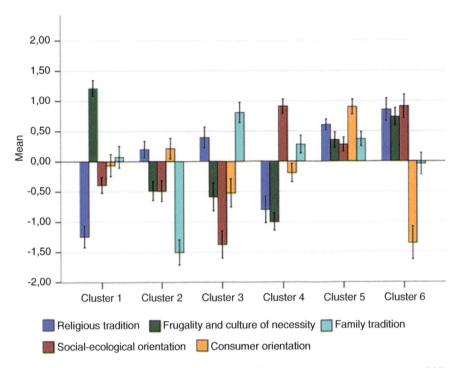

Fig. 5.9 Grouped bar chart of cluster centres (mean) for each cluster. Error bars represent 95% confidence interval levels

(two-sided test). With this preparation of data at hand, the author then examined the clusters in detail, based on the "active" variables. The underlying factors thematically cover a relevant array of characteristic orientation dimensions, which are specific for the contemporary Indian urban context (see Sect. 5.2.3).

After this first interpretational step, additional "passive" variables have been drawn upon, in order to further describe and analyse the clusters. There are both categorical and metric variables, some of which were categorised for better interpretation (e.g. income, education) and some of which were computed through principal component analysis, such as for wealth (specialised procedure for building an index) and self-reported shopping, leisure, and holiday practices (see Sect. 4.1.3.3). The final and core element of the cluster analysis was the examination of personal carbon footprints. Overall and sector-specific distributions of personal GHG emissions are therefore also part of the following cluster descriptions.

Tables 5.20, 5.21, 5.22, and 5.23 give an overview of all "passive" variables that were drawn upon for the further analysis of the clusters. Tables 5.20, 5.21, and 5.22 show all variables with a metric level of measurement.[3] It shows the median values

---

[3] The ordinal level of education, as it was reported by the respondent, was exceptionally treated as a metric variable in supplement to the actual categorical analysis as shown in Table 5.20.

Table 5.20 Overview of medians and results of H-test after Kruskal-Wallis for all "passive" metric variables of social demography

Variable Social-Demography	Cluster Median						Overall	H-test statistics after Kruskal / Wallis
	1	2	3	4	5	6		
Age	31⁻	35⁺⁺⁺	30*	28⁻	33⁺⁺	30*	30,0	H-test (K.-W): Var. sign. affected by cluster affiliation; H(5) = 11.3, p < .05
Education	4*	4*	5⁻	5⁻	4*	4*	4,0	H-test (K.-W): Var. sign. affected by cluster affiliation; H(5) = 13.5, p < .05
Wealthindex	16,0⁻	20,9⁻	24,2⁺⁺⁺	22,6⁻	19,9⁻	18,3⁻	19,8	H-test (K.-W): Var. sign. affected by cluster affiliation; H(5) = 19.7, p < .01
Household Size	4⁻	4,5*	3⁻	6⁻	4,5*	4⁻	4,5	H-test (K.-W): Var. sign. affected by cluster affiliation; H(5) = 25.7, p < .01

Values above overall median are marked red, values below overall median are marked blue, and plus (high), minus (low), and star (same as overall median) mark clusters' relative position for each variable

Table 5.21 Overview of medians and results of H-test after Kruskal-Wallis for all "passive" metric variables of social demography

Variable Preferences and Behavior	Cluster Median						Overall	H-test statistics after Kruskal / Wallis
	1	2	3	4	5	6		
Shopping in Centralized Locations (Malls / Supermarkets)	-0,67⁻	-0,13⁻	-0,10⁻	0,51⁺	0,52⁺⁺⁺	-0,20⁻	-0,07	H-test (K.-W): Var. sign. affected by cluster affiliation; H(5) = 39.8, p < .001
Shopping Local (Kiranas / Street Vendors)	0,52⁺⁺⁺	-0,18⁻	-0,29⁻	0,65⁺⁺⁺	0,07⁻	-0,03⁻	0,07	H-test (K.-W): Var. sign. affected by cluster affiliation; H(5) = 40.4, p < .001
Leisure - Action and Consumption Oriented	-0,97⁻	-0,70⁻	-0,05⁻	1,08⁺⁺⁺	0,20⁺	-0,18⁻	-0,14	H-test (K.-W): Var. sign. affected by cluster affiliation; H(5) = 101.0, p < .001
Leisure - Homely Oriented	0,60⁺⁺⁺	-0,50⁻	0,02*	0,00⁻	-0,03⁻	0,50⁻	0,08	H-test (K.-W): Var. sign. affected by cluster affiliation; H(5) = 25.0, p < .001
Holidays - Action and Consumption Oriented	-0,39⁻	-0,34*	-0,36⁻	-0,34*	-0,07⁺	-0,54⁻	-0,34	H-test (K.-W): Var. sign. affected by cluster affiliation; H(5) = 19.8, p < .01
Holidays - Traditionally Oriented	-0,22⁻	-0,33⁻	0,01*	0,33⁺	0,40⁺	-0,26⁻	0,01	H-test (K.-W): Var. sign. affected by cluster affiliation; H(5) = 35.8, p < .001
Media Use - Sophisticated / English	-0,82⁻	-0,29*	-0,04⁻	0,09⁺⁺⁺	-0,05⁻	-0,40⁻	-0,29	H-test (K.-W): Var. sign. affected by cluster affiliation; H(5) = 30.5, p < .001
Media Use - Red Top	-0,18⁺	-0,36⁻	-0,40⁻	0,37⁻	0,84⁺⁺⁺	-0,40⁻	0,00	H-test (K.-W): Var. sign. affected by cluster affiliation; H(5) = 51.1, p < .001
Expenditure on Internet	250*	200⁻	400⁻	100⁻	250⁻	500⁻	200	H-test (K.-W): Var. sign. affected by cluster affiliation; H(5) = 22.0, p < .01

Values above overall median are marked red, values below overall median are marked blue, and plus (high), minus (low), and star (same as overall median) mark clusters' relative position for each variable

of each variable in each cluster and also depicts the test results of the H-test with degrees of freedom, H-value, and asymptotic significance level (p-value). The highest and the lowest median values are shown in colour for better illustration.

5.2 Construction and Descriptive Analysis of Latent Variables and Components...

Table 5.22 Overview of medians and results of H-test after Kruskal-Wallis for all GHG emission variables (metric)

Variable	Cluster Median							H-test statistics after
GHG Emissions	1	2	3	4	5	6	Overall	Kruskal / Wallis
Emissions Electricity	703++	685+	1235+++	400--	575-	619*	619	H-test (K.-W): Var. sign. affected by cluster affiliation; H(5) = 34.9, p < .01
Emissions Cooking	135++	92--	180+++	86---	90-	120*	108	H-test (K.-W): Var. sign. affected by cluster affiliation; H(5) = 33.2, p < .01
Emissions Food	270++	197---	274+++	261*	229-	226--	238	H-test (K.-W): Var. sign. affected by cluster affiliation; H(5) = 23.1, p < .01
Emissions Long Distance Travel	0--	0--	10-	16+++	14++	12*	3	H-test (K.-W): Var. sign. affected by cluster affiliation; H(5) = 61.8, p < .01
Emissions Air Travel	0	0	0	0	0	0	0	H-test (K.-W): not significant
Emissions Public Transport	62*	62*	52---	55*	80+	54--	62	H-test (K.-W): not significant
Emissions IMT	53--	53--	165+++	121++	119*	95-	103	H-test (K.-W): Var. sign. affected by cluster affiliation; H(5) = 15.0, p < .05

Values above overall median were marked red, values below overall median were marked blue, and plus (high), minus (low), and star (same as overall median) mark clusters' relative position for each variable

For all "passive" categorical data (e.g. caste, gender, and marriage status), Table 5.23 gives an overview of the results from cross tabulation of these variables in regard to the respective cluster distribution. The table depicts the residuals with highlighted significance levels as well as Pearson's chi-square test results.

The above-given overview of passive variables is the basis for the second step of the cluster analysis process, namely, the description of the clusters. Section 6.4 will give a comprehensive description of all six clusters, first in respect to their inherent structure of values and attitudes based on the active variables. This first step is foundational also for the naming of each cluster. This basic description is supported with the analysis of the clusters based on the passive variables.

Table 5.23 Overview of results from cross tabulation for all "passive" categorical variables

Variable	Category	Cluster 1	2	3	4	5	6	Pearson's chi-square
Income	Deprived	0,96	0,33	-0,09	-1,47	0,31	0,02	χ2 (15) = 29.74
	Aspirers	-0,49	-0,90	-2,57*	3,22***	0,49	-1,09	p < .05
	Seekers	0,74	0,76	0,72	-1,86	-1,23	1,58	[2 cells (8,3%) expected count < 5]
	Strivers and Globals	-1,51	-0,12	3,07***	-1,27	0,67	-0,65	
Wealth Index (Quintiles)	Poorest	4,44***	-1,03	0,64	-2,22*	-1,04	-0,34	χ2 (20) = 42.74
	Second	-0,16	0,24	-2,03*	0,32	0,47	1,09	p < .01
	Third	0,12	0,10	-1,28	-0,78	1,29	0,32	
	Fourth	-2,23*	-0,49	-0,17	2,43*	0,52	-0,42	
	Richest	-2,09*	1,15	2,75**	0,24	-1,21	-0,62	
Employment	Housewife	6,20***	-0,53	-1,69	-0,44	-2,03*	-1,45	
	Retired	0,58	0,49	0,08	-0,48	-0,86	0,45	
	Unemployed	-1,23	-0,65	-0,31	-1,51	1,52	2,41*	
	Student	-1,79	2,45*	2,09*	-0,83	-2,67**	1,60	χ2 (50) = 109.3
	Semi- and Un-Skilled Worker	0,93	-1,19	0,06	-1,30	0,29	1,55	p < .001
	Skilled Manual Worker	0,21	0,56	0,13	-1,23	1,50	-1,57	[36 cells (54,5%) expected count < 5]
	Lower-Grade Technicians	-0,73	-1,57	1,18	-0,23	1,46	-0,27	
	Small Proprietors (No Employees)	-2,55*	-0,67	1,73	0,75	0,89	-0,27	
	Small Proprietors (Employees)	-0,16	-1,87	-1,55	-0,61	2,13*	2,01*	
	Routine Non-Manual Employees	0,05	1,10	0,23	-0,16	-0,42	-0,85	
	Professionals	-1,79	-0,59	-1,17	3,18**	0,06	-1,44	
Education	No Degree	1,97*	-1,05	1,15	-1,76	-0,52	0,77	χ2 (10) = 20.30
	Up to 12th class	1,64	0,53	-1,71	-1,52	0,73	0,24	p < .05
	Graduation and Above	-3,05***	0,26	0,82	2,78**	-0,33	-0,80	
Education Father	No Degree	1,57	0,83	-1,05	-5,40***	2,01*	2,26*	χ2 (10) = 45.3
	Up to 12th class	-1,43	-2,21*	-0,29	4,91***	-0,50	-0,76	p < .001
	Graduation and Above	-0,72	0,75	1,45	-2,49*	-1,96*	-2,06*	
Visual Analogue Class Rating	Lower Class	2,43*	0,04	-1,23	2,52*	-2,49*	-1,11	χ2 (10) = 24.40
	Middle Class	-0,08	0,08	0,11	-0,72	-0,32	1,21	p < .01
	Upper Class	-2,25*	-0,13	1,06	-1,60	2,78**	-0,33	
Gender	Female	4,14***	-0,04	-1,93	0,43	-1,39	-1,17	χ2 (5) = 20.68, p < .01
Religion	Hindu	1,39	-1,04	-5,97***	1,35	2,98**	0,33	χ2 (10) = 42.81
	Christian	-0,97	0,46	2,99**	-1,31	-0,97	0,28	p < .01
	Muslim	-0,95	0,87	4,78***	-0,70	-2,70**	-0,53	[5 cells (27,8%) expected count < 5]
Caste	Scheduled Caste OBC	3,46***	-0,97	-0,15	-2,56*	0,03	0,51	χ2 (5) = 16.56, p < .01
Marriage	Single	-3,00**	0,64	2,06*	1,98*	-2,56*	1,52	χ2 (5) = 21.9
	Married	3,00**	-0,64	-2,06*	-1,98*	2,56*	-1,52	p < .001
Daughters Unmarried	Yes	4,03***	-0,96	-1,56	0,01	-1,24	0,01	χ2 (5) = 17.81, p < .01
Children in Same Household	No Kids Below 13 Living in Household	1,92	-0,58	2,20*	-4,40***	-1,13	3,27***	χ2 (5) = 33.8, p < .001
Meat Consumption Level	Vegetarian	-1,53	1,14	-2,04*	-0,82	2,41*	0,28	χ2 (10) = 32.4
	Ocassionally (< 1 / week)	3,78***	-0,41	0,11	-2,22*	-1,68	1,21	p < .001
	Regularly (> 2 / week)	-2,88**	-0,45	1,53	3,11**	-0,12	-1,55	
Dominant Mode of Transport	Walking and Cycling	2,49*	0,09	0,54	-0,18	-1,07	-1,84	χ2 (15) = 27.75
	Public Transport	-1,40	2,98**	0,00	-0,09	-1,98*	0,85	p < .05
	Two-Wheeler	0,20	-2,92**	-0,54	0,08	2,74**	-0,01	[2 cells (8,3%) expected count < 5]
	Four-Wheeler	-1,59	-0,49	0,11	0,28	0,48	1,27	
Emission Classes	Low Emissions (< Indian pca)	0,37	1,42	-4,08***	2,42*	-1,47	1,05	χ2 (10) = 38.05
	Mid Level Emissions (< Dev. Cntrs. avg.)	1,03	-0,84	0,91	-1,24	1,31	-1,39	p < .01
	High Emissions (> Dev. Cntrs. avg.)	-1,90	-0,99	4,82***	-2,02*	0,46	0,28	

The table shows residuals and asymptotic significance levels (p-values). Red marks positive, blue marks correlation, and residuals outside ±1.96 are significant at $p < 0.05$ (*), outside ±2.58 are significant at $p < 0.01$ (**), and outside ±3.29 are significant at $p < 0.001$ (***). Underlined values represent residuals based on expected counts <5

Chapter 6
Results Part II: Income, Practice, and Lifestyle-Oriented Analysis of Personal-Level GHG Emissions

Keywords Income · Wealth index · Investive consumption · Consumption practices · Lifestyle segmentation · Hedonism · Family tradition · Religious tradition · Ethical consumption

6.1 The Structure of Consumption-Based Personal-Level GHG Emissions (Carbon Footprints) in Hyderabad

The core element of the secondary lifestyle analysis is the analysis of personal GHG emissions. The carbon footprint calculations comprise emissions from household-based electricity consumption, transportation, household-based cooking, and the most important domains of food, namely, milk, meat, and rice consumption. Resulting in yearly per person GHG emissions measured in tonnes of CO_2 equivalents (CO_2e), the calculator adjusts household-based consumption (e.g. electricity and cooking fuels) to the equivalised number of household members (adults = 1, children = 0.5).[1] Table 6.1 depicts the results of the carbon footprint calculation ($n = 605$).

While there is a substantial range between the largest (16.6 tonnes of CO_2e/cap/year) and the smallest footprint (0.27 tonnes of CO_2e/cap/year) with noticeable variations expressed by the high standard deviation, the mean of the overall emissions for this sample is 1.88 tonnes of CO_2e/cap/year. However, as data of emissions is not symmetrically distributed, it is more accurate to refer to the median, which is 1.38 tonnes of CO_2e/cap/year.

With a mean of almost half of the overall emissions' structure, electricity carries the highest share in this sample at 47% (890 kg CO_2e/cap/year), followed by individual motorised transport at more than 16% (310 kg CO_2e/cap/year) and food consumption at almost 15% (274 kg CO_2e/cap/year). While public transportation accounts for 9% (170 kg CO_2e/cap/year), the use of fossil fuels for cooking is 8% (151 kg CO_2e/cap/year) high. The smallest share of all sectors is carried by

---
[1] This standard is also used for calculation of equivalised income.

Table 6.1 Sector-specific distribution of individual GHG emissions in Hyderabad

	N	Mean (kg CO$_2$e/cap/year)	Median (kg CO$_2$/cap/year)	Percent share based on mean	Standard deviation	Minimum (kg CO$_2$e/cap/year)	Maximum (kg CO$_2$e/cap/year)
Overall emissions	590	1881	1383,5	100	1571	267	16,568
Electricity	590	890	640,5	473	796	0	6366
Individual motorised transport (IMT)	605	310	94,9	16,5	706	0	5074
Food	590	274	233,4	14,6	163	25	1779
Public transportation	605	170	62,4	9,1	336	0	2990
Cooking	590	151	108,5	8	169	0	2739
Long-distance travelling	605	86	5,6	4,6	533	0	8920

long-distance travel[2] with less than 5% (86 kg CO_2e/cap/year; includes bus and train as well as air travel).

Table 6.2 gives a more specific picture of the carbon footprint calculation, with a breakdown of the contributing consumption modes for each sector. After electricity, individual motorised transport (IMT) represents the second most relevant sector: even though two-wheelers (includes moped, scooter, and motorbike) are less carbon-intensive, they have a share of almost 35% of all CO_2 emissions in this sector. On the other hand, the use of the car is less widespread (not even 12% of all IMT users), but this mode contributes significantly to the overall emissions from IMT at more than 65%.

Meat consumption is the most relevant domain in the food sector, as the emission factor for meat is almost ten times that of milk and rice (Table 6.2). Compared to the proportion of vegetarians reported in the *Meat Atlas* (between one third to one fourth for all India; 2014: 36), this study shows relatively high shares of respondents eating meat (84%).

Concerning public transport, more than half of all emissions in this sector go back to the use of auto-rickshaws. Used by one third of the respondents, it is a frequently used mode of transport. However, the most important public transport mode is the city bus with 57% of all respondents relying on it. Nevertheless, only one fourth of all emissions in the public transport sector can be traced back to the use of city buses (without air-conditioning). The Multimodal Transport System (MMTS), which is a suburban rail system, is used by around 4% of all respondents but contributes only a little more than 2% to the emissions in this sector. All other modes are rather negligible, both from their contribution to the overall emissions and also with regard to their dissemination.

The use of fossil fuel-based cooking stoves is quite common in the Indian context, with less than 7% relying on electricity for cooking. Most of the interviewed households (95.4%) use liquefied petroleum gas (LPG) for cooking. LPG is a relatively clean fuel with regard to indoor air pollution compared to kerosene and biomass, and it also has a relatively good climate balance. With respect to biomass, respondents were asked to report their use of firewood, charcoal, and cow dung, but only firewood was reported to be used, with more than 5% of all respondents using it for cooking, mainly as a supplement to LPG or kerosene.

The share of long-distance travel emissions was relatively small, mainly due to the low amount of air travel within this sample, which is very carbon-intensive compared to other modes of transport: only 5% (not given in the table) of all respondents reported at least one flight in the past 1 year, of whom a little more than 2% took only domestic flights and 1.8% used also international flights. Most of the long-distance travel was made with bus (60% of all trips) and train (40% of all trips, figures not given in table). In all, a little more than 45% of the respondents (not given in table) did no long-distance travel in 2010.

---

[2] The term long-distance travel is used here only for bus, train, and air travel. In the Indian context, the car or motorbike is not frequently used for longer distances, though this may change soon.

Table 6.2 Sector- and mode-specific structure of the overall carbon footprint of Hyderabad ($n = 590$)

Sector	Consumption mode	Percent share based on mean for each mode	Percent share of respondents using respective mode	Emission factor
Individual motorised transport	Four-wheeler	65,1	11,6	4,20 kg CO_2e/h
	Two-wheeler	34,9	60,7	0,65 kg CO_2e/h
	Overall sector	*16,5*	*61,8*	n.a.
Food	Rice	38,1	100	0,92 kg CO_2e/kg
	Meat	31,2	84	9,66 kg CO_2e/kg[a]
	Milk	30,7	92,1	0,83 kg CO_2e/l
	Overall sector	*14,6*	*100*	n.a.
Public transport	Auto-rickshaw	52,2	33,6	1,80 kg CO_2e/h
	Regular city bus	24,2	57,2	0,20 kg CO_2e/h
	Shared auto-rickshaw	8,8	15,9	0,49 kg CO_2e/h
	AC taxi	5,3	1,2	4,66 kg CO_2e/h
	Taxi non-AC	3,3	1,3	3,30 kgCO_2e/h
	MMTS local train	2,2	4,1	0,60 kg CO_2e/h
	Chartered office/school bus	2,1	1,8	0,22 kg CO_2e/h
	AC city bus	1,4	2,1	0,40 kg CO_2e/h
	Car sharing	0.5	0,5	1,77 kg CO_2e/h
	Overall sector	*9,1*	*77,5*	n.a.
Cooking	LPG	83,2	95,4	3,10 kg CO_2e/kg
	Kerosene	16,8	13,4	2,46 kg CO_2e/l
	Overall sector	*8*	*99,2*	n.a.
Long-distance travel	Train	11,2	38,5	0,70 kg CO_2e/pass/h
	Bus	10,9	34,4	0,60 kg CO_2e/pass/h
	International flights	69,6	1,8	666,33 kg CO_2e/pass./flight[b]
	Domestic flights	8,4	2.1	99,67 kg CO_2e/pass/flight[b]
	Overall sector	*4,6*	*53,1*	n.a.

[a]This emission factor is an average factor for all different kinds of meat; for the actual calculations, the type of meat and the specific factor were considered, with mutton (12,69 kg CO_2e/kg), chicken (4,48 kg CO_2e/kg), beef (kg CO_2e/kg), and pork (kg CO_2e/kg pork)

[b]This emission factor is an average for different flight length: for the actual calculations, the length of the flight and the specific factor were considered, with flights of less than 4 h and 4–8 h and flight lengths above 8 h

6.2 Income-Class-Specific Analysis of Personal GHG Emissions

Electricity is not included in Table 6.2. Its calculation is only based on reported electricity bills, and therefore a direct specification is not possible. However, electricity data show a substantial increase of power consumption during the four summer months (March–June). On average, the use of electricity climbs by 30% during summer. Moreover, Sect. 6.3 will give an overview of practice-specific per capita emission effects, i.e. average annual per capita carbon footprints of certain consumption practices. For electricity, the author depicts the average carbon footprints from using different household appliances. This approach facilitates a more differentiated perspective on household electricity consumption.

6.2 Income-Class-Specific Analysis of Personal GHG Emissions

Income is understood as a necessary, but not a sufficient, condition for higher levels of consumption and carbon footprints. Figure 6.1 depicts an illustration of the structure of sector-specific emission profiles within the equivalised income groups.

As stated above with respect to the overall distribution, electricity and IMT make out the highest share compared to all other sectors. It is remarkable that in both of these sectors, the income effect is most substantial. Figure 6.1 can only serve as first illustration. It depicts the sector-specific distribution in means without giving evidence of whether the differences between the selected groups are significant. The data of this sample in respect to sector-specific GHG emissions is not normally distributed. Therefore, to compare groups in respect to the selected variables, nonparametric tests have to be conducted in order to show whether the groups differ significantly. For the comparison of more than two independent groups, the nonparametric H-test after Kruskal-Wallis is recommended (Bühl 2012, pp. 381, 395). Along with bar charts with means and error bars (95% confidence interval), significance levels, as well as the medians, are reported for each variable and group (Figs. 6.2, 6.3, 6.4, 6.5, 6.6, 6.7, 6.8, 6.9, 6.10, and 6.11).

Figure 6.2 gives a more precise picture of the personal overall GHG emissions. The H-test for all sectoral variables of GHG emissions is significant, i.e. for every variable (overall emissions, electricity, food, cooking, long-distance travel, air travel, public transport, and IMT), the income classes differ significantly. Chi-square values, degree of freedom, and significance levels are reported below the median table for each examined variable, respectively (Figs. 6.2, 6.3, 6.4, 6.5, 6.6, 6.7, 6.8, 6.9, 6.10, and 6.11). The integrated table indicates the rank based on the median, and the bar chart depicts the mean carbon footprint for each income group with error bars.

For the overall distribution of GHG emissions, Fig. 6.2 shows a strong linear positive relation between income and personal carbon footprints. The steepest increases can be observed between the four lower-income segments (deprived, aspirers, seekers, and strivers). In the last two higher-income groups (strivers and

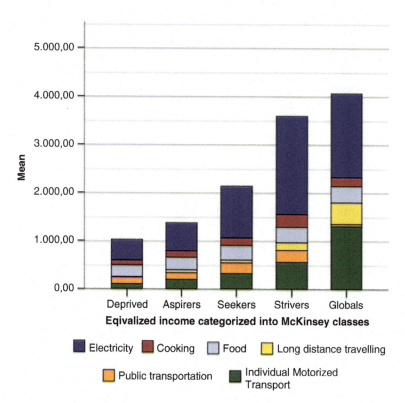

Fig. 6.1 Stacked bar chart of sector-specific distribution of personal GHG emissions (CO_2e/cap/year) per equivalised income class

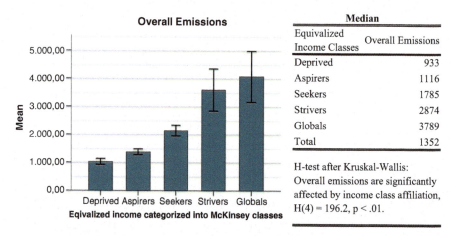

Fig. 6.2 Bar chart with mean levels of overall personal GHG emissions (CO_2e/cap/year) per equivalised income class; table with median levels for overall emissions, as well as results of H-test after Kruskal-Wallis. Error bars show 95% confidence intervals

6.2 Income-Class-Specific Analysis of Personal GHG Emissions

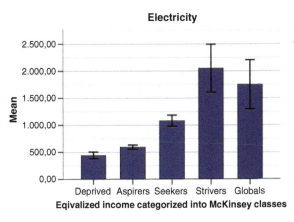

Fig. 6.3 Bar chart with mean levels of GHG emissions (CO_2e/cap/year) from electricity consumption per equivalised income class; table with median levels for electricity emissions, as well as results of H-test after Kruskal-Wallis. Error bars show 95% confidence intervals

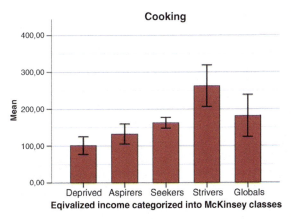

Fig. 6.4 Bar chart with mean levels of personal GHG emissions (CO_2e/cap/year) from cooking fuel consumption per equivalised income class; table with median levels for cooking emissions, as well as results of H-test after Kruskal-Wallis. Error bars show 95% confidence intervals

globals), overall emissions vary quite substantially, and the level of increase is less steep compared to all other groups. Also, the size of both groups is relatively small. Therefore, differences between these two income groups should be handled with caution. This will become more obvious in the following, as the linear relation changes, when we look at the disaggregated sectoral emissions.

Similarly, in respect to electricity (Fig. 6.3), there is a very strong positive linear effect on emissions up to the second highest-income level, the strivers. Quite

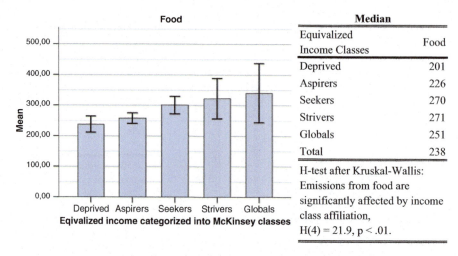

Fig. 6.5 Bar chart with mean levels of personal GHG emissions (CO_2e/cap/year) from food consumption per equivalised income class; table with median levels for food emissions, as well as results of H-test after Kruskal-Wallis. Error bars show 95% confidence intervals

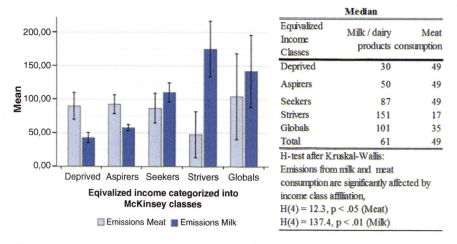

Fig. 6.6 Grouped bar chart with mean levels of personal GHG emissions specified for meat and dairy consumption per equivalised income class; table with median levels for meat and dairy consumption, as well as results of H-test after Kruskal-Wallis. Error bars show 95% confidence intervals

unexpected, however, is the apparent decline in electricity emissions among the highest-income segment, the globals. Both the mean level and the median are lower among the globals compared to the strivers. Interestingly, the same effect can be identified with respect to the use of cooking fuels (Fig. 6.4): a rather clear slope for all lower segments with a substantial hike from the seekers to the strivers and then a drop in emissions from the strivers to the last segment, the globals.

6.2 Income-Class-Specific Analysis of Personal GHG Emissions 183

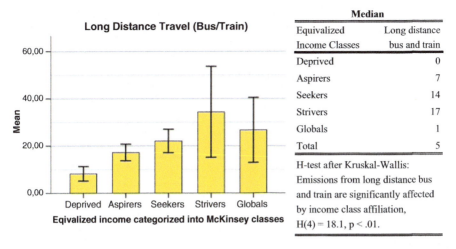

Fig. 6.7 Bar chart with mean levels of personal GHG emissions (CO_2e/cap/year) from long-distance bus and train travel per equivalised income class; table with median levels for long-distance bus and train travel emissions, as well as results of H-test after Kruskal-Wallis. Error bars show 95% confidence intervals

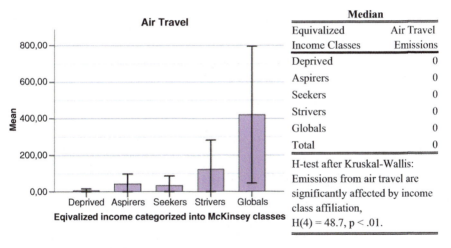

Fig. 6.8 Bar chart with mean levels of personal GHG emissions (CO_2e/cap/year) from air travel per equivalised income class; table with median levels for air travel emissions, as well as results of H-test after Kruskal-Wallis. Error bars show 95% confidence intervals

According to the wealth index, which measures the level of household and personal amenities (see Sect. 5.2.1), there are more households with high levels of wealth among the strivers than among the globals. That means the income group of strivers tends to have more household appliances that may lead to higher consumption of electricity. Furthermore, the income group of globals tends to have larger household sizes, with a median of 3 among the globals (household size is adjusted

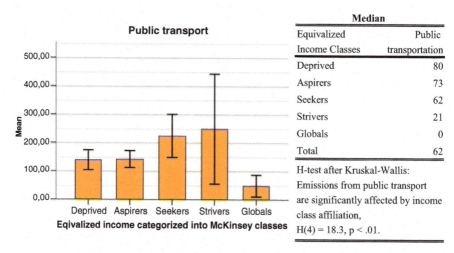

Fig. 6.9 Bar chart with mean levels of personal GHG emissions (CO_2e/cap/year) from public transport per equivalised income class; table with median levels for public transport emissions, as well as results of H-test after Kruskal-Wallis. Error bars show 95% confidence intervals

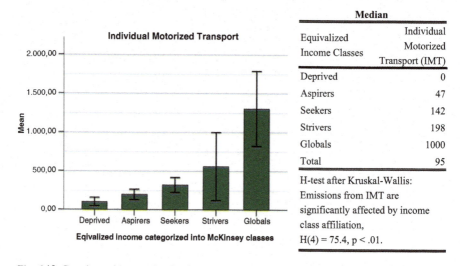

Fig. 6.10 Bar chart with mean levels of personal GHG emissions (CO_2e/cap/year) from individual motorised transport (IMT) per equivalised income class; table with median levels for IMT emissions, as well as results of H-test after Kruskal-Wallis. Error bars show 95% confidence intervals

using a factor 1 for adults and a factor 0.5 for children below 13 years) and a median of 2.25 members among the strivers. With larger household size, there tends to be a scale effect for the use of household energy, e.g. for space cooling, use of refrigerators, and cooking (similar amounts of energy divided among larger number of household members).

6.2 Income-Class-Specific Analysis of Personal GHG Emissions

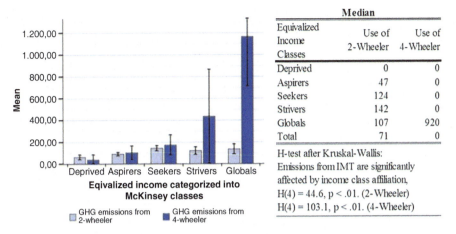

Fig. 6.11 Grouped bar chart with mean levels of personal GHG emissions specified for both two-wheeler and four-wheeler per equivalised income class; table with median levels for both use of two-wheeler and four-wheeler, as well as results of H-test after Kruskal-Wallis. Error bars show 95% confidence intervals

With respect to food, differentials between income segments are less clear than with regard to other domains of consumption, even though the H-test is significant. There is a small hike between the second (aspirers) and third (seekers) segment. Interestingly, as depicted in a more specific overview in Fig. 6.6, income effects on meat consumption differ considerably different compared to the effects on dairy consumption.

The medians for emissions from meat remain on the same level in the first three income segments until there is a considerable drop among the strivers. Meat consumption among the globals remains on a lower level but with quite some variance. In respect to dairy, emission levels increase substantially with higher income up to the group of strivers and fall back again to quite a moderate level among the globals.

Concerning long-distance travel by bus and train, it is striking in general how low the personal carbon footprints in this sector are. The maximum amount of personal emissions in this sector is 210 kg of CO_2e per year. This is because with an emission factor of 0.6 kg (bus) and 0.7 kg (train) CO_2e per person per hour travel, carbon intensity of bus and train travel in India is extremely low, due to the extremely high degree of capacity utilisation.

Also, income effects are less obvious in this sector. While the H-test is significant, the bar chart shows high levels of variance especially in the two highest segments. Only the poorest, the deprived, and the richest, the globals, differ substantially on a very low level from all other groups. Between the other groups, there may be a slight positive trend with rising income up to the strivers. Remarkably, the median among the globals is very low, almost negligible, just as the zero median value among the deprived (see table in Fig. 6.7).

Taking a look at the air travel emissions in Fig. 6.8 and even more specifically at the actual travel practices of the different income groups, the above-stated results become more plausible. While 27% (significant with $p < 0.001$) of all globals travelled by air in 2010, only 10% (not significant, due to small numbers) of the strivers and only 2.5% (not significant, due to small numbers) of the seekers used an aeroplane during the same time span of 1 year.[3]

With regard to public transportation, there is a remarkable negative income effect on personal GHG emissions, which does not, however, appear from the distribution of mean values among income segments. As this data is not normally distributed and there is huge variance within the higher-income segments, the mean values have to be treated with caution. But as the H-test after Kruskal-Wallis is significant, inter-group differentials based on the median can be well interpreted. While public transport seems to play an important role for the lower-income groups, the importance of it decreases very steadily with growing income until it reaches a zero median in the highest segment. This finding should be examined more closely in conjunction with the analysis of individual motorised transport, as the conclusions from both domains go together very well.

While public transport still plays a vital role for the lower-income groups, there is evidence for the fact that with rising income, respondents switch from public transport to individual motorised transport (IMT). Among the deprived, use of IMT is very limited with a mean of about 60 kg of CO_2e per person per year (Fig. 6.9).[4] This very low level of GHG emissions is in spite of about 40% of all households in this class having a motorbike or a scooter, but the use of public transport seems still more important in this lowest-income class. This pattern changes with rising income: as the use of public transport and associated personal GHG emissions decreases, so does the use of IMT rise with growing levels of income. Figure 6.10 shows the mean and median levels of IMT emissions with a very clear income effect. Interestingly, however, there is also a steep hike in the last segment, the globals, which calls for a closer look at the distribution of emissions from two-wheelers and four-wheelers.

Figure 6.11 allows a more specified examination of this domain. It shows a split in emissions between use of two-wheelers and four-wheelers. Up to the strivers, the median values of emissions in this domain (table in Fig. 6.11.) rise steadily. However, among the globals, emissions from two-wheelers drop substantially, while emissions from the use of four-wheelers reach to a level far beyond that of all other income classes: a median of almost 1 ton CO_2e per person per year is a huge amount in relation to a median of 1.3 tonnes of overall emissions across all income groups (Fig. 6.11).

The diffusion patterns of vehicle ownership support the above-given findings: the ownership of motorbikes and scooters is well disseminated among the aspirers (54%), seekers (56%), and strivers (50%). Interestingly, the class of globals has a

---

[3] The Pearson's chi-square test was significant with $\chi 2 (4) = 48.3$ and $p < 0.001$. However, three cells (30%) have an expected count <5

[4] The median is zero, because less than 50% of this class own a vehicle.

6.3 Consumption-Practice-Specific Analysis of Personal GHG Emissions... 187

much smaller share of motorbikes and scooters than the lower-income groups, showing less than 17% of vehicle ownership. Instead of two-wheelers, however, the globals have a substantial share of car ownership. With 73.3% of the globals having a car, this mode of transport seems to have almost replaced both public transport (median of emissions is zero) and the use of two-wheelers in this group. This is compared to not even 40% of all strivers having a car in their household. The fact that on average, cars emit six times more emissions than a two-wheeler (Table 6.2) makes it clear that people who can afford to own a car will commonly face substantial hikes in their personal GHG emissions.

The next section will delineate a rather new approach to the nexus of income dynamics, consumption practices, and related GHG emissions. It attempts to quantify and compare the carbon-footprint-related effects of different consumption practices and calculates their dissemination among the overall sample of respondents. This approach is meant to suggest a bridge between income-related GHG accounting methods and the lifestyle-related approach to GHG emission differentials.

6.3 Consumption-Practice-Specific Analysis of Personal GHG Emissions and Associated Key Points of Intervention

It has been indicated above that personal GHG emissions tend to increase with rising income levels. However, there is also evidence that income is not the only determining factor for differences in personal carbon footprints. The author has also been able to show that investive consumption decisions are highly relevant for the understanding of lifestyle, consumption practices, and related personal carbon footprints. It is assumed that investive consumption decisions tend to create path dependencies that have long-term effects on social practices of consumption and their related GHG emissions.

To get closer to an understanding of such effects and related differences in consumption-based personal GHG emissions, the author has developed a quite simple and straightforward approach that specifically highlights the effects of certain consumption practices that are usually related to investive consumption decisions or the decision for a certain routinised pattern of consumption. Based on results of the carbon calculator, the author has computed the average per capita amount of GHG emissions associated with different consumption practices.

Table 6.3 gives an overview of the sector-specific per capita averages of all measured consumption domains. It quantifies each domain in its per capita average GHG emission effect, i.e. personal GHG emissions that accumulate on average over a year, if a person starts following such a practice of consumption. These figures are able to differentiate the impact of various practices (e.g. using an AC) along the line of certain sectors or functional areas of consumption (e.g. space cooling). The figures of domain-specific GHG emission effects have to be taken as estimates, but

Table 6.3 Overview of domain- and sector-specific personal average emission effects (based on mean and median) and percent shares of respondents associated with domain

Sector	Consumption mode	Percent share of respondents using respective mode	Per capita effect of appliance dissemination [kg CO$_2$e/year] based on mean	Per capita effect of appliance dissemination [kg CO$_2$e/year] based on median
Electricity	Incandescent bulb	78,7	n.a.	66,7
	CFL bulb	51,1	n.a.	15,4
	Tubelight	95,2	n.a.	72,0
	Fan	89,6	n.a.	180,3
	Air-conditioner	10,7	n.a.	464,4
	Refrigerator	67,4	n.a.	165,3
	Air cooler	35,2	n.a.	51,3
	Geyser	21,3	n.a.	332,9
	Washing machine	31,1	n.a.	31,8
	TV	96,0	n.a.	124,3
	Computer	29,6	n.a.	38,4
Private motorised transport	Two-wheeler	60,7	178,7	165,1
	Four-wheeler	11,6	1745,3	1594,3
Public transport	AC taxi	1,2	784,7	242,3
	Taxi non-AC	1,3	421,9	286,0
	Auto-rickshaw	33,6	265,1	140,4
	AC city bus	2,1	107,2	62,4
	Regular city bus	57,2	71,9	34,7
	Shared auto-rickshaw	15,9	94,7	52,0
	MMTS local train	4,1	91,5	31,2
	Car sharing	0,5	163,6	184,1
	Chartered office/school bus	1,8	200,7	183,0
Long-distance travel	Train	38,5	25,0	16,8
	Bus	34,4	27,3	16,8
	Domestic flights	2,1	335,1	284,0
	International flights	1,8	3296,7	2500,0
Food	Milk	92,1	88,8	67,3
	Meat	84,0	100,0	49,5
	Rice	100,0	101,5	100,7
Cooking	LPG	95,4	126,7	108,3
	Kerosene	13,4	199,9	109,6

these estimations allow one to evaluate the structure of personal and household-based emissions and to quantify the effects of particular consumption practices for much more targeted interventions and lifestyle policies.

Such key points become obvious if one takes a closer look at the per capita consumption effects, which are highest in regard to international air travel (2.500 kg CO_2e/year/cap), followed by the use of cars (1.600 kg CO_2e/year/cap), the use of air-conditioners (464 kg CO_2e/year/cap), the use of electric geysers for heating water (333 kg CO_2e/year/cap), and the practice of using domestic flights for transport (284 kg CO_2e/year/cap) (Table 6.3).

It is instructive to also look at the dissemination levels of particular consumption practices. Based on the findings of this approach, income dynamics and social-cultural behavioural change will have the greatest effects concerning international air travel. Less than 2% of respondents followed this practice in 2010 (Table 6.3). A recent projection on the Indian market for international air travel suggests that international passenger traffic is projected to grow from 50.8 million in 2014–2015 at the rate of 7–8% and will reach 176.11 million by 2031–2032 (Singh et al. 2016, p. 223). According to the estimations of this study in terms of personal-level GHG emissions associated with the use of air travel, the expected growth in demand for international air travel will lead to an estimated increase of related yearly emissions from 127 million tonnes in 2014–2015 to 440.3 million tonnes of CO_2e per year in 2031–2032.

Mobility based on using cars is also highly relevant against the growth dynamics of car ownership in India, especially in cities. Verma et al. (2015, p. 113) highlighted this issue, referring to an increase of car ownership in Indian cities by 23–75% in 4 years between 2007 and 2011. The highest growth rate has been found in Bangalore, a city that is similar to Hyderabad in terms of its dynamics concerning economic development and service-sector orientation. The baseline of dissemination on the level of 12% in 2010–2011 (Table 6.3) still bears high potential for future growth in this sector.

Air-conditioners are used by only 10% of all surveyed households, while fans are much more common with a dissemination rate of almost 90%. According to "India Air Conditioners Market Forecast & Opportunities, 2020", the market for air-conditioners in India is projected to grow at a compound annual growth rate (CAGR) of over 10% during the period 2015–2020 (ReportLinker 2015). This projected steep rise in sales of ACs indicates a major change in practices related to maintaining indoor thermal comfort, with ACs playing a key role here.

The use of electric warm water geysers is another good example of a consumption mode with a high per capita emission effect and a still low level of dissemination at around 20% of all surveyed households. It has the second highest emission effect just after the AC, but it has a very cost-efficient and near to zero emission alternative, the solar water heater. Solar water heating systems are a great alternative in most regions in India, but this technology has not yet become widely accepted (in this study only one household had a solar water heating system). The household practice of warm water heating could therefore become an effective key point for driving sustainable consumption forward in the urban Indian context.

Moreover, Table 6.3 depicts other key points such as the practice of domestic air travel, which is expected to grow at the rate of 8–9% (which is a faster rate than for international air travel, see above) reaching a level of 741 million passengers by the end of 2031–2032 (Singh et al. 2016, p. 223). Air traffic in general is expected to grow very fast: according to the "IATA Air Passenger Forecasts Global Report", "the total Indian air passenger market is forecast to climb from 9th largest in the world in 2014 to 6th largest over the next five years, and to continue on into the top-three in the 2030ies" (IATA 2014, p. 7).

Apart from those practices with yet low levels of dissemination and high future potential, there are other consumption practices that are already followed on a large scale. Some of these realms of consumption have a great reduction potential, despite having a rather low emission effect. Because of the large absolute number of followers, reductions by means of efficiency improvements or by means of replacement through low-carbon options bear huge potential for prevention of future GHG emissions. For instance, one such low-hanging fruit can be found in the practice of using auto-rickshaws. Auto-rickshaws are widely used in Hyderabad, with a dissemination rate of more than 30% (Table 6.3). Even though related emissions are not as high as in other domains, alternative less carbon-intensive modes of transport would bring considerable reduction effects, e.g. by reconsidering the effectiveness and potential of cycle rickshaws for short-distance transport in cities or by improving the quality and density of integrated lower-carbon public transport systems. Similar examples can be found with the use of incandescent light bulbs, which could be replaced by low-carbon options such as CFL and LED lights. Moreover, the use of conventional fans can be addressed in terms of efficiency improvements, with quite a scope to make possible reductions in absolute terms.

The following chapters will present the results from the cluster analysis with the factor analysis. The resulting lifestyle segmentation will provide the basis for a very different perspective on the issue of consumption and related GHG emissions. It will also point to the challenges that occur with such an approach, and it will critically address the question of whether this approach actually meets the requirements of modern environmental-science-related typification strategies (Fig. 6.12).

6.4 Interpretive Analysis of Lifestyle Clusters

6.4.1 *Cluster 1: Aspiring Involuntary Economisers (n = 62)*

6.4.1.1 Values

This cluster's value structure appears to be very interesting with respect to the dynamics of recent economic development processes, especially in the urban Indian context. The cluster has a size of 62 members, which is quite average compared to the other clusters. First of all, respondents in this group show the highest levels of agreement in the factor of frugality. This criterion, along with a very strong rejection of religious

6.4 Interpretive Analysis of Lifestyle Clusters

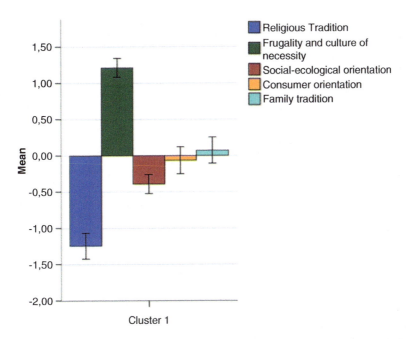

Fig. 6.12 Bar chart with the cluster-specific (cluster 1) agreement levels on the underlying "active" attitudinal and value dimensions. Error bars show 95% confidence intervals

tradition, makes this cluster unique and very characteristic. Less pronounced, but to be mentioned, is a tendency to disagree with prosocial and ecological norms and ideals. In respect to hedonism and family tradition, respondents in this group rather range along the overall sample's average. A neutral position towards hedonistic conspicuous consumption combines well with the positive attitude towards frugality and thrift. And, as thrift seemingly represents a determining factor and guiding principle with respect to consumption and the spending of money, concerns of quality, eco-friendliness, or social issues do not play much of a role with regard to purchase. Obviously, the objective to vote for the cheaper price tends to challenge pro-environmental consumption, which can be costlier and also costs more time it takes to research the right product. Social and environmental injustice can be understood as a rather distant topic for people in this segment, as personal achievements are more basic and urgent.

Indicated by the high levels of disagreement to religious tradition, people in this group tend to rather have an internal locus of control for many situations in their life. They may be assumed to be more optimistic and indeterministic than other respondents, and they rather tend to look forward and aspire to a wealthier life. With this approach people in this segment can be seen much more than others as architects of their own future, tending to be more active and more enthusiastic. With respect to religious orientations, they are not necessarily atheistic. However, they do not leave things to God or destiny as they do not passively accept every situation. Respondents in this group rather strive to ameliorate a problematic situation for their own benefit.

The description reads well, and initially, the author was surprised by the structural consistency of this value orientation. Already by looking at the value structure (active variables), it can be assumed that this group is socially positioned at the lower end of society. However, for the interpretation of consumption preferences expressed through the strong agreement to frugality, it makes sense to take a closer look at the social-demographic data. This is important, because it is not absolutely clear whether agreement to the factor of frugality describes a choice only of necessity due to financial constraints or whether it has deeper, more virtue-related reasons.

Respondents in this group have the lowest median in regard to income, and they are less likely than other respondents to belong to the highest-income group (strivers and globals). However, the analysis of income distribution among clusters does not allow a final conclusion, as the results of this cluster for income both in its categorical and its metric form are not significant (see Table 5.23 and Sect. 5.2.4). Subsidiary to income, the educational structure of this cluster helps to draw a clearer picture: cluster 1 has the highest share of people without any educational qualification ($p < 0.05$) and the smallest number of respondents having graduated ($p < 0.01$; see Table 5.23). Moreover, respondents in this cluster tend to have a father with lower levels of education (not significant; residual 1.57, see Table 5.23). Also, far above the average, cluster 1 shows the highest share of members of the scheduled caste/tribes ($p < 0.001$). These results clearly indicate a lower social position of this cluster suggesting that the high level of agreement to frugality is more strongly related to financial reasons than driven by moral concerns.

In addition to these findings, it was seen that a very large share of almost 55% in this cluster is represented by women. As the overall gender balance is substantially biased against the share of women (see Sect. 5.1.1), this result is far beyond the average at a significance level of $p < 0.001$. Moreover, among these women, over 82% are housewives and not otherwise employed ($p < 0.001$). It is worthwhile to examine whether the characteristic value structure of this cluster is related to its distinctive gender balance.

Taking a look at the overall sample shows that women in general and housewives in particular tend to rather agree to an orientation towards frugality and thrift. Especially being a housewife highly correlates with frugality with a Pearson coefficient of $r = 0.18$ significant at the 0.01 level. This effect is even more obvious as regards disagreement to the value of religious tradition (Pearson correlation coefficient $r = -0.21$ significant at the level of 0.01).

In addition to this striking result, it was even more surprising to find this cluster and its pronounced orientation towards thriftiness being strongly influenced by the familial structure of large shares of this segment. Compared to all other clusters, it has by far the highest share of respondents with daughters who are as yet unmarried. Almost 60% of all respondents in this group still expect their daughter to get married in the near or more distant future. Against the background of a widely accepted social-cultural norm of dowry (see Sect. 5.1.2), it is quite obvious that the high agreement levels of frugality are in many cases associated with this family-related specificity. For the bride and her family, the financial implications and the social

6.4 Interpretive Analysis of Lifestyle Clusters

pressure that arises from this social practice is tremendous. The following interview sequence taken from a female respondent who was just about to get married illustrates the financial burden associated with the dowry, and it somewhat shows how it is taken as a matter of course:

> He works in the airport. My photo was shown to him and his family liked me and so further talks went on and they decided to take me as their daughter in law. [...] My parents have to give dowry in marriage; it is like INR 100,000 in furniture and crockery, 50 grams of gold, a [motor]bike, and about 21 sarees; total expenditure comes to INR 500.000. I am taking a loan of INR 100.000.

The interviewed woman lives in a slum pocket quite close to a major sewage drain in Banjara Hills – an upper class area – where she works as a beautician, earning most of the household's combined income, which was reported with a total annual amount of INR 130.000. That means the dowry is almost four times the *annual* combined household income of this family. Her father is blind and not able to carry on his work as a labourer. For the marriage proposal, the family invested in renovating their slum house (kitchen, roof, areaway, furniture), and for this they had to draw a loan of INR 200.000, which again exceeds the yearly household income by far and leads to substantial debts.

Interestingly, the correlation between having unmarried daughters, the related expectation of dowry that has to be paid, and the positive attitude towards the value of frugality are limited only to this cluster. There is no such correlation in the overall sample, nor in any of the other clusters. Quite obviously, the system of dowry only affects respondents (given that they have an unmarried daughter) in the lower social segments; higher-income groups seem to be unaffected. The above interview sequence depicts an exemplary case in which marriage functions quite fundamentally as means of gaining social status. With the lack of financial resources, it may be felt important to invest the limited resources on effective means for upward social mobility: a good marriage match for the offspring apparently is able to lift a family inter-generationally into higher strata of the society.

For this purpose, scarce financial resources and even loans are painfully mobilised in order to invest extensively into improved household infrastructure and dowry gifts for the groom's family. Seemingly, all resources are mobilised for this event, and it becomes quite clear why frugality and thrift play a crucial role for a family with the above-delineated features.

Against this background, the characteristic wealth structure of this cluster increasingly makes sense: respondents in cluster 1 clearly show very low levels of wealth, especially in regard to household infrastructure. It is the cluster with the smallest median of the wealth index, and more than 40% of all respondents belong to the poorest quintile ($p < 0.001$). The cluster also contains most of all those households that have to live without their own water supply connection ($p < 0.001$) and who are required to share a tap with other households publicly ($p < 0.001$). Moreover, there is a much smaller share of respondents with landline connection, computer/Internet, washing machine, air cooler, and grinder.

6.4.1.2 Activities

With respect to shopping, cluster 1 is quite characteristic as respondents in this group tend to prefer local kirana stores and street vendors over more centralised facilities such as supermarkets and malls. This preference pattern coheres closely with the factor of homely oriented leisure: cluster 1 has the highest median in this dimension, and therefore, activities such as watching TV at home, chatting with neighbours on the street, or doing pending house chores are quite typical for this group. Conversely, the other factor with more of an action and consumption orientation has the smallest median in this cluster, i.e. activities such as visiting malls, cinemas, and cafes or hanging out in parks are less likely to be found among members of this cluster.

This characteristic shopping and leisure orientation combines well with the fact that people in cluster 1 are more likely to report their dominant mode of transport to be based on walking and cycling ($p < 0.05$). Members of cluster 1 are also less mobile than other respondents; they have the lowest level of emissions from individual motorised transport, and they are less likely than other respondents to have a car (not significant, residual 1.59).

For both dimensions of holidays, cluster 1 shows a rather low level of dissemination. It is, however, more unlikely to find a member of cluster 1 taking an action-oriented holiday ($p < 0.05$) such as hiking or going to the beach than going on a more traditional vacation to religious sights of interest or visiting relatives (not significant, residual − 1.1).

In respect to the use of information media, cluster 1 is most unlikely among all clusters to use more differentiated and more sophisticated media, such as an English newspaper. This is more characteristic than the use of red-top media with an average level of dissemination. Also the use of the Internet is on an average level with quite insignificant levels of expenditure. Last but not least, respondents from cluster 1 tend to only eat meat occasionally. With more than 90% of all respondents in this group eating "non-vegetarian" food, it is not very likely to find a full-fledged vegetarian among members of cluster 1. However, as people in this group tend to eat meat only occasionally and if then only in small amounts, they can be classed as not morally but financially driven vegetarians.

6.4.1.3 Personal GHG Emissions

As to be expected, the overall emissions in this cluster range on a low level, and people in this segment are less likely to be found in the highest emission class (not significant; residual −1.90). The cluster however has the second highest emissions from using cooking fuel, and they have the lowest median together with cluster 2 in respect to individual motorised transport.

6.4 Interpretive Analysis of Lifestyle Clusters

6.4.1.4 Summary and Discussion

People in this cluster tend to look forward with an aspiring attitude to the future. As architects of their own future, they are more optimistic and indeterministic and aspire to achieve a better life. In terms of social position, the aspiring involuntary economisers range at the lower level of the pyramid in terms of income levels and wealth as well as with regard to education, employment, and caste. Thrift and frugality play out as the most important guiding principles for this cluster, both more generally, driven by financial constraints, and on the other hand for large parts of this cluster due to an expectation of paying out a dowry in the coming future. Social-ecological concern therefore ranges on a very low level of priority. Self-enhancement and individual-level improvements of one's situation in life are guiding principles and weigh much more than moral principles, hedonistic practices, and societal or ecological concern.

A large share of the cluster is female and many of them are housewives. Conduct of life is rather traditionally oriented with a preference for local shopping and homely oriented leisure and with the dominant modes of transport being walking and cycling. Holidays seem to play a minor role, and also dietary patterns (financially driven vegetarians) are accommodated within the very limited scope of action that is due to financial constraints and the priority to save for the marriage of one's offspring.

For this cluster, a combined analysis of values, behavioural patterns, and social determinants based on the concept of lifestyle allows for an improved understanding of the constraints, drivers, and motives of consumption. Thrift and frugality translate into low levels of GHG emissions. This is not only because of limited financial resources and low levels of income but also due to the important priority to save money for a dowry. The attitudinal and value structure also indicates the fact that this social-cultural segment aspires to climb the social-economic ladder. Changes in behavioural and consumption patterns are likely to occur once children have been married off. Before that is achieved, income increases may not have so much impact on patterns of consumption, and the value orientation of frugality may persist even without the need to save money. Moreover, societal and ecological concern is quite unlikely to play a role in the near future, even if awareness levels are raised (Fig. 6.13).

6.4.2 Cluster 2: Liberal Pragmatists (n = 72)

6.4.2.1 Values

The second cluster is less clear in its structure and interpretation. With a size of 72 respondents, it is already a slightly larger group. Very obvious is a strong disagreement with the paradigm of family tradition, followed by a tendency to reject frugality and social-ecological values. Somewhat indicative is also the small but significant

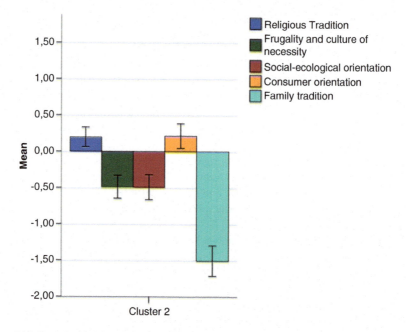

Fig. 6.13 Bar chart with the cluster-specific (cluster 2) agreement levels on the underlying "active" attitudinal and value dimensions. Error bars show 95% confidence intervals

positive orientation towards the paradigms of religious tradition and hedonist conspicuous consumption. Based on this structure of agreement, the author decided to name this segment of people the liberal pragmatists.

The most important for the interpretation of this cluster is the complete rejection of the paradigm of family tradition, which is quite striking here. Respondents belonging to this group disclaim a too strict sense of family tradition, which includes the inherent ideal of hierarchy and propriety. People in this group tend to regard this paradigm as too constrictive. Liberal pragmatists rather share an open-minded and liberal world view with a clear here and now attitude. While obedience and hierarchic order are not important for people in this cluster, a disagreement with family traditional values does not imply a neglect of the family. It rather proposes a more enlightened and individualistic attitude towards peers, kin group members, or authorities. This implicit liberal attitude is somewhat supported by the fact that religious tradition tends to be rated low but slightly positive. This minor tendency towards fatalism corresponds with the rejection of a too-strict interpretation of moral norms and rules. They serve as guidelines, not as restrictions, because fate is seen as a God-given fact.

Rejection of frugality and social-ecological values coheres well with this rather pragmatic and liberal world view. In respect to income, people in cluster 2 are not more likely to belong to the lower strata, nor do they represent larger shares of richer people (Table 5.23). This indicates – different than in cluster 1 – that people in this group tend to interpret the paradigm of frugality as a moral issue rather than a financially driven necessity. Austerity for ethical reasons is rather peculiar for a liberal pragmatist. Similarly, they are reluctant to constrict themselves with prosocial-ecological norms, because such action-guiding principles are not understood to be meaningful for a liberal pragmatist due to the lack of perceived self-efficacy. Environmental problems and questions of social justice are not to be addressed by individual actions, because such issues are far beyond their control and result from a God-given structure.

Another interesting aspect is the small but significant position towards the paradigm of hedonistic conspicuous consumption and consumer culture. People in this group may perceive consumption still rather pragmatically and functionally as a fulfilment of material needs. But it also represents a level of expressive freedom from too many restrictions. Quality concerns may only play a minor role, more oriented towards practicality. Social practices and consumption should not become an issue of concern in any regard.

6.4.2.2 Social Demography

In general, this cluster is characteristic as it does not stand out in terms of any abnormal results in respect to social demography, preference patterns, and emissions. Many of the passive describing variables rather show average results. For instance, income ranges along the average and so does material wealth. Taking a more specific look at the distribution of asset ownership, cluster 2 does not exhibit very characteristic features in terms of particular amenities. Remarkable only is the fact that the level of dissemination of flat-screen TV sets is the lowest among all clusters ($p < 0.05$). On the other hand, households in this cluster are less likely to share a tap with other households instead of having their one ($p < 0.05$), and they are more likely to live in a pucca house, which is a more solid and more permanent house (not significant; residual 1.56). These features may give a slight indication of a lived pragmatism and the importance of functionality.

In respect to education, cluster 2 is difficult to describe, as there is no characteristic feature. Looking at the education level of the parental house, it is significant to find a smaller share of respondents with fathers having a 12th class qualification ($p < 0.05$). Considering employment, it is remarkable that relatively more students are found in this cluster than in others ($p < 0.05$). This observation somewhat contradicts the fact that cluster 2 is the oldest cluster with a median age of 35 years (overall sample median is 30 years). Together with cluster 1 and cluster 4, people in this segment are more likely than other clusters to have their own children (not significant; residual 1.66). Moreover, cluster 2 members tend to have household sizes that range slightly above the average.

6.4.2.3 Behavioural and Consumption Preferences and Patterns

With regard to general preferences and behavioural patterns, cluster 2 does not depict a specific structure. It shows average values in the dimension of centralised shopping, while it ranges in the lower segment with respect to locally oriented shopping in the neighbourhood. People in this group tend to reject consumption- and action-oriented leisure activities, but they even more disclaim home-oriented leisure, such as watching TV and chatting with neighbours. Respondents in this cluster are also most likely to reject traditionally oriented holidays. They tend to prefer action- and consumption-oriented holidays with a median ranging along the average. Moreover, members of cluster 2 are more likely to use public transport ($p < 0.01$) than a two-wheeler ($p < 0.01$) as their dominant mode of transport.

Last but not least, there are no peculiar observations in respect to the use of media and expenditures on internet. While the usages of different media range along the average, both for red-top as well as more differentiated media, the expenditure on the Internet is slightly below the average.

6.4.2.4 Personal GHG Emissions

With regard to personal GHG emissions, cluster 2 ranges at the lower end with the second lowest median in overall emissions. Electricity emissions are well positioned at the mid-level, and so are emissions from the use of cooking fuel. Remarkable is the low level of emissions from food. Together with cluster 4, it has the lowest median of its carbon footprint from milk consumption and the lowest footprint from meat consumption. However, there is no tendency of higher levels of vegetarianism. Hence, there is no obvious explanation for the small amounts of meat consumed in this cluster. Remarkable are the low-level emissions in the domain of mobility. Cluster 2 has the lowest median value from long-distance travel with train and bus, and it ranges at the lowest level of emissions from individual motorised transport together with cluster 1. This result is not surprising, as cluster 2 has the highest share of people for whom public transport is the dominant mode of transport.

6.4.2.5 Summary and Discussion

Liberal pragmatists reject a too strict interpretation of moral norms and rules. They tend to have a more enlightened and more individualistic attitude towards peers, kin-group members or authorities. Environmental problems and questions of social justice are rather perceived as beyond individual control and therefore not worth engaging in. Frugality and thrift for moral reasons also seem meaningless for this rather open-minded and liberal world view. In regard to consumption, liberal pragmatists orient their choices based on pragmatism, practicality, and functionality. For

6.4 Interpretive Analysis of Lifestyle Clusters

them, social practices and consumption should not become an issue of concern in any regard.

Liberal pragmatists form some share of the new urban middle classes, with quite reasonable levels of income and a decent educational background. The character of a typical liberal pragmatist household can be described as rather practical and functional. Members of this group usually do not stand out as luxurious or conspicuous. This is true also for their patterns of consumption, leisure, and holiday. For instance, extravagant leisure activities are rather disapproved in the same way as religiously oriented holidays. However, members of this group typically prefer action- and consumption-oriented holidays. Interestingly, liberal pragmatists tend to rather use public transport than individual motorised transport.

It is mainly patterns of mobility with a preference for public transport that lead to low levels of personal GHG emissions in this cluster. In terms of electricity and cooking fuels, liberal pragmatists show quite average results, while for mobility this cluster has the lowest per capita emissions due to the typical use of public transport. Also in regard to dietary patterns, the cluster shows rather low levels of personal GHG emissions. However, liberal pragmatists do not typically follow a vegetarian diet, nor do they avoid dairy products for either ethical or religious reasons (Fig. 6.14).

6.4.3 Cluster 3: Well-Established Traditionalists (n = 57)

6.4.3.1 Values

The third cluster is very characteristic in regard to its very strong rejection of social-ecological values. This attitude coheres well with a high level of agreement to family tradition and religious tradition. The private sphere tends to be more important than concerns about societal problems, politics, environment, and social issues. Maybe also due to the rather higher social position and higher levels of emissions (see below), people in this segment push away their responsibility to the government and other actors. Apparently, it is not society or the environment which call for a moral instance, but the family and the related traditional moral values, such as kin-based hierarchies. This moral standpoint may function as an ethical compensation to substantial changes in ways of living and consuming.

Also remarkable is the low-level disagreement with frugality on the one hand and consumer orientation on the other hand, with both variables seemingly contradicting each other. Together with the strong agreement with the paradigm of family tradition, this cluster indicates an interesting aspect of modern urban consumption in the Indian context. By tendency, there seems to be an inherent ambivalence towards modern consumption patterns: While frugality is rejected, which is an aspect that indicates to a positive orientation towards mass consumption, the paradigm of hedonistic conspicuous consumption is rejected too. This subtlety is very interesting against the background of a high degree of agreement towards both reli-

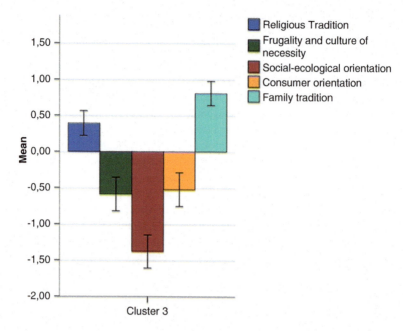

Fig. 6.14 Bar chart with the cluster-specific (cluster 3) agreement levels on the underlying "active" attitudinal and value dimensions. Error bars show 95% confidence intervals

gious family tradition. Tradition, religious and mythological thought, and fatalism seem to play an important role for people in this cluster, but probably, such attitudinal position is rather pretentiously claimed due to social expectancy and as a moral stance against substantial changes in the ways of living and consuming.

6.4.3.2 Social Demography

A look at the social-structural analysis, such as the average income and wealth level of this cluster, also shows that this segment and its attitudinal structure are to some extent contradictory. People in this group feature the highest levels of income as well as wealth, and the latter characteristic at least indicates that on average, people in this group tend to have very high levels of consumption.

Cluster 3 is well-positioned social-economically compared to all other clusters. Income-wise it has the least share of aspirers ($p < 0.05$), which is the second lowest-income segment, and the highest shares of strivers and globals ($p < 0.01$), i.e. the highest-income groups. This indicates that the financial situation of most respondents in this group allows for a relaxed way in dealing with money and consumption. And it explains why this group shows the highest levels of disagreement with the paradigm of frugality. It also coheres with the highest levels of material wealth, i.e. ownership of consumer goods and other assets: respondents of cluster 3 are most likely to belong to the richest wealth segment compared to all other clusters

($p < 0.01$). The cluster also has the smallest share of the second lowest wealth class ($p < 0.05$). Taking a more specific look at the wealth distribution, cluster 3 members are most likely to have a washing machine ($p < 0.001$), they are most likely to have a credit card ($p < 0.001$), they are more likely than other segments to have a radio ($p < 0.05$), and the cluster depicts a high share of respondents that have a flat-screen TV ($p < 0.05$). Not significant, but relevant to mention is the higher share of respondents having a full-time maid (H-test not significant; residual 2.64) and also that more respondents tend to have a pucca house (not significant; residual 1.87). These quite high levels of consumption do not necessarily lead to an orientation towards conspicuous consumption and lavish lifestyle, as indicated by the disagreement with factor 4.

In respect to employment, the most characteristic feature is the relatively higher share of students in this cluster ($p < 0.05$), however, less significant than in cluster 2. With regard to education, cluster 3 does not depict any significant structural features. It tends to have a smaller share of respondents with a 12th class qualification (not significant; residual −1.71), and it has more respondents with either no educational qualification (not significant; residual 1.15) or who are graduates (not significant; residual 0.82). However, results only indicate very subtle tendencies.

Considering the education of the father of the respondent, results are not significant, but there is a tendency towards higher qualifications (graduation and above, residual 1.45) also in the parental house of cluster 3 respondents. Cluster 3 also shows the highest shares of male respondents, compared to all other clusters (not significant; residual 1.93).

Remarkable in this segment is the religious affiliation. There is a highly significant share of Muslims ($p < 0.001$) and Christians ($p < 0.001$), and respondents in this segment are – compared to all other clusters – less likely to be Hindus ($p < 0.001$): while the overall sample has a share of more than 75% of Hindus, cluster 3 only has a little more than 40% of Hindus, compared to more than 40% of Muslims (19% in the overall sample). This social-cultural feature also has a noticeable effect on the value orientation. There is a tendency among Muslims to disagree more strongly with the paradigm of frugality (Pearson correlation coefficient $r = -0.12$ significant at the level of 0.05) and especially with the paradigm of social-ecological orientation (Pearson correlation coefficient $r = -0.16$ significant at the level of 0.01). Consequently, this cluster's value structure is to quite some extent influenced by its much larger share of Muslims, who tend to rate the paradigm of frugality and social-ecology quite differently than Hindus. This result indicates the relevance of contextual variables that determine respondents' value orientations. Clusters emerge based on attitudes on similar attitudinal orientations, but the contextual conditions of these attitudes may be quite heterogeneous and in many cases not visible in the data.

The age structure of this cluster ranges along the average. With respect to the household size, people are more likely to live in smaller households, with the smallest median among all other clusters. Similarly to cluster 4, this group has a higher share of singles than all other clusters ($p < 0.05$). Consequently, respondents in this segment are less likely to have own children ($p < 0.05$), and it is also significant that there are less children living in the same household with the respondent than in other clusters ($p < 0.05$; except cluster 6).

With respect to dietary habits, respondents in cluster 3 are the least likely to be vegetarians ($p < 0.05$). Cluster 3 does not show any significant effect in respect to the most dominant mode of transport.

6.4.3.3 Behavioural and Consumption Preferences and Patterns

While cluster 3 has the lowest level of preference for locally oriented shopping, e.g. in kirana grocery stores, respondents have an average preference pattern towards centralised shopping facilities. In respect to leisure activities, cluster 3 ranges along the average, both for action- and consumption-oriented leisure. Similarly, holiday preferences are very much positioned on the mid-level as well, both action holidays and traditionally oriented holidays. Considering media use, cluster 3 members quite typically consume sophisticated and more differentiated media and show the lowest level of usage of red-top media. Moreover, expenditure patterns on the Internet use are slightly above the average.

6.4.3.4 Personal GHG Emissions

Cluster 3 clearly has the highest levels of GHG emissions among all clusters, and this is the case for all domains except for long-distance travel and public transport – both on mid-level – and for food consumption, where the cluster shows the second highest emission levels. Consequently, respondents in this cluster are most likely to be found in the highest emission class with very high significance ($p < 0.001$).

6.4.3.5 Summary and Discussion

Very characteristic for well-established traditionalists is the obvious rejection of social-ecological values. This feature goes along with an agreement to both paradigms of tradition, and here it is more family-oriented traditional values than religious tradition. This attitudinal structure indicates to emphasis of the private sphere that is perceived more relevant than concerns about societal or ecological problems. For the solution of such problems, well-established traditionalists tend to have more trust in formal institutions and the government, while self-efficacy is perceived as rather low. Together with a low but significant rejection of frugality on the one hand, and consumer orientation on the other, there seems to be also a moral ambivalence towards consumer culture. While members of this group embrace modern consumption – indicated by the very high levels of investive consumption and the rejection of frugality – they still reflect carefully about implications that modern consumption may have in terms of morality and tradition. Family traditional values therefore are held high in distinction to "Western" influences that are perceived to come along with new ways of life and consumption. This is seen as an explanation also for rejection of hedonistic consumer values – a tendency that is not reflected in the actual consumption patterns (highest wealth index among all clusters).

Well-established traditionalists show the highest average levels of disposable per capita income and of investive consumption. Their financial situation allows for a relaxed way of dealing with money and consumption. In regard to employment and education, this segment rather shows average results, but in terms of religious affiliation, there is a highly significant share of Muslims and to a minor degree of Christians compared to other clusters. The high share of Muslims has an effect also on the value orientation, as Muslims in general tend to disagree with the paradigm of frugality and especially with the paradigm of social-ecological orientation. Also typical for this cluster are smaller households and a higher share of singles, with a tendency against respondents having their own children in the household. Moreover, cluster members are least likely to be vegetarians. Well-established traditionalists prefer shopping in centralised facilities rather than locally oriented shopping. In regard to all other behavioural profiles, this cluster ranges along the average.

The high levels of investive consumption already indicate high levels of GHG emissions, which are in this group at the highest level among all clusters. Except for long-distance travel and public transport (both average), all sectoral emissions range at the highest (electricity, cooking, IMT) or second highest level (food) (Fig. 6.15).

6.4.4 Cluster 4: Aspiring Modern Achievers (n = 76)

6.4.4.1 Values

Among all clusters, this group disagrees the most with the dimension of frugality. With regard to social-ecological orientation, this cluster shows the highest level of agreement together with cluster 6. Both groups are on par with each other, being the only clusters with significantly high levels of agreement to this dimension. Another apparent feature is the rejection of religious traditional values. This goes together with a very low level of agreement to family tradition.

As discussed in Sect. 4.1.2, the dimension of social-ecology draws mainly on consumption-related issues. Only one item in this factor states a rather general attitude towards social and environmental responsibility. It makes sense that social-ecologically responsible consumption does not combine well with a rather frugal orientation that does not allow paying a higher price, just in order to support a fairer or more environmentally friendly product. Also, a rejection of the religious traditional paradigm makes sense together with social-ecology, as the latter orientation stands for higher levels of self-efficacy and a rather internal locus of control. Religious tradition contrasts this feature by its rather fatalist perception of issues related to society or ecological problems. Even though it ranges on a very low level, the agreement to family tradition also fits well with the overall attitudinal structure.

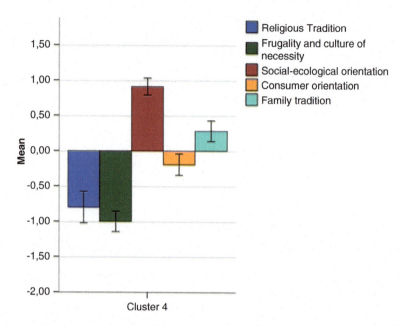

Fig. 6.15 Bar chart with the cluster-specific (cluster 4) agreement levels on the underlying "active" attitudinal and value dimensions. Error bars show 95% confidence intervals

The paradigm of family tradition represents a more moderate and worldlier tinged perspective and better combines with a more modern-oriented way of life. The following analysis of social-demographic and behavioural aspects in this cluster will make the inherent attitudinal structure of this segment clearer.

6.4.4.2 Social Demography

Cluster 4 is remarkable with respect to its social determinants. Respondents in this group do not typically report high levels of income: it is the cluster with the highest share of aspirers, i.e. the second lowest-income class (highly significant with $p < 0.001$). People in this cluster also tend to live in households that share the smallest number of rooms, with a median of half a room per household member. This household membership density goes well together with the larger household size: it is the cluster with the highest median of people living together. Interestingly, more than 90% of all respondents in this cluster have indicated to living together with children (below 13 years). As almost 40% of these respondents are singles, quite some of them tend to live together with their parents. Moreover, cluster 4 represents the youngest cluster among all six segments, with a median of 28 years (overall sample median is 31 years). Despite the lower age structure and despite the larger share of singles, respondents in this segment are – together with respondents of clusters 1 and 2 – more likely to have their own children (not significant, residual 1.67) compared to clusters 3, 5, and 6.

With this apparent rather low social profile (low income and large households), it is surprising that other variables indicate a much higher social position. For instance, the wealth distribution is quite high, with the second highest median of wealth (household and personal assets) just behind cluster 3. Also in respect to education, this cluster shows the highest proportion of respondents with an academic qualification (60% are at least graduates, $p < 0.01$). This result is supported by data on respondents' parents' highest educational qualification. People in this segment are much more likely to have a father with final qualification at least the 12th class ($p < 0.001$), with many of those even having graduated ($p < 0.05$). The tendency of high levels of education, both in respect to the parental house and the respondent, also explains the very high share of professional service employees ($p < 0.001$) in this cluster at around about 38%. Quite some of these professionally employed people are working in the IT sector. Last but not least, cluster 4 has the lowest proportion ($p < 0.05$) of people with a scheduled caste or tribal background.

6.4.4.3 Behavioural and Consumption Preferences and Patterns

Members of cluster 4 are much more likely to purchase their day-to-day groceries in local mom-and-pop grocery stores (kiranas). This cluster shows the highest proportion of people using these local shops ($p < 0.001$), but shopping in more centralised shopping localities, such as supermarkets and malls, is also apparently important for this segment ($p < 0.01$). The latter affinity to more centralised shopping locations also coheres with the general orientation of this cluster towards activities outside the house and the neighbourhood. Leisure activities are very likely to be action- and consumption-oriented activities, such as shopping as well as visiting cafes, cinemas, and parks and enjoying malls as entertainment ($p < 0.001$). With regard to home-oriented leisure, there is no significantly higher affinity compared to other clusters. Holidays play a role in this segment, but with more of a traditional orientation like visiting relatives and religious-related travelling. Action-oriented holidays, such as sightseeing, beach, or trekking holidays are not more widespread than in other clusters. In respect to the use of media, both categories are relevant, but media use is much more likely in the realm of more differentiated and more sophisticated media, such as daily English newspapers, books, and the Internet.

Interestingly, respondents in this cluster have the lowest level of expenditure on Internet use, and apparently, this is not due to the scale effect from generally larger household sizes that may in some cases push down the costs for the Internet. Respondents in this cluster are not more likely to live in households with an Internet connection. Therefore, the use of Internet media as a source of information is less likely to take place at home than in the office, where it means no additional costs for the respondent.

6.4.4.4 Personal GHG Emissions

For respondents of this cluster, all emissions associated with household-based consumption tend to be very low: the cluster shows the lowest level of electricity emissions, it has the lowest emissions from the use of cooking fuels, and it ranges among those clusters with the least amount of milk consumption. For all these household-based emissions, the respective amount per household is passed on to the number of people living in the household.

All other domains, which are based on the amounts reported in respect to the individual consumption of respondents, tend to be somewhat higher: accordingly, respondents of cluster 4 are likely to have the highest emissions from long-distance travel with train and bus. They also range among the three clusters with the highest emissions from individual motorised transport, which consists – in the case of cluster 4 – of emissions mostly from use of two-wheelers. This explains the rather low level of emissions from public transport, which is among the lowest three clusters. With regard to food consumption, it is interesting to find cluster 4 having the highest levels of meat consumption (individual consumption), while it ranges among the lower end with regard to the consumption of milk (household based assessment; see above). With respect to air travel, there were only four respondents (<5% of cluster 4; $n = 76$) who reported to have travelled by air during the reference year 2010, but air travel in general is very low within the overall sample.

Overall, the emissions in this cluster show the lowest mean among all clusters. Also, the categorised variable for overall emissions shows the highest proportions in the lowest emission segment (below Indian per capita average). There is an apparent difference between household-level emissions and emissions from outside activities. This difference coheres well with the action- and consumption-oriented behavioural profile outlined above.

6.4.4.5 Summary and Discussion

Very characteristic for an aspiring modern achiever is her or his rejection of religious tradition and frugality and a positive attitude towards social-ecological values. While hedonistic consumer attitudes tend to the average, there is a minor inclination towards family tradition. Social-ecological values are perceived more relevant than financial concerns. The rejection of the religious traditional paradigm with its inherent fatalism is a logical consequence of a social-ecological orientation that perceives high levels of self-efficacy and emphasises the individual responsibility in regard to consumption. Family tradition with its more moderate and worldlier tone combines well with this rather life-affirming positive value-orientation structure.

Aspiring modern achievers range at the lower end of the income ladder, and they tend to use less living space than members of other clusters. Surprisingly, contrary to the income levels, levels of investive consumption are very high on second position just after cluster 3. Also the high levels of education and the above-average

6.4 Interpretive Analysis of Lifestyle Clusters

share of professional service employees are at variance with the reported income. This group also shows the lowest proportion of scheduled caste or tribal background.

It is the youngest cluster with high shares of singles and a close to average rate of having own children. However, 90% of all respondents have children living together with them in the same household. This segment also shows the largest household sizes among all clusters. The author interprets this structural feature as an indication to higher shares of very young and well-educated entry-level employees, who still live together with their parents (and siblings). While equivalisation of household income based on household size directly affects the level of income, such effects are minor in regard to levels of investive consumption.[5] These young aspiring achievers may have reached the entry point of their career, but their way of life is still strongly influenced by their parents and the structural features of their household. It is possible that people in this group change to more carbon-intensive lifestyles once they move out of their parents' houses and lead their own life. It would be productive for a better understanding of lifestyle and sustainable consumption to get hold of such dynamics and interactions between values and behaviour that are related to such shifts (e.g. Thøgersen 2012).

In terms of behavioural and consumption preferences, aspiring modern achievers show a significant tendency towards centralised shopping in malls and supermarkets in coherence with a strong orientation towards action- and consumption-oriented leisure activities outside the house and the neighbourhood. This feature fits in well with the observations made above that life at home is still prestructured by the way of life of the parents, while individual behavioural patterns rather follow the logic of a more modern consumer orientation (low levels of frugality), with individuals social-ecological consciousness. In regard to holidays, members of this group are still more traditionally oriented. In regard to media use, cluster 4 shows a tendency towards the use of more differentiated and more sophisticated media.

Similarly, all emissions associated with household-based consumption tend to be at the lowest level, i.e. with regard to electricity emissions, the use of cooking fuels, and emissions from milk consumption. Individual-level consumption tends to be higher, e.g. in regard to emissions from long-distance travel and individual motorised transport (mostly from use of two-wheelers). Public transport emissions are significantly low, which indicates that people in this cluster prefer individual motorised transport. Also in regard to meat consumption, which is measured on the basis of individual consumption, members of this cluster show the highest levels of emissions. However, in regard to the overall level of emissions, the cluster shows the lowest median of GHG emissions among all clusters. Hence, also the emission structure of this cluster indicates an apparent difference between household-level emissions and emissions from outside activities, which coheres well with the action- and consumption-oriented behavioural profile (Fig. 6.16).

---

[5] The wealth index (investive consumption) partly builds on personal investments (own mobile phone, credit card) and to a large extent does not equivalise for household size effects, except for counted items (e.g. number of rooms, fans, air-conditioners, or cars).

6.4.5 Cluster 5: Hedonic Ostentatious Consumers (n = 105)

6.4.5.1 Values

The value structure of cluster 5 is highly dominated by an agreement to the consumer culture paradigm. It is the only cluster with a significant level of agreement to this dimension. This attitude is combined with a strong orientation towards religious tradition. These two high loading factors are supported by lower levels of agreement with frugality, family tradition, and social-ecological values. It is somewhat striking that respondents in this cluster agree to a consumer culture and at the same time hold high the values of frugality. Even though frugality ranges on a rather low level, it contradicts an approach that requires spending and consumption as means of identity management. Moreover, the ideals and morals of religious tradition may create some conflict with the social-ecological paradigm but also with the concept of consumer culture.

A closer examination of the social-demographic variables and the analysis of behavioural and preference patterns will shed more light on this cluster and may help to gain a better understanding of this ambivalent value structure.

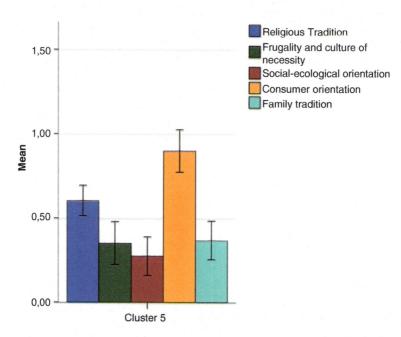

Fig. 6.16 Bar chart with the cluster-specific (cluster 5) agreement levels on the underlying "active" attitudinal and value dimensions. Error bars show 95% confidence intervals

6.4.5.2 Social Demography

Cluster 5 does not show very characteristic features in regard to social position and there is no significant differentiation from the overall sample in terms of income. Taking a more specific look at the distribution of material wealth, it is this cluster least likely to own an air cooler, but it is – just behind cluster 3 – most likely to have a credit card. There is a higher share of people with landline telephones, and the cluster has the highest share of people with the Internet. It is striking that by far the most respondents with a well or a handpump are gathered in this cluster ($p < 0.001$), in all 22% of respondents. However, this can lead one to drawing wrong conclusions, as of those respondents in this cluster who have a handpump, almost 75% also have their own tap. Another puzzling result is that 47% of all respondents who do live in a katccha or semi-pucca house are gathered in cluster 5. However, this is not even 10% of the overall cluster. Nevertheless, cluster 5 remains a rather heterogeneous cluster in respect to the social position of its members.

The employment structure shows the least share of housewives ($p < 0.05$) and students ($p <= 0.01$), compared to all other clusters. Moreover, it has the highest share of small proprietors with employees ($p < 0.05$). Even though it is not significant, the cluster seems to have a higher share of unemployed respondents as well as skilled manual workers and lower grade technicians.

Considering education, there is a tendency of more respondents having a qualification of up to 12th class (not significant; residual 0.73). However, the familial background in terms of education is remarkable, as it is very likely to find respondents with fathers having no qualification and least likely to find respondents with graduate fathers having graduated (for both categories, $p < 0.05$). Another very significant feature is the highest share of Hindus in this cluster and that it is least likely to find any Muslims as member of this group (both categories; $p < 0.01$).

In terms of familial structure, members of this segment tend to be more likely to be married than single ($p < 0.05$), but there is a larger share of respondents who do not have their own children (not significant; residual 1.79). And it is the second oldest cluster, just behind cluster 2. The household size of members of cluster 5 ranges along the average.

In respect to food habits, cluster 5 is more typical for being vegetarian ($p < 0.05$). In regard to mobility, respondents tend to use a two-wheeler ($p < 0.01$) much more than taking a bus or the MMTS ($p < 0.05$).

6.4.5.3 Behavioural and Consumption Preferences and Patterns

It is a striking result that the consumer orientation measured by active variables above coheres very well with the preference pattern in terms of shopping. Cluster 5 members prefer to shop in centralised locations, such as malls and supermarkets. In respect to leisure, however, people in this segment tend to behave more moderately: both in action- and consumption-oriented leisure and home-oriented leisure, the cluster ranges along the average. Also considering holiday preferences, respondents in this cluster are more likely to go on a traditionally oriented vacation. Action- and consumption-oriented holidays take place rather on an average level. And, while this cluster has the second highest median in the use of red-top media, the use of sophisticated, more differentiated media is less disseminated. In terms of expenditure on Internet use, respondents rather show average levels.

6.4.5.4 Personal GHG Emissions

In terms of emissions, this cluster takes a mid-level position with the third highest GHG emission profile. Electricity consumption and emissions from cooking are both located on a lower level, as the cluster in both these domains ranges among the second lowest positions. Individual motorised transport (third highest), long-distance travel with bus and train (second highest), and emissions from public transport are a bit higher (highest, but not significant), but rather along the average among all clusters.

6.4.5.5 Summary and Discussion

Hedonic ostentatious consumers stand out clearly with their agreement to the consumer culture paradigm. It is the only cluster with such a strong agreement to this dimension. The cluster is also characteristic in its strong agreement towards religious tradition. These two paradigms do not necessarily contradict one another: the inherent fatalistic attitude of the religious tradition paradigm allows for light-heartedness in regard to consumption. It is assumed that the attitudinal items have been interpreted by these respondents in a worldlier sense emphasising the role of karma and liberation. The modest but still significant positive attitude towards frugality, however, opposes the author's interpretation. This ambivalence is also reflected by an inclination towards social-ecologically conscious consumption. However, as agreement levels towards factors 2, 3, and 5 range on a significantly lower level, it is the combination of the consumer culture paradigm with religious tradition that differentiates this cluster from the others.

Difficulties also apply to the interpretation of social demography and behavioural and consumption preference patterns. There is no significant differentiation based on either income or wealth. Coherent with the rather modern consumption orientation reflected in the cluster's value orientation, respondents of this group are more likely to have a credit card and have the Internet at home. The cluster comprises fewer housewives and students and has higher shares of small proprietors. Quite striking is the high share of respondents' fathers who have no qualification and the least share of respondents with fathers having graduated. The cluster also has the highest share of Hindus and contains more vegetarians than other clusters. In regard to mobility, hedonic consumers prefer private two-wheelers over public transport, and they show a considerable preference for shopping in centralised locations, such as malls. This coheres strongly with hedonic consumer values. In terms

6.4 Interpretive Analysis of Lifestyle Clusters

of leisure and holidays, the cluster does not stand out as clearly: both leisure dimensions range along the sample's average, while in regard to holiday, there is a moderate inclination towards traditionally oriented vacation. The use of red-top media is preferred over sophisticated and more differentiated media.

There is no clear relationship between cluster membership GHG emission levels. The cluster ranges on a mid-level position. While electricity and cooking range on a lower level, all other sectors tend to range along the average (Fig. 6.17).

6.4.6 Cluster 6: Religious- and Nature-Oriented Traditionalists (n = 45)

6.4.6.1 Values

The most characteristic about cluster 6 is the pronounced rejection of the hedonic and ostentatious paradigm of consumer culture, the highest level of disagreement among all clusters. Content specifically, this attitude suits well with its agreement to frugality. Religious tradition also has a prominent positive influence on respondents in this cluster; overall, it is the cluster with the highest degree of agreement to this factor. This theme somehow contrasts with the concurrent positive attitude towards social-ecological values. The paradigm of religious tradition with its fatalistic pattern of reasoning may stand in conflict with the rather positive value of

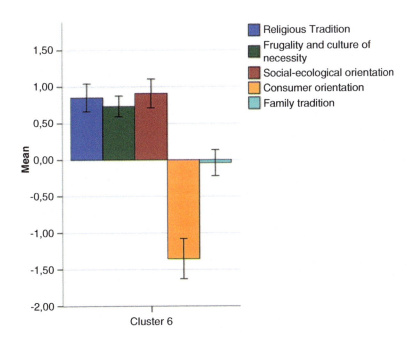

Fig. 6.17 Bar chart with the cluster-specific (cluster 6) agreement levels on the underlying "active" attitudinal and value dimensions. Error bars show 95% confidence intervals

social-ecology, with its high levels of self-efficacy and its inherent social-ecologically-driven attitude of societal responsibility. Social-ecology as framed in this context has a consumption momentum, which rather values quality and consumer responsibility over cheap prices. This may also indicate ambivalence with the agreement to frugality. The value of family tradition for respondents in this cluster ranges around the overall mean and does not say much about this cluster.

6.4.6.2 Social Demography

With only 45 respondents, cluster 6 is the smallest of all clusters. The small size is a great challenge for the analysis, especially as regards drawing conclusions from the examination of passive categorical variables such as income, employment, education, dominant mode of transport, and even religion. However, the analysis is able to indicate certain trends. Without being statistically significant, there tends to be a higher share of seekers, the income segment which belongs – according to McKinsey Global Institute (McKinsey Global Institute 2007) – to the Indian middle class. Moreover, there are higher shares of respondents found in the second poorest wealth class (not significant; residual 1.09). More specifically in terms of the wealth distribution in this cluster, respondents of cluster 6 are least likely to have an air cooler and also least likely to have a pressure cooker. Interpretation of these results, however, is difficult, as both these amenities are widespread among the middle classes. Likewise puzzling is the fact that cluster 6 members are more likely to live in a semi-pucca house ($p < 0.01$) and less likely to live in a pucca house ($p < 0.05$).

In respect to employment, respondents in this cluster are more likely to be unemployed ($p < 0.05$) compared to other clusters, and there is a larger share of people with the employment status small proprietors with employees, but this tendency is less obvious than in cluster 5 ($p < 0.05$). In regard to education, there is no obvious distribution pattern. However, as in cluster 5, but even more pronounced, is the higher share of respondents' fathers having no qualification and being least likely to have graduated (both $p < 0.05$).

Cluster 6 is the second youngest cluster with a median of 29.5 years. There is the tendency for respondents to be single rather than married (not significant; residual 1.52), and there is a highly significant share of people living in households without any children below 13 ($p < 0.01$). Additionally, the cluster has a rather below-average tendency of household sizes. Respondents of this cluster are also more likely to eat meat but only occasionally (not significant; residual 1.21). They are also less likely to walk or cycle as their dominant mode of transport (not significant; residual 1.84).

6.4.6.3 Behavioural and Consumption Preferences and Patterns

Preferences for shopping are not at variance with the overall sample. Respondents of cluster 6 range along the lower preference level in respect to centralised shopping. Likewise, decentralised shopping on neighbourhood level shows average results. Action- and consumption-oriented leisure activities also range on the average level, while home-oriented leisure, such as watching TV or chatting with friends on the street, is located on the second highest level. Respondents in this cluster are more likely to have a traditional kind of vacation than an action-oriented holiday; however, both types range on a very low preference level.

In regard to the usage of information media, cluster 6 respondents tend to prefer more differentiated and more sophisticated media over red-top media, but both dimensions are not placed above the overall average. Interestingly, cluster 6 has the highest level of expenditure on the use of the Internet.

6.4.6.4 Personal GHG Emissions

Concerning emissions cluster 6 shows average carbon footprints, being located on the third lowest level among all six clusters. Taking a more specific look at the single consumption domains, there are no characteristic amplitudes in any of those sectors. All consumption domains range along the average level: while cluster 6 is the third lowest group in regard to electricity emissions, it ranks among the three highest clusters in respect to the use of cooking fuels. Cluster 6 has the second lowest food emissions, and there are no characteristic features in regard to the amount of meat and milk consumption, which both range along the average level. The same is true for individual motorised transport, which locates on the average level.

6.4.6.5 Summary and Discussion

Religious- and nature-oriented traditionalists form a value opposition to cluster 5, the hedonic consumers. Religious- and nature-oriented traditionalists show the highest levels of rejection of the hedonic consumer culture paradigm. This combines very well with positive attitudes towards religious tradition and frugality. However, frugality and the fatalistic paradigm of religious tradition ambivalently oppose social-ecological concern. As disagreement towards consumer culture reflects the strongest attitude in this cluster, this dimension is being considered most seriously for the interpretation. However, cluster 6 is also the smallest cluster with only 45 respondents, and together with the ambivalences mentioned above, interpretation involves huge challenges. This applies not only to the interpretation of the value structure but also to drawing conclusions from the examination of passive categorical variables, such as income, employment, education, dominant mode of transport, and even religion. Therefore, analysis and interpretation of this cluster is quite limited to a small set of variables, and conclusions need to be made very cautiously.

Like cluster 5, religious- and nature-oriented traditionalists have higher shares of respondents' fathers with no qualification, and they are least likely to have a graduate father. It is the second youngest cluster and has a significant share of people living in households without any children below 13. This cluster has a lower tendency to enjoy centralised shopping and greater preference for home-oriented leisure. Traditional vacation is preferred over action-oriented holiday. Respondents in this cluster reported the highest average levels in regard to expenditure on internet.

Cluster membership of this cluster does not allow for predictions in regard to GHG emission levels. All surveyed emission sectors range along the overall sample's average.

6.5 Cluster-Specific Analysis of Personal GHG Emissions

The following sections will give a summarising overview of the cluster-specific GHG emission accounts. It will also show how effects of cluster membership are largely masked by income and household size. Figure 6.18 illustrates the sector-specific distribution of GHG emissions among the clusters indicating the mean values. Obvious differences between clusters can be observed concerning modes of transport and the use of electricity. Differences in emissions from food and from cooking fuels are less recognisable. To test whether the samples (in this case clusters) originate from the same distribution, analysis of variance has to be conducted. As argued in Sect. 6.2 concerning not normally distributed samples, the non-parametric H-test after Kruskal-Wallis was conducted (Bühl 2012, pp. 381, 395).

Figure 6.19 gives a more precise picture of the personal overall GHG emissions. The H-test is significant for almost all sectoral variables of GHG emissions. It is highly significant at a level of $p < 0.001$ for the variables "overall emissions", "elec-

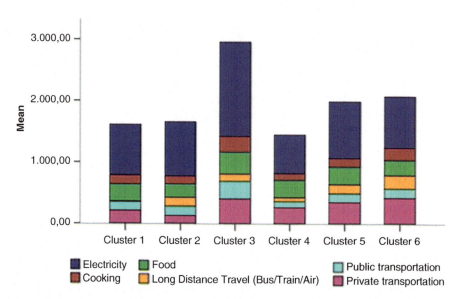

Fig. 6.18 Stacked bar chart of sector-specific distribution of personal GHG emissions (kg CO_2e/cap/year) for each lifestyle cluster

6.5 Cluster-Specific Analysis of Personal GHG Emissions

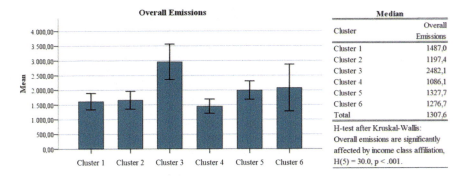

Fig. 6.19 Bar chart with mean levels of overall personal GHG emissions (kg CO_2e/cap/year) per lifestyle cluster; table with median levels for overall emissions, as well as results of H-test after Kruskal-Wallis. Error bars show 95% confidence intervals

tricity", "food", "cooking", and "long-distance travel". It is significant at a level of $p < 0.05$ for "individual motorised transport", but it is not significant for "air travel" and "public transport". For all significant tests, chi-square values, degrees of freedom, and significance levels are reported below the median table for each examined variable respectively (Fig. 6.19). The median table indicates the rank based on the median, and the bar chart depicts the mean carbon footprint for each lifestyle group with error bars.

Overall GHG emissions vary significantly between the six clusters, however, with quite some intra-cluster variance in cluster 6 as well as in cluster 3 (Fig. 6.19). The highest carbon footprints can be found in cluster 3 with a median of almost 2.5 tonnes of CO_2e/cap/year. The drop down to the second highest level of overall emissions is quite substantial. Its median with about 1.5 tonnes of CO_2e/cap/year ranges a tonne below the highest level. From this level, the other clusters show rather modest steps down to the lowest level of about 1.1 tonnes of cluster 4. A more detailed and sector-specific analysis of the lifestyle clusters can be found in Annex V.

Section 6.2 has analysed the influence of income on personal carbon footprints. The findings show that income has a determining effect on personal GHG emissions, and this influence tends to be highest in the most carbon-intensive domains of consumption. Apart from income, there are other factors leading to differences in carbon footprints. By applying the lifestyle concept, the author has attempted to get hold of these other factors with a focus on the role of social-cultural determinants. The above-given description of clusters shows that this study was able to meaningfully differentiate respondents according to their underlying value orientations. The clusters' orientational patterns are inherently consistent, and underlying values largely cohere with behavioural and consumption patterns. Likewise, average emissions within these clusters are largely explicable based on the determinants of lifestyle, i.e. values and preferences, behavioural patterns, and social demography. Accordingly, lifestyle as conceptualised here is a valid approach to comprehensively describe and explain social-cultural differences among the urban population of Greater Hyderabad.

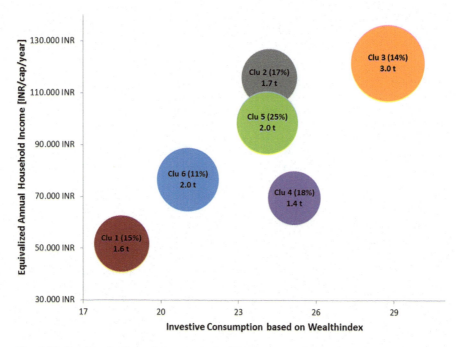

Fig. 6.20 Combined effects of cluster membership, income, and investive consumption on personal carbon footprints. The size of the "bubbles" indicates the mean of CO_2e footprints (also given in numbers – in tonnes of CO_2e/cap/year). Percentages in brackets designate the share of the cluster in the overall sample

However, a comparative analysis of the clusters with respect to income and investive consumption indicates how the "soft" or subjective criteria (lifestyle) and social-economic determinants interact quite substantially with each other. Figure 6.20 illustrates this interesting relationship between lifestyles, investive consumption (wealth index), and income quite strikingly.

Cluster 1 has the smallest income average, and in terms of investive consumption, the cluster shows the lowest results among all clusters. The characteristic social-economic situation of large parts of this cluster matches well with a very characteristic and coherent structure of values. In particular, low levels of income and the prevalent feature of expected dowry payments translate into a frugal lifestyle consistently underlined by the values of simplicity and thrift. The very low levels of investive consumption as well as the routines and behavioural patterns clearly underpin this observation. These characteristic features of low income and frugal lifestyle let one anticipate the lowest average of carbon footprints. However, the cluster has a slightly higher mean of emissions than cluster 4, despite obvious income and wealth differences.

Especially the outside orientation of cluster 4, with higher emissions from the use of two-wheelers and from travelling long distances with bus and train, indicates a less frugal and more action- and consumption-oriented lifestyle. The structure of values supports this finding, as cluster 4 rejects frugality. The pronounced social-ecological orientation, however, appears not to play a very great role in respect to investive consumption and conduct of life.

Cluster 2 ranks at a higher social position with the second highest income among all clusters. In this light, the rather low levels of emissions are surprising and even more so because these low levels can largely be traced back to a preference towards public transport against IMT. Household-related emissions such as electricity range along the average for the overall sample. A comparison with cluster 3 reveals striking differences. High income in both these clusters leads to very different levels of emissions as well as investive consumption. Cluster 3 has the highest emissions and the highest investive consumption among all clusters. The wealth index shows that cluster 3 is exceptionally well endowed with household amenities, which includes electric appliances. Moreover, cluster 3 prefers IMT to public transport, quite contrary to cluster 2. Hence, the differences between these two clusters in terms of GHG emission levels are less an effect of structural factors but related to the choices people take. And this observation is well underlined by the structure of values measured from these two clusters. While liberal pragmatists tend to being guided by pragmatism and functionality, members of cluster 3 show a rather relaxed way of dealing with money and spending it. Moreover, cluster 3 rejects the social-ecological paradigm much more than cluster 2. Cluster 3 also shows a tendency of being morally ambivalent with "modern" ways of consumption, but this attitudinal feature cannot be found to have an effect on consumption patterns (see Sect. 4.1.2.1).

Interesting to compare are also cluster 5 and cluster 6. Both clusters show the same levels of average carbon footprints, but they differ in terms of income and wealth (Fig. 6.20). In part, both clusters have a similar structure of values and attitudes, in particular with regard to religious tradition, frugality, and social-ecological orientation. The most significant difference between the clusters can be found concerning consumer culture, with cluster 5 showing the highest agreement levels. This attitudinal feature contrasts with cluster 6 that largely rejects this paradigm. This difference in attitudes towards the paradigm of modern consumer culture does in fact translate into different patterns and levels of consumption (wealth index), but it does not lead to a difference in the average carbon footprint between the two clusters.

Chapter 7
Discussion

Keywords Carbon footprint · Income-related emissions · Meat consumption · Air-travel · Consumption expenditure · Path dependencies · Consumption practices

7.1 Restating the Research Questions

7.1.1 General and Income-Related Analysis of Personal GHG Emissions

7.1.1.1 Per Capita Carbon Footprint

The first research question deals with the average level and structure of consumption-based personal-level GHG emissions (carbon footprint) in Hyderabad: "What is the average consumption-based personal carbon footprint in Hyderabad and what is the sector-specific structure of personal-level GHG emissions being emitted in Greater Hyderabad?"

The average carbon footprint in this study is 1.38 tonnes of CO_2e per capita per year. With emissions only being estimated based on personal consumption, this average for Greater Hyderabad ranges below the national average of 1.7 tonnes of CO_2 (without land use, land-use change, and forestry, LULUCF) (Indian Network for Climate Change Assessment 2010, p. 48). However, this national per capita average includes production-based as well as public sector emissions. It therefore cannot serve as a basis for comparison. However, there are only very few studies that are based on (primary) household survey data and a consumption-based footprinting approach. Comparisons between studies should be taken with caution, because the underlying emission factors and considered domains vary quite substantially.

A recent study, for instance, issued by IIT Bombay (Bhoyar et al. 2014) estimates a per capita carbon footprint of 2.5 tonnes of CO_2e in urban areas (mainly different localities in Mumbai) and 0.85 tonnes of CO_2e in rural areas. The study is quite comprehensive as it considers consumption of fruits and vegetables, municipal waste, as well as the use of mobile phones (not considered in this study). The emission factor (EF) for cellphone operation, however, seems quite high with a factor of 57 g of CO2e per min. of talktime, resulting in a share of 15% of the total footprint in urban areas and a share of 9% of all emissions in rural areas only from the use of cellphones (Bhoyar et al. 2014, p. 56).

The study conducted by Greenpeace in 2007 arrives at a figure of 0.501 tonnes of CO2/cap/year taking into account household energy consumption (electricity and cooking) and private motorised transport (Ananthapadmanabhan et al. 2007). This study only covers consumption-based emissions from electricity use, cooking fuels, and transport, and it does not differentiate between urban and rural consumption.

7.1.1.2 Sector-Specific Structure

The sector-specific structure of personal GHG emissions in Hyderabad is heavily influenced by the consumption of *electricity*. At almost 50% of overall personal-level emissions, the use of electricity has to be seen as one of the most relevant sectors for climate change mitigation in urban India. This is in terms of the amounts of household-level consumption but also concerning the carbon intensity of this sector. On average, respondents in Hyderabad cause electricity-related emissions of 890 kg CO2e per capita per year. Other studies show quite different results in terms of electricity-related emissions: the Greenpeace study has an average of 326 kg CO2e/cap/year from electricity consumption (Ananthapadmanabhan et al. 2007, p. 9), while the IIT study arrives at an average of 120 kg CO2e/cap/year in rural areas and 821 kg CO2e/cap/year in urban areas (Bhoyar et al. 2014, p. 57). The latter case, based on data from Mumbai, is at the same level of emissions as estimated in this study.

However, comparisons between studies are problematic due to differing emission factors (EF). Especially concerning emissions from electricity, many studies ignore the relevance of transmission and distribution (T&D) losses, which are substantial in the Indian context and amount to an additional factor of 0.3–0.4 kg CO2e/kWh (Brander et al. 2011, p. 6; cf. cBalance 2010). While this study uses a conversion factor of 1.33 kg of CO2e/kWh (see Brander et al. 2011, p. 6) to account for T&D losses, both other studies use a much lower conversion factor without losses (0.87 kg CO2e/kWh in the Greenpeace study and 0.82 kg CO2e/kWh in the IIT study). Consideration of T&D losses however is very crucial, because they occur due to the *consumption* of electricity (Brander et al. 2011, p. 2). The difference between these two approaches is substantial and weighs heavily in terms of the overall structure of carbon footprints in India. Thus it is all the more surprising that the IIT study features very similar results. Under consideration of T&D losses, electricity emissions in Mumbai would arrive at a value of about 1429 kg CO2e/cap/year (based on own calculations), which is about 1.6 times higher than that of this study. The author assumes that emissions in Mumbai tend to be higher because of higher consumption levels but also due to a bias towards wealthier households. The study has a sample size of just 97 households for both rural and urban locations and does not indicate the sampling approach.

7.1 Restating the Research Questions

Moreover, most studies (also the two above-mentioned studies) ignore seasonal effects of consumption. The findings of this study clearly indicate that households' electricity emissions significantly increase during the summer months: on average, the use of electricity climbs by 30% during summertime. The largest share of this seasonal hike can be traced back to air-conditioning, i.e. the use of fans, air coolers, and ACs.

Altogether, the electricity sector offers huge potential for avoiding future GHG emissions, both in terms of generation and consumption. Concerning generation, the electricity sector in India suffers from very low-quality coal and a high level of inefficiency in power plants, which makes Indian electricity in general very carbon-intensive with an emission factor of 1.33 kg CO_2e/kWh (Ananthapadmanabhan et al. 2007, p. 13; Brander et al. 2011, p. 6). In respect to consumption, there is huge potential for increasing energy savings and efficiency at the household level, especially concerning air-conditioning.

Individual motorised transport is the second most relevant sector at 310 kg CO_2e/cap/year, which is 16% of overall personal GHG emissions. Large shares of IMT users still rely on two-wheelers (60% of all respondents), which are much less carbon-intensive than cars. An expected increase in the dissemination of cars however will substantially change this picture. While four-wheelers are used by only 12% of all respondents, they are responsible for more than 65% of all GHG emissions in this sector. Other studies are less specific about the structure of emissions in this sector and are problematic due to sectoral overlaps. Bhoyar et al. (2014) aggregate all considered transport emissions within the category of transport (air travel, local and long-distance trains, autorickshaw) and arrive at a figure of 872 kg CO_2e/cap/year in urban areas and 100 kg CO_2e/cap/year in rural areas (long-distance bus travel and local city buses were not considered). Also, Ananthapadmanabhan et al. (2007, p. 9) subsumed all transport domains into a single category and figured out that this sector emits an average level of 70 kg CO_2e/cap/year across India. Due to the highly aggregated character of the figures, comparisons are problematic, and it is impossible to figure out which of the included domains weigh most.

Based on the findings of this study, however, a structural shift from the use of two-wheelers to the use of cars will substantially raise the overall share of emissions from this sector in competition with electricity and food-related emissions.

Food consumption makes up an overall share of almost 15%, which is an average of food-related emissions of 274 kg CO_2e/cap/year. This relatively high share of emissions in the food sector can be traced back to the consumption of quite large amounts of rice and milk. Rice is consumed in almost all households and dairy products by more than 90% of all respondents. While it can be viewed as a cultural feature to some extent, especially concerning rice consumption, it is the dynamics related to the increasing amounts of milk consumption in the Indian context that will become a matter of concern in respect to climate change mitigation. Even though they have a relatively low carbon intensity compared to meat, milk products will increasingly play a role in GHG emission accounting in the Indian context.

Most relevant in this sector, however, is the consumption of meat and its related dynamics. With an emission factor that is ten times higher than that of milk and rice and with the high dissemination rate of meat consumption in the urban context (84%), the relatively low share of meat-related emissions in this sector (31% of all food emissions) is surprising. The author argues that this low share of meat-related emissions is due to the still very low level of consumption. An average of 8.6 kg of meat is consumed per year in Hyderabad according to this study, i.e. about one-tenth of the average of almost 80 kg of meat consumed in the Global North according to the *Meat Atlas* (Chemnitz and Becheva 2014, p. 10). In the light of a likely continuation in upwards social mobility and the associated lifestyle dynamics, a significant increase in meat consumption can be expected in the coming years. According to the *Meat Atlas*, an almost tenfold increase in demand for meat can be expected up to 2050 in India, and most of this increase will be due to increased per capita consumption rather than to an increase in population (Chemnitz and Becheva 2014, p. 40).

Comparing these findings with other studies, it is surprising that Bhoyar et al. (2014, p. 57) estimated food-related emissions to be slightly higher in rural areas (145 kg CO2e/cap/year) than in urban areas (115 kg CO2e/cap/year). The authors attribute this difference to the much higher levels of rice consumption and the less diverse character of diets in rural areas. This finding, however, contrasts with the observation that meat consumption carries more weight in terms of emission intensity and that the prevalence and amounts of meat consumption in urban areas are substantially higher than in rural areas (nearly double, Devi et al. 2014, p. 509). Unfortunately, the study does not further discuss this interesting finding and fails to give a more specified account of the composition of the sector (which includes also vegetables, fruit, potatoes, and cereals).

Public transport contributes with a share of 9% to the overall emissions in Hyderabad. The most carbon-intensive public mode of transport is the use of autorickshaws. More than 50% of all public transport emissions go back to this domain, which is used only by about one-third of all respondents. The use of city buses is more prevalent (57%), but even with its old technology, its contribution of 24% of GHG emissions to the public transport sector is still quite low. The high levels of dissemination in the bus sector have to be seen as an asset for climate change mitigation. However, the growth trends in the IMT sector have to be taken very seriously and need to be understood also as an answer to far-reaching problems in the public transport sector. Improvements in regard to reliability, quality, comfort, and affordability may disincentivise people from switching to IMT.

The use of fossil fuel-based *cooking* stoves makes up a share of 8% (i.e. 170 kg CO2e/cap/year) of the overall emissions and has a dissemination rate of more than 95% of all interviewed households. LPG is relatively clean, and it has a relatively good climate balance. Biomass has been shown to be used only to a small extent in Hyderabad with only about 5% of all respondents using it for cooking, mainly as supplement to LPG or kerosene. This is a very small number compared to findings from other studies. The National Sample Survey Organization's (NSSO) survey in 2009–2010 found that 22.5% of the urban population in India use traditional fuels

7.1 Restating the Research Questions

such as firewood, chips, dung cake, and other locally available biomass (National Sample Survey Organisation 2012). This huge difference can be attributed to the characteristic definition of urban areas (cf. NSSO 2001, pp. 5–6) in the NSSO surveys and the substantial differences between different urban areas. The findings of this study are quite consistent with the study from Mumbai that arrives at a figure of 103 kg CO2e/cap/year (Bhoyar et al. 2014, p. 57). Interestingly, Bhoyar et al. (2014, p. 57) also show that cooking fuel-related GHG emissions are far higher in rural areas due to the common use of less efficient biomass-based cooking stoves in rural areas. Worldwide burning of biomass and coal in inefficient household stoves represents roughly 15% of global energy use and is associated with the release of black carbon (apart from GHG emissions). Black carbon is a product of incomplete combustion and has a higher global warming potential than, e.g., carbon dioxide (Jeuland and Pattanayak 2012; Venkataraman et al. 2005).

The smallest share of all GHG emissions is based on *long-distance travel* (bus, train, air travel) less than 5%. As this sector includes air travel, the author was surprised about this low level of emissions in this sector. However, air travel is still only being used to a very limited degree, with about 5% of all respondents having reported any trip on an aircraft over the reporting period of 1 year. Nevertheless, the trends and future projections indicate a very different direction, with steep increases in air travel in competition to buses and trains (Drews 2011, p. 14f). This fact is interesting against the light of 45% of all respondents not having done any long-distance travel over the reporting period. It is therefore important that policy measures support growth in long-distance travel by train and bus and disincentivise the highly carbon-intensive air travel. Unfortunately, there are no disaggregated findings from other studies that measure overall personal-level carbon footprints in India. The two studies cited above (Ananthapadmanabhan et al. 2007; Bhoyar et al. 2014) do not differentiate between IMT, public transport, and long-distance travel (see above). National inventory studies show that the transportation sector contributed 8.2% to the total national carbon emissions in 2007, with shares of 87.3% from road activities, 7.3% from domestic air travel, and 4.3% from railways. The same study shows that emissions from aviation have more than trebled since 1994. Taking into account international air travel, India shows an exceptional increment of 165.7% in international air travel between 1990 and 2008, compared to the world average of 76.1% (Drews 2011, p. 14f).

7.1.1.3 Income and Personal-Level Carbon Footprints

The second research question raises the issue of social-economic differences in personal-level carbon footprints. It is assumed that with higher levels of disposable income, an individual or household is able to consume greater volumes of consumer goods and services. As an effect, carbon footprints increase due to higher levels of consumption-based GHG emissions. The question is "What is the influence of income on personal-level carbon footprints?" The first sub-question (No. 2.a.) reads: "Do higher levels of income lead to higher levels of personal GHG emissions?" The second sub-question (No. 2.b.) is "Which domains of consumption show stronger effects on income?"

The results of this study have shown that income is very critical for the analysis of GHG emissions and that it is *conditio* sine qua non for most realms of urban consumption in India. Certain consumption patterns are only possible at all with the available financial resources. Nevertheless, the findings of this study indicate that variance levels of personal carbon footprints increase with rising incomes and that consumption and related emissions do not necessarily have to rise with higher levels of income. For instance, disposable income can be saved for other purposes in the future, or it can be used for the consumption of higher-quality products. Moreover, the findings of this study have shown that income effects vary significantly between sectors. The most obvious effects have been identified within the emission sectors *IMT, public transport,* and *long-distance travel*. Also within the sectors electricity, cooking fuels, and food, income effects are measurable, but with a few minor restrictions.

Very interestingly, the sectors *public transport* and *IMT* closely correspond with each other in terms of income-class-specific carbon footprints. While IMT-related emissions rise substantially, emissions from public transportation remarkably drop with increasing levels of income. Public transport obviously plays an important role for the lower income groups, but this importance clearly diminishes with growing income levels, to as low as zero median in the highest segment. In contrast, IMT-related emissions show the highest positive income effects among all sectors. A specific look at the income-specific distribution between car and two-wheeler-related emissions underlines an important trend, which becomes obvious from these findings. Among the lower income classes – with rising incomes – people tend to switch from public transport to the use of two-wheelers. However, within these lower segments, two-wheelers still tend to serve as a supplement to public transport. It is only within the higher income groups that the car gains relevance, and it is only in the highest income class in which car-related emissions start to play an important role. In parallel, two-wheeler emissions drop substantially, and public transport-related emissions become irrelevant. The analysis of diffusion patterns of vehicle ownership clearly supports this picture (Sect. 6.2).

Concerning *air travel* and *long-distance travel*, there are also obvious income-class-specific effects that seem to correspond between these two sectors. The use of air travel in India is still restricted to the higher income groups with the globals standing out most clearly: while 27% of the globals reported using air travel during 2010, it was only 10% of the strivers and 2.5% of the seekers. These figures also show that air travel is still very exceptional. However, among the higher income segments, there seems to be a tendency of switching from the more common mode of bus and train travel to the faster but more expensive mode of air travel. Bus and train-related emissions tend to rise with income, but for the highest income group, the globals, train and bus-related emissions play a negligible role compared to the lower income segments. With an emission factor of 0.6 kg (bus) and 0.7 kg (train) CO2e per person per hour travel, long-distance travel (bus and train)-related GHG emissions are almost negligible for the overall emission sector analysis as well as

7.1 Restating the Research Questions

for climate change mitigation. In India, carbon intensity of bus and train travel is extremely low, due to the very high degree of capacity utilisation.

For *electricity, cooking, and milk consumption*-related emissions, income shows a significant positive influence, too, but there are unexpected restrictions to this observation. There is a clear positive effect from the lowest income segment up to the second highest class (deprived to strivers), and there is a negative slope from the second highest to the highest income class (from strivers to the globals). Therefore, strivers clearly stand out as having the highest personal carbon footprints in all these three sectors. These findings certainly need to be considered with caution due to the small sample size of both income classes, with $n = 28$ for the strivers and $n = 30$ for the globals. However, other results of this study somewhat explain this interesting decline of emissions from the second highest to the highest class.

One explanation for this finding is the home-based character of these emissions. For instance, milk consumption has been measured on the household level, while milk being consumed elsewhere is not accounted for. The author argues that more than 55% of all globals in this sample either work as professionals (e.g. specialised IT sector) or are self-employed as proprietors. In consequence, lower levels of household energy use (electricity and cooking fuel) and milk consumption could be due to long office hours (less amount of cooling energy during the day) and the use of canteens or outside food (fast food, restaurants, milk-tea stalls), with maybe both partners working. As in most other footprinting studies, such outsourced emissions have not been considered within the framework of this study. Moreover, globals tend to have larger household sizes that lead to scale effects for the use of household energy, e.g. for space cooling, use of refrigerators, and cooking (similar amounts of energy divided among a larger number of household members).

A very striking feature of this dataset is the tendency among strivers to have the highest levels of wealth, measured by means of the wealth index. The higher wealth index levels among the strivers are in part an effect of the smaller averages of household size in this class (see above). However, to some extent, this result may be in fact attributed to different patterns of consumption. Grunewald et al. (2012, p. 19) show that increases in income tend to affect consumption patterns quite differently, depending on the current level of consumption. The authors show that households with currently low carbon footprints tend to invest income increments into more carbon-intensive domains, such as transport or household-energy-intensive household appliances. In contrast, households with currently higher levels of emissions show a saturation effect, i.e. households show higher levels of spending on luxury goods and services (jewellery, medical goods, or entertainment), which the authors argue to be less carbon-intensive (Grunewald et al. 2012, p. 19). That means the consumption basket of carbon-intensive goods and services tends to saturate at a certain expenditure level with then fewer carbon-intensive luxury goods coming in and adding to the expenditure level, but less to the carbon footprint.

This finding could be one explanation for the lower carbon footprints among people of the highest income class (globals) in this study. It is conceivable that people from the highest income class tend to have quite different patterns of consumption, which is not only indicated by the lower carbon footprint average but also by the lower average of the wealth index. However, such conclusions are merely speculative. To give more insight on this matter, a larger sample size is needed (representative esp. for the higher income groups), and the patterns of consumption need to be looked at more comprehensively.

Moreover, as shown in Sect. 6.2, certain conclusions from an income-related analysis of carbon footprints based on expenditure data are problematic. With such an approach, income is not measured directly, but approximated based on household consumption expenditure data. Hence, strictly speaking, these classes are *not* income classes, but consumption expenditure classes. These classes are formed based on different aggregate levels of spending, i.e. people within the same class may have different incomes, depending on the level of saving. Moreover, people within the same class may also have very differently structured consumption baskets, which may contain a huge number of luxury goods with less carbon intensity as well as other goods with high carbon intensity.

Very surprisingly, there is no clear income effect on meat-related emissions: *emissions from meat* range on the same level in the first three income segments, followed by a considerable drop among the strivers. The globals have higher meat-related emissions than the strivers, but range below the overall sample's average, however, with quite some variance. Self-reported meat consumption therefore is less prevalent in the higher income segments, while consumption of dairy products increases. To some extent, a social expectancy effect is conceivable with regard to reporting levels of meat consumption: as meat consumption is still stigmatised from a traditional-religious perspective, it is possible that respondents tended to under-report their meat consumption especially among the higher classes. On the other hand, rice consumption and associated GHG emissions have no income effect.

In sum, aggregated to the overall carbon footprint, the income-related structure of total emissions follows the pattern delineated for electricity, cooking, and dairy: it shows positive income effects up to the strivers level and a negative jump from the second highest to the highest income segment. On the whole, mobility-related sectors show the most significant income effects. Among these sectors, it is IMT and air travel which are most carbon-intensive. Income effects are also highly relevant concerning electricity use, because half of the overall emissions stem from this sector. Hence, the greatest future dynamics due to social mobility can be expected within the sectors IMT, air travel, and electricity.

Other studies largely support these findings. Most studies conclude that income is the strongest predictor for consumption-based personal GHG emissions. A similar study that compares German, Czech, and UK households in terms of the determinants of personal carbon footprints found income to be the most relevant factor across the three study areas (Peters et al. 2013, p. 243). However, the study omits to differentiate income effects according to emission sectors.

Bhoyar et al. (2014, p. 56) analyse income effects based on three income categories, the poor with a yearly household income below 400.000 INR, the middle class with income of between 400.000 and 1.500.000 INR, and the rich with household incomes above 1.500.000 INR per year. Their findings highlight the greatest income effects in the sectors electricity and travel (aggregated from public transport, IMT,

7.1 Restating the Research Questions

and long-distance travel). The sectors food, waste, and cooking fuels depict only minor income effects. However, in terms of cellphone use, income tends to play quite a role.

Similarly, the Greenpeace study (Ananthapadmanabhan et al. 2007, p. 9), which classifies seven income segments (from 3.000 INR up to 30.000 INR per household per month), gives clear evidence for the relevance of income. While the effects are small for the consumption of cooking fuels, increases are very pronounced for electricity consumption and transport.

Concerning electricity, the Greenpeace study (Ananthapadmanabhan et al. 2007, p. 9) nicely shows appliance-specific peaks of emissions in relation to the income classes. Fans and lighting reach a peak of emissions at an income level of 5.000 INR per month, while washing machines start playing a role only from this segment onwards. Interestingly, washing machine-related emissions peak at a household income level of 15.000 INR per month, indicating that the upper income class tends to prefer using laundry services. This finding partly explains the unexpected lower levels of electricity-related emissions in the highest income class of this study. Such outsourcing of services leads to different underestimations within classes (Ananthapadmanabhan et al. 2007, p. 9). Based on consideration of these findings, the author of this study developed the practice-specific analysis of personal carbon footprints (Sect. 4.1.4). This analysis estimates the average carbon footprint of single practices, such as using a washing machine or a car. Together with an estimation of the diffusion level of single practices, this analysis delivers highly relevant insights for the identification of key points for climate policy intervention. Moreover, practice-specific marginal income thresholds can be calculated based on income data, i.e. from which income level onwards practices start proliferating. This approach has been shown in the Greenpeace study (Ananthapadmanabhan et al. 2007, p. 9).

Another interesting finding of the Greenpeace study is the income-specific use of air-conditioners (AC). ACs become relevant only starting at a monthly income of 10.000 INR, however, at a quite minor level of below 65 kg CO_2/cap/year (Ananthapadmanabhan et al. 2007, p. 9). Moreover, emissions from electric geysers increase gradually with rising income, without indicating any income-related peak of emissions. The Greenpeace study also shows that electricity-based emissions from the bulk of *other* household appliances ("small electronic devices") depend highly on income, and, in terms of overall relevance, the effect of these other appliances is substantial. Within the highest income class, these small devices account for an average of 534 kg CO_2/cap/year (Ananthapadmanabhan et al. 2007, p. 10).

The appliance and income-specific analysis of electricity-related GHG emissions signifies that certain domains of consumption start becoming relevant only at certain levels of income. The authors of the Greenpeace study (Ananthapadmanabhan et al. 2007, p. 10) show this phenomenon also for the transport sector. While two-wheeler-related emissions increase gradually with every step on the income ladder, emissions from the use of cars start carrying weight at an income of more than 10.000 INR. Just as in this study, air travel emissions massively increase at the highest income level of above 30.000 INR, while it remains negligible in the lower income segments (Ananthapadmanabhan et al. 2007, p. 10).

The given results closely correspond with the findings of this study. The most carbon-intensive domains of consumption are also those domains that show the greatest positive income effects, partly in competition with low-carbon options as shown with IMT against public transport. The next section delineates the findings from practice-specific footprint analysis, indicating carbon intensities of practices as well as associated diffusion levels.

7.1.2 Consumption-Practice-Oriented GHG Emissions and Related Key Points of Intervention

7.1.2.1 GHG Emission-Related Evidence for Consumption Practice-Related Key Points of Intervention

Due to a multiplicity of driving factors, consumers show significant differences in consumption patterns and investments (investive consumption) with the effect of major differences in personal carbon footprints. This study aims to shed light on these driving factors, but it also hopes to contribute to a better understanding of the structure of personal and household-based GHG emissions. It moreover aims to facilitate exploration and development of new methodological approaches to analyse consumption-related emissions that may become easy-to-handle tools for applied studies in this field. Looking at consumption practices and their associated social-technical systems is such an approach, which has been developed by the author. A sector-specific analysis (see above) does not allow conclusions to be drawn concerning specific lifestyle practices that are most relevant for climate change mitigation. A social-practice approach is much better able to identify relevant key points of intervention. This theme is addressed by the third research question: "What are consumption practice-related key points of intervention?" The first sub-question is (No. 3.a.) "What is the average contribution of specific consumption practices to the overall personal GHG emission balance of a person?" The second sub-question (No. 3.b.) is "What is the overall share of people following each of these consumption practices?" Based on these results in regard to contribution and share of users, the third sub-question (No. 3.c.) addresses the issue of identifying key points of intervention: "Which of the everyday consumption practices are most relevant for climate-policy intervention?"

The analysis of consumption-practice-related personal GHG emissions attempts to quantify and compare different consumption practices in relation to their carbon footprint effect and reckons their dissemination among the overall sample of respondents. This approach is meant to suggest a bridge between income-related GHG accounting methods and the lifestyle-related approach to GHG emission differences. Routines or patterns of everyday behaviour are not fixed. Individuals adopt or choose such practices under the influence of a multiplicity of factors. Investments

7.1 Restating the Research Questions

in household infrastructure and personal assets (investive consumption) play an important role in determining the structure of everyday behaviour. Some of the investments tend to create path dependencies for individuals and households that have long-term effects on social practices of consumption and their related GHG emissions. Therefore, it is productive to analyse the GHG emission effects of such rather long-term consumption decisions.

For this purpose, the author has developed a simple and straightforward approach that specifically highlights these effects. Based on results of the carbon calculator, the author has computed the average per capita amount of GHG emissions associated with different consumption practices. The number of relevant practices is manageable and the approach to data and analysis straightforward. In many cases, the consumption practice is associated with a social-technical system, the material and technical basis of the consumption practice. With regard to food consumption only, the material dimension is non-technical. The emission effects of consumption practices are quantified in terms of yearly per capita GHG emissions, estimated based on averages of consumption. Only those respondents are considered for this calculation who actually follow a respective practice.

Some of these practices have been identified as most relevant for the structure of GHG emissions associated with personal consumption. These practices are conceptualised as key points of consumption concerning emission reduction strategies. The approach not only focuses on the emission effects of specific practices but also on the level of dissemination of these practices. A highly carbon-intensive practice which is followed by only a few respondents has little relevance for climate change mitigation – unless it has a dynamic potential for much wider dissemination in the future. In the Indian context of dynamic economic and social-cultural change, it is duly important to consider such practices as, e.g., shown by the use of air-conditioners, the use of cars, or the practice of air travel. Such domains are still relatively small, but the rate of growth is considerable, and their attractiveness to consumers means adoption of such practices is very high.

Key points of intervention can be easily identified based on associated GHG emission effects. The largest effect is found in international air travel (2.5 tonnes CO_2e/year/cap.), followed by the use of cars (1.6 tonnes CO_2e/year/cap.), the use of air-conditioners (0.464 tonnes CO_2e/year/cap.), the use of electric geysers for heating water (0.333 tonnes CO_2e/year/cap.), and the practice of using domestic flights for transport (0.284 tonnes CO_2e/year/cap.). Concerning dissemination, all mentioned practices have a very high potential for future growth.

Income dynamics and social-cultural behavioural change will have the greatest emission effects among these practices. Most relevant is international air travel with today less than 2% of respondents following this practice and a rate of growth of 7–8% over the next 15 years (Singh et al. 2016, p. 223). With the very high emission effect of 2.5 tonnes of CO_2e per year per user and the projected growth in absolute numbers of users, emissions will rise substantially in this domain.

Similarly, with an emission effect of 1.6 tonnes CO_2e per year per user and a dissemination rate of around 12%, the practice of automobility already weighs considerably, but bears even more potential for the future growth of associated GHG emissions. In fact, car ownership in Indian cities has increased by 23–75% within just 4 years between 2007 and 2011 (Verma et al. 2015, p. 113).

Air-conditioners represent another social-technical system with cooling practices and related carbon emissions averaging at 0.464 tonnes of CO_2e per year per user. While fans still represent the lion's share in regard to dissemination of space cooling appliances (90%), the market for air-conditioners shows the greatest potential for future growth (compound annual growth rate of over 10% in the coming 5 years according to ReportLinker 2015).

Also, electric warm-water geysers have a high per capita emission effect and a still rather low level of dissemination at around 20% of all surveyed households. Solar water-heating systems are a highly cost-efficient alternative to electric geysers with near to zero emissions. Against the light of rising incomes and general changes concerning preferences towards increased comfort and quality of life, rising demands for warm-water-heating systems are very likely. Therefore, economic incentives and the promotion of solar water-heating systems bear huge potential for the prevention of future carbon emissions in household energy use.

The results of this study have shown two major aspects of consumption practices, those realms with high emission effects, yet low levels of dissemination and a great future potential, and those realms that commonly have low levels of emission effects, but a huge dissemination rate and in turn huge reduction potential due to scale effects. In this latter field, wide-reaching small reductions have considerable effects due to the scale effects associated with the large absolute number that can be reached. The use of autorickshaws has been shown as relevant example. Also, the promotion of alternatives to incandescent light bulbs or a regulation of the market can have substantial effects.

In reference to the findings concerning income effects on consumption (Sect. 6.2) and in reference to the study issued by Greenpeace (Ananthapadmanabhan et al. 2007, p. 10), it is striking that those practices with the highest emission effects are also those with the greatest income effects. Most of these practices are characteristic concerning income effects as they depict obvious income thresholds, i.e. these practices start proliferating only from an identifiable income threshold (see Sect. 6.2). This is the case with air travel, the use of cars, and the use of ACs (Ananthapadmanabhan et al. 2007, p. 10). These practices are therefore duly relevant in a context of rapid economic growth and social mobility.

The theoretical approach to identifying key points of climate policy intervention has been introduced by Michael Bilharz (2009). However, Bilharz only estimates the differences of household energy consumption (in kWh) based on variance levels of certain domains of consumption. He identifies heating (corresponding to the use of ACS in India), the use of cars, air travel, and other unspecified diverse consumption (Bilharz 2009, p. 246). These findings obviously correspond with the findings of this study. However, a calculation of emission effects allows for a more comprehensive analysis, as it specifically and directly translates consumption into carbon footprints. This simple approach is better able to support climate change communication and raise public awareness, as the results are straightforward for inter-practice comparisons and for an understanding of individual carbon footprints. Therefore, this study suggests an adaptation of the approach outlined by Bilharz (see Sect. 6.3).

7.1 Restating the Research Questions

To sum up, the shown quantification of average per capita emission effects of various consumption practices is straightforward and flexible, and it allows key points for personal and household-based interventions to be easily defined. For environment-related social practice research, such quantification also provides an interesting vantage point. The approach builds on defining the material basis or the social-technical system that is associated with or foundational for certain climate-relevant practices. The approach therefore offers a flexible framework that allows for a specific analysis of other dimensions, such as practice-related symbolism, meanings, and value systems, as well as contextual or structural barriers or facilitating features of a practice.

Moreover, this simple approach of quantifying practice-specific personal-level carbon footprints (emission effects of following a practice) is able to practically inform targeted interventions and policies. Studies that highlight context-specific key points of intervention can build the foundation for more creative, integrated, and participative approaches to sustainable and inclusive urban planning and architecture. Other measures such as tax incentives or regulative measures can be explored based on the knowledge of such key points that are promisingly most effective in respect to reducing consumption-based GHG emissions.

7.1.3 Lifestyle-Related Analysis of Personal GHG Emissions

Values and attitudinal patterns are difficult to measure, and linear interpretation patterns are not viable to adequately reproduce these complex patterns. The study aims to get hold of these social-cultural factors that influence behavioural and consumption patterns and related GHG emissions. One of the most crucial questions of this research therefore deals with the issue of identifying interrelation and structural features of value orientations that tend to better explain behavioural and consumption-related specifics of everyday life. The study attempts to reveal lifestyle-specific structures or patterns of ideal-typical value orientations that build the foundation of a particular lifestyle. The aim of this basic research is to better understand whether and how certain structural configurations of values and attitudes structure behavioural and consumption patterns. These revealed patterns represent ideal types of lifestyles as basis for an analysis of lifestyle-specific personal GHG emissions. The fourth research question therefore builds on the concept of lifestyle and asks "Does the concept of lifestyle contribute to an improved understanding of differences in personal carbon footprints?" The question follows the rationale that linear interpretation pathways fail to adequately reflect important interrelational aspects of values and attitudes. A combined analysis of several dimensions based on typification is better able to shed light on an interdependent set of value orientations with their inherent ambivalences and contradictions. Hence, the first sub-question (sub-question No. 4.a.) is "Does the concept of lifestyle that draws on ideal-typical patterns of value orientation apply to the urban Indian context?" Assuming that such ideal patterns and structures of value orientation exist, it is unclear whether they are a meaningful basis for the conceptualisation of lifestyle and whether they are able to explain differences in personal-level GHG emissions. Sub-question No. 4.b. therefore asks: "Does this concept explain group-specific differences in behavioural and consumption patterns and related personal carbon footprints?"

The aim of the cluster analysis was to merge the overall sample of respondents into homogenous segments to meaningfully structure the dataset based on the selected attitudinal characteristics. The factor analysis provided a good foundation for this analysis, because it reduced the number of variables to a manageable and more interpretable number of latent dimensions based on the response pattern in the dataset.

The segmentation is based on a k-means algorithm, which rests on a preceding centroid-based algorithm of hierarchical cluster analysis (Ward) that provided the start values. To formally examine different cluster solutions statistically, a number of statistical tests were conducted such as Eta^2, PRE-value, F-max (cf. Schendera 2010, p. 119), and the F-values test (Backhaus 2011, p. 446; Fromm 2010, p. 214). All conducted tests pointed to an optimum being achieved with a six-cluster solution. Also, the evaluation of intra-cluster homogeneity by means of the F-values test shows very good levels of homogeneity within the clusters and heterogeneity between clusters. However, the most relevant quality criterion for the evaluation of a cluster solution is the content-based interpretation of the clusters. Interpretability of the cluster solution has priority over formal criteria. As the cluster analysis is solely based on attitudes and values, each cluster represents a quite homogenous group of individuals who share common beliefs and attitudes on the selected issues and areas. The interpretation was to lay out a description of the clusters based on the underlying factor variables. The given structure of each cluster can be understood as a representation of the underlying attitudes and values of people who belong to this group. The further description of the segments based on passive variables aimed to arrive at a typology of value-structure-specific lifestyles.

The content-related interpretation of the six-cluster solution was straightforward for most of the segments. It was facilitated by the reduced number of value orientation dimensions, which was achieved through PCA. The most meaningful interpretation and differentiation based on values was accomplished for cluster 1, cluster 2, cluster 3, and cluster 4. These value clusters were also found to be consistent in respect to social demography, behaviour, and consumption patterns, as well as to personal-level carbon footprints. Less clear was the interpretation of cluster 5 and cluster 6 which contained in part ambivalent structural features of value orientation and exhibited less coherent patterns of behaviour and social-demographic variables. For both clusters, a minimum interpretation was achieved that also is able to interpret some of the inherent ambivalences and incoherent structural features of the clusters.

The most interesting features are found in *cluster 1, cluster 4, and cluster 3*. *Cluster 1* (aspiring involuntary economisers) stands out in regard to its inclination towards frugality and thrift, which is in part driven by social-cultural and social-demographic factors. Both financial constraints and the precondition of expected dowry influences attitudinal features of thrift and in consequence behavioural structural features of frugality. Characteristic also is the life-affirming and aspirational orientation within this cluster with dismissive attitudes towards religious tradition and a fatalistic life view. Members of this cluster optimistically aspire for individual-level improvements of their life situation. Societal and ecological concerns play a negligible role for this cluster.

Cluster 4 (aspiring modern achievers) can be seen somewhat in opposition to cluster 1, with its rejection of frugality and thrift and its high level of agreement with social-ecological values. For this cluster, social-ecological values are highly relevant. The religious traditional paradigm with its inherent fatalism is a logical consequence of high levels of perceived self-efficacy and the emphasis on individual responsibility concerning consumption. The cluster is characteristic in terms of its high shares of very young and well-educated entry-level employees, who still live together with their parents (and siblings). Young aspiring achievers tend to have reached the entry point of their career, but their way of life is still strongly influenced by their parents and the structural features of their household. In terms of outside-oriented activities, members of this group have a very clear orientation towards centralised shopping and action and consumption-oriented leisure with a clear preference for IMT (mostly two-wheelers). This is coherent with the rejection of frugality. Social-ecological concern seems to translate into very low levels of personal GHG emissions, but this is mainly driven by household structural factors: while household-based emissions are very low (electricity, cooking fuels, rice, dairy), individual-level consumption is more carbon-intensive, with high levels of emissions from IMT and meat consumption and very low levels of public transport emissions.

Concerning emissions, *cluster 3* (well-established traditionalists) shows the most considerable results. Except for long-distance travel and public transport (both average), all sectoral emissions range at the highest (electricity, cooking, IMT) or second highest level (food). The cluster is very characteristic in this regard also due to the obvious rejection of social-ecological values. While there tends to be concern with issues related to the private sphere (high levels of agreement with family tradition), societal or ecological issues are beyond scope and not worth to engage for. In combination with an inclination towards religious tradition, people in this cluster tend to have more trust in formal institutions and the government, while self-efficacy is perceived rather low. Moreover, the cluster shows highest levels of investive consumption in coherence with a rejection of frugality. However, a concurrent rejection of the consumer culture paradigm indicates to a moral ambivalence with and critical reflection of modern ways of consumption. It is also indicative that larger shares of this cluster are Muslims. Across the overall sample, Muslims tend to reject social-ecological values as well as frugality and thrift. Muslims are also more likely for a non-vegetarian orientation. To a certain extent, the value and behavioural structure of this cluster is therefore influenced by religious-cultural features.

Cluster 2 (liberal pragmatists) is less clear in regard to interpretation, but the interpretation still makes sense to a large extent. Respondents of cluster 2 reject a too strict interpretation of moral norms and rules, and they tend to have a more enlightened and more individualistic attitude towards peers, kin-group members, or authorities. Environmental concern and questions of social justice are not very important for this segment, and so a morally driven orientation towards frugality and thrift is seen as rather meaningless for an open-minded and liberal worldview. Consumption and household structural features are based on pragmatism, practicality, and functionality. For the respondents of this cluster, social practices and consumption should not become an issue of concern in any regard. Liberal pragmatists form some share of the new urban middle classes, with quite reasonable levels of income and a decent educational background. Both in regard to behaviour and consumption as well as associated emissions, it is striking only that liberal pragmatists tend to use public transport rather than individual motorised transport, which translates into lower levels of carbon footprints. In all other realms, the cluster tends to range along the average.

Cluster 5 and *cluster 6* are more difficult to analyse in terms of interpretability and consistency in behavioural patterns and social demography. *Cluster 5* (hedonic ostentatious consumers) stands out with a high agreement to the consumer culture paradigm. It is the only cluster with a strong agreement to this dimension. Puzzling in relation to this consumer culture orientation is the additional strong alignment towards religious tradition. However, these two paradigms are not necessarily contradictory. A worldlier interpretation of religious tradition with an emphasis on the role of fatalism allows for light-heartedness concerning consumption. But the modest inclination towards frugality and the social-ecological paradigm contradicts the consumer culture paradigm. As agreement levels towards factors 2, 3, and 5 range on a much lower level, the interpretation has been focused on the combination of the consumer culture paradigm with religious tradition. In terms of social-demographic features, cluster 5 does not show any specificity, except for its highest share of Hindus and its larger shares of vegetarians. Hedonic ostentatious consumers are more likely to have a credit card and Internet at home. They show a preference for private two-wheelers and a preference for shopping in centralised locations such as malls, which both cohere strongly with hedonic consumer values. In terms of emissions, cluster 5 does not stand out, except for its lower levels in the use of electricity and cooking fuels.

The last and smallest cluster 6 (religious and nature-oriented traditionalists) somewhat opposes cluster 5. Both clusters have a somewhat similar value orientation structure, except for the dimension of consumer culture. Other than cluster 5, cluster 6 strictly rejects the paradigm of hedonic consumption. This coheres with positive attitudes towards religious tradition and frugality. As cluster 6 is quite small with $n = 45$, descriptive analysis with passive variables is a challenge. The quite unspecific features in terms of social demography and behaviour add to this problem. The only feature that stands out is that respondents' fathers are more likely to have lower levels of education. Cluster 5 has a lower tendency towards centralised shopping and greater preference towards homely oriented leisure. Traditional vacation is preferred over action-oriented holidays. Cluster membership of this cluster does not allow for predictions concerning GHG emission levels.

7.1.3.1 Discussion of These Results

The character of this lifestyle study is explorative in different respects: first, the approach of this study is based on a rather new method, which was not applied and tested elsewhere so far. Second, the typology builds on rather newly developed value dimensions with underlying questionnaire items that were in part specifically conceptualised for this study. Except for a few of these items, most of them have not been tested in India up to now. Third, it is the first time that the lifestyle concept has been applied to an Indian context – a context in which poverty still largely prevails and where the application of the concept therefore represents a conceptual and methodological challenge (see Sect. 2.3.2). Regardless of the explorative character of this study, the findings are quite striking. Yet, these findings need to be handled with proper caution. This research should be understood as vantage point from which the concept of lifestyle can be further developed for future research in the Indian context.

So far, to the author's knowledge, there is only one study which analyses lifestyle-related differences in personal carbon footprints. This outstanding comparative study on Germany, the UK, and the Czech Republic, issued by Peters et al. (2013), takes a slightly different segmentation approach by combining values and preferences together with income in a two-step cluster analysis. The segmentation was done separately for each country and includes values of traditionality, hedonism/materialism, self-fulfillment, and environmental awareness, as well as consumption preferences such as thriftiness or materialistic, hedonistic, and sustainable preferences (Peters et al. 2013, p. 232).[1]

Similarly as in this study, values and preference patterns vary significantly between clusters despite similar economic resources. In the same way, similar value orientations are found between clusters of very different social-economic position. The German case study identifies two segments with very high social-ecological values, but very different levels of income: the high income group "combines ecological orientation with hedonistic interests and a focus on political correct consumption, the less affluent and older lifestyle group […] embeds ecological values in thriftiness and traditional values" (Peters et al. 2013, p. 233). This is, while the rich segment depicts the biggest carbon footprint among the German sample, while the poor cluster shows the smallest footprint. Similarly, based on a review of different lifestyle studies in Germany, Poferl (1998, p. 307) concludes that ecological orientation and behaviour tends to unfold both within a broader paradigm of self-fulfillment and accompanying traditional and conservative values. This feature was also identified in the lifestyle segmentation of this study between cluster 4 (aspiring modern achievers) and cluster 6 (religious and nature-oriented traditionalists). Both clusters are strongly inclined towards social-ecological values, but while cluster 4

---

[1] The author of this study has borrowed some of the questionnaire items and adapted these to the Indian context (see Sect. 4.2.4.1).

tends to reject traditionality and values of frugality, cluster 6 holds traditional values and thriftiness high. This feature is very interesting, as it shows a similar tendency of value orientation patterns in India and Germany.

Moreover, the results from the UK exhibit a similarly well-differentiated structure of values, income, and carbon emissions, but as in Germany, differences in carbon footprints cannot be clearly traced back to value orientations or preference patterns. Lastly, the Czech lifestyle segmentation did not deliver such differentiated results in terms of carbon footprints in association with values and preferences. All clusters in the Czech case also have very similar income levels (Peters et al. 2013, p. 233).

Overall, the study's results reflect the problem of traceability delineated in Sect. 5.2.3. Lifestyle segmentations based on a combination of different dimensions, such as social-economic variables and value orientations, bear the problem that effects of cluster membership cannot be clearly traced back to a single dimension. In the case of the mentioned study (Peters et al. 2013), it is not clear whether levels of carbon footprints depend on income or rather on the structure of values. This is especially the case for the two most outstanding clusters in the German sample, with a similarly strong social-ecological orientation, but highly differing income levels and highly differing general values. In this study, the problem of traceability is less pronounced, because the author has deliberately separated the underlying lifestyle dimensions and has only included variables of value orientation in the cluster analysis.

Multivariate analysis, as suggested by Peters et al. (2013, p. 233), does not solve this problem, as the targeted variables of value orientation can only be analysed in terms of linear effects and not based on the combined structure of values. Thus, the multivariate hierarchical regression analysis only reveals linear effects from social-demographic variables (mainly income), while it shows no influence from the singled-out value and preference variables (Peters et al. 2013, p. 234). This is because value orientations tend to predict behaviour, consumption, and related emissions much less linearly, but rather in their specific combination and in a specific structure or pattern. This problem was also identified in an explorative multivariate regression analysis in the Hyderabad case study: only the dimension of frugality predicts lower levels of carbon footprints, while all other dimensions show no effect. This problem was identified as one of the most relevant arguments for the approach of building a typology of lifestyles based on the combination of different value dimensions and the resulting patterned structure of values and attitudes (de Haan et al. 2001, p. 9f).

7.2 Challenges and Critical Reflection on Methodology and Applied Methods

This research project has its origins in a broader project context of the BMBF-funded Future Megacities Programme with the project title "Hyderabad as a Megacity of Tomorrow: Climate and Energy in a Complex Transition Process

7.2 Challenges and Critical Reflection on Methodology and Applied Methods

Towards Sustainable Hyderabad" (in the following the short name "Sustainable Hyderabad Project" is used). The project design emerged from the question of how (local) lifestyle dynamics contribute to climate change and how lifestyle changes can help to reduce local emissions and the vulnerability to global climate change. The Sustainable Hyderabad Project was – both from its funding and from its participant structure – a German undertaking.

Therefore, working in a post-colonial country such as India requires the researchers to reflect on their own role, cultural background, perspectives, theoretical approach, and other aspects that can influence the research process and the outcome of the study. The author attempted to make this reflexive process become an integrative part in the overall research process of this study. Long periods of fieldwork in Hyderabad, during which the author lived in the city, allowed him to develop a more critical way of thinking about his own perception and the learned theoretical concepts. Facilitative was also the close and long-term interaction with other scholars from India and Germany in the context of the overall Sustainable Hyderabad Project.

Likewise, the underlying concepts, theories, methodologies, and normative implications are fundamentally based on a European or politically northern way of thinking. For instance, the notion of sustainability as interpreted in the context of the Sustainable Hyderabad Project is based on a rationale fundamentally built on "Western" thinking. It may convey quite different meanings in a different social-cultural context, such as India. This concern is even more relevant for the lifestyle concept as it has emerged in response to highly political- and cultural-specific transformation processes in Europe. The transfer to a different cultural environment bears huge challenges in respect to translating the underlying components and dimensions into an applicable framework that does justice to the very specific conditions in urban India. Particularly, the most important dimension of the concept is based on values and attitudes, which build on social-cultural determinants that are known by the researcher only based on an outsider's perspective. The new conceptualisation of lifestyle and the very explorative character of the study in this new context added to these challenges.

However, it must be stated as well that an outsider's perspective tends to be in fact more sensitive to the respective research context and its subjects. Kluckhohn argues:

> *Ordinarily we are unaware of the special lens through which we look at life. It would hardly be fish who discovered the existence of water. Students who had not yet gone beyond the horizon of their own society could not be expected to perceive custom which was the stuff of their own thinking. Anthropology holds up a great mirror to man and lets him look at himself in this infinite variety.* (Kluckhohn 1944, *p. 16 his emphasis*)

Moreover, most of the above stated challenges were addressed methodologically, based on the author's experience in qualitative ethnographic research. Especially for the identification and understanding of major societal divisions or cleavages in respect to general themes of everyday life, such as globalisation-related issues, tradition, conduct of life, consumption, environment, etc., preliminary qualitative research was fundamentally important. These insights were then used for the culturally adapted conceptualisation of lifestyle and the preparation of the questionnaire.

Poverty and social-economic uncertainty were a critical issue for the transfer of the concept, because for a large share of the overall population, choices of leisure activities and choices of consumption largely still depend on economic boundary conditions. The assumption that the social-economic position also tends to structure peoples' value orientation patterns was derived to some extent also from this preliminary research. In sum, based on this qualitative research, the author was able to make a well-informed effort to translate the underlying theoretical concepts into a foreign social-cultural context.

Apart from these quite general and methodological challenges, there were other more concrete problems and difficulties related to language and translation, sampling, employment of field assistants, and data analysis. The problem of language is an important aspect of intercultural research. Many of the interviews and interactions with people in the field required a good knowledge of Telugu or Hindi. With the researcher's quite basic level of Hindi and lack of knowledge in Telugu, the assistance of a reliable interpreter was indispensable. For the qualitative interviews, an experienced field worker was found with experience in geographic field research. The process of translation doubles the risk of an expectancy effect, i.e. the tendency of the researcher as well as of the interpreter to obtain results that they expected (Bernard 1994, p. 233). Moreover, respondents tend to react and may adapt their response to the social-cultural background of the researcher and the field assistant (e.g. caste or income), which potentially leads to an exacerbation of social expectancy effects (Sect. 4.1.2.1). Also an increased risk of a language-driven bias was given by the questionnaire, which was initially prepared in English and then translated into Telugu. This problem was addressed through comparing and discussing both questionnaire versions among the team of research assistants and with two native and well-experienced field researchers (Sect. 4.2.4.1).

One of the most critical errors in this research is related to the sampling process of the standardised survey and its implementation through the field investigators. Because of the lack of basic social-economic data disaggregated to the municipal level of Hyderabad, the selection of respondents involved quite a number of important methodical steps to be considered by the field investigators with the effect of an increased potential of sampling bias. Particularly at the household level (selection of a household member based on birthday method; see Sect. 4.2.4.2), implementation of the sampling rule was challenged through respondents forging ahead. Despite very explicit and proper guidelines and training on the methodical steps of the sampling process, it was followed with equal consistency among all researchers, which produced some avoidable sampling bias. The greatest effects have been revealed in respect to gender imbalances (see Sect. 5.1.1). However, this study is an explorative study that aimed to develop an applicable approach for transferring the lifestyle concept, to test this new methodological approach in the Indian context, and to explore its practicality for environmental social science research. Hence, the representativeness of the results is not the most important quality criterion.

Last but most importantly, quantitative research and analysis is often suggestive of being maximally precise and reliable as it tends to convey a sense of being able to reconstruct social "reality" through precise use of concrete numbers. The author of this study argues that the researcher's perspective, her or his underlying knowledge and assumptions, as well as the particular context of the situation of enquiry cannot be seen as detached from the research process. Moreover, social reality is always in part a construction based on our perceptions, knowledge, and observations, which means that this subjective view on "reality" structures or influences the measurement (cf. Creswell 2003, p. 9).

In this study in particular, methodological assumptions and decisions on, e.g., the choice of questionnaire items have largely been subjective, supported by a chosen set of research literature on the topic. The author constructed the concept and defined relevant criteria for differentiation, based on which meaningful groups were typified. However, group members do not necessarily perceive themselves as a group, because the group-defining criteria remain unknown to them. Groups become meaningful based on the subjectively set criteria and the author's interpretive decoding effort. However, the criteria have in some sense been understood and rated by the respondents, and hence they measure their subjective evaluation of everyday relevant issues. Thereby, group-specific characteristics reflect an ideal-typical differentiation pattern of a population, seen, described, interpreted, and documented – through the lens of the researcher.

Furthermore, the group members' value orientations and the described behavioural patterns do not equally resemble the ideal-typical pattern of the value cluster. In terms of certain criteria, a respondent may be closer to an orientational pattern of another cluster. In some ways he or she may not agree with the author's given descriptions and interpretations. Nevertheless, this approach is able to delineate a simplified model of social interaction patterns that helps highlight subtle or deeper divisions between groups and facilitates understanding of how such orientational patterns translate into certain patterns of behaviour and consumption.

In this light, it would be desirable in a further research step to introduce some of the cluster descriptions to different groups of similar people (focus groups) – independent of their potential cluster affiliation. Focus group discussions (Flick 2006, p. 198) could serve as an approach to evaluate the validity of the identified groups and help to detect improved criteria for a further or adapted differentiation.[2] This validation and conceptual improvement would be an adequate response to the critique stated by Poferl (1998). She argues that cluster analytical approaches are in fact able to capture subjective evaluations based on attitudes and values, but this measurement remains stuck in its limited access to the matter because of its standardised research approach. Such an approach – according to her relevant argument – was not able to fully reconstruct the *emic* elements of meaning located within the concrete common-sense world ("Alltagswelt") of the people (Poferl 1998, p. 307).

In fact, this explorative research was not meant to deliver a ready-made and transferable concept of lifestyle for further application in an urban Indian context. The concept and the underlying criteria of differentiation need to be further devel-

---

[2] Initially, this research step was considered by the author, but due to financial and time constraints, it was discarded in the end.

oped based on the findings of this research. With such an improved and much more simplified methodological approach, the concept of lifestyle could serve as relevant perspective for environment-related social science research in urban India. The next section will demarcate the most relevant research steps for the further development of this urban-Indian-specific lifestyle concept.

7.3 Implications for Future Application

As stated in the section above, the author was bound to take an explorative approach for the objective of typification of lifestyle in urban India. Based on this extensive and time-consuming effort, the author was able to show that the application of the concept can help to identify and explain group-specific evaluations of commonly known issues of everyday life. Based on these specific evaluation patterns, underlying social values can be inferred, and typical patterns of behaviour and consumption thereby become clearer. However, this typification can be refined through further research.

7.3.1 Refinement of Value Dimensions

First, this research has made a large step towards identifying those dimensions of value orientations that are most relevant for lifestyle and consumption in a context of rapid change as found in Hyderabad. Through explorative principal component analysis (PCA) of a large set of attitudinal items that form the basis for measuring a variety of underlying value dimensions, the five most relevant and most differential value dimensions have been identified. Both the selection of items and the identification of value dimensions were informed by statistical analysis of the patterns of rated agreement on the underlying items. The final factor solution was able to structure the rated agreement levels of respondents, and this structure involves the strongest and most relevant divisions and cleavages measured among the people in Hyderabad. These dimensions have been found to be useful and comprehensive for the analysis of subjective social-cultural inequalities.

Nevertheless, for further application, these dimensions can be further optimised in terms of precision and scope. For instance, the measurement of attitudes towards frugality and thrift could be refined to allow discrimination between frugality as a reaction to one's financial situation (culture of necessity) and frugality as a morally driven virtue (see Sect. 5.2.3.2). Also, for evaluating social-ecological orientation, the author suggests a differentiation of measurement in terms of consumption-related attitudes (as formed with the factor of social-ecological consumption in this study) and more general environment and climate-related values. For the latter, one could draw a lot more on issues related to environmental movements[3] in India and related public discourses, as delineated, for instance, by Gadgil and Guha (1995, p. 101f) or Mawdsley (2004, p. 80; 2006, p. 382). Finally, in fact, the two paradigms of tradition are highly relevant in the Indian context. For a better interpretation of the two, items measuring these should be able to delineate a clearer picture of what people actually think of, when they rate high or low on one or the other of the two paradigms. The items should therefore be conceptualised as being more rooted within commonly discussed themes of societal change.

For a well-founded understanding, such refinements can build on existing research literature, but it could also be informed by further qualitative research that specifically targets the mentioned themes of value orientation. Furthermore, social-economic or other structural factors that closely and specifically interact with certain value orientations should be identified and explained on the basis of qualitative social research. Such research can serve as a basis for the conceptualisation, interpretation, and communication of advanced lifestyle research in India.

7.3.2 *"Ground Truthing" of Identified Lifestyles*

Similarly, as with measuring value orientations, further research is suggested to develop and test a simplified tool for measuring lifestyle, which builds on the identified components and dimensions and which benefits from a more targeted and more specific conceptualisation without being bound to have extensive criteria for exploration as in this study. A refined follow-up study would have to adopt a shorter and more concise questionnaire, and the data analysis would be more straightforward in such a second step. Moreover, further lifestyle research would profit from a qualitative evaluation and validation of the lifestyle groups. Such a "ground truthing" of the identified clusters could aim to get hold of how people evaluate the interpretation of the clusters and how they explain relational aspects between these groups.

One methodological approach to this validation would be to have focus group discussions with the objective of allowing people to evaluate the most relevant and/ or the most ambivalent features of the lifestyle groups. This not only includes the explanation of value orientation patterns but also their related patterns of behaviour and consumption and the way in which these are structured by social-demographic and external factors. By means of discussion among a group of rather socially homogeneous people, such research would allow to validate, adapt, or reject the respective cluster interpretations. Such research would in fact contribute to developing a more simplified standardised approach to measuring lifestyles.

---

[3] It may not be commonly known that India is regarded as having the world's largest and probably most diverse environmental movement (cf. Mawdsley 2006, p. 382; Nelson 2000, p. 4).

7.3.3 Extending the Scope and Depth of Practice-Related Analysis of Personal GHG Emissions

The main objective of this research was to explore the feasibility of applying the lifestyle concept to an urban Indian context with the aim of identifying and explaining patterns in conduct of life and related personal GHG emissions. To support this analysis, the author has measured social-economic and household-structure-related factors that influence lifestyles. More than income, the level of investive consumption – measured by the wealth index – has been found to have the greatest effect on personal carbon footprints (see Annex VIII). To analyse this effect more specifically, the author has then identified and looked at different consumption practices and measured their direct average effect on personal-level carbon footprints.

This simple approach can be helpful in identifying the most relevant consumption practices as key points for policy interventions regarding climate change mitigation. These key points can moreover serve as a vantage point for more detailed research on the underlying social practices, their meanings and functions for carriers of the practice, and the related structural and infrastructural factors influencing these practices. For example, the use of cars largely depends on social-economic factors, but it also involves multiple other motives and features that play a distinctive role for choices for or against this practice.

Social practice theory has made substantial methodological advance over the last two decades, and some scholars have started to take a practice perspective to sustainable consumption. With the objective of challenging common framings in current sustainability policies and finding new ways in analysing problems of environmental justice, climate change, and sustainability in relation to consumption and demand, an innovative strand of research has emerged. Especially Elizabeth Shove and her colleagues (Shove 2003; e.g. Shove et al. 2012; Walker et al. 2014) have further developed methodology in social practice theory for the analysis of the relationship between consumption, everyday practices, and ordinary technology, with a focus on sustainability and climate change mitigation. One of the strengths of social practice theory is its framework that combines objective criteria (material) with subjective realms of competences and meanings. The element of material involves architecture, infrastructures, and urban planning issues, which closely relate to urban governance and policies on different levels. This social-technical system embeds and influences individual and household consumption decisions. This is while consumption decisions in turn feed back to the system and influence its structure. Consumption decisions also tend to structure routines and behavioural patterns on individual and household level. Some of these decisions tend to have impacts that go as far as being able to create path dependencies for future routines and consumption, which applies, e.g., to the purchase of a car.

Meanings and functions of practices are relevant for the understanding of how unsustainable practices (key points) are embedded socially and culturally. Shedding light on the material and subjective aspects of practices is likely to deliver a differentiated basis of knowledge that could better inform targeted sustainability policies

7.3 Implications for Future Application

(cf. Climate Change Leadership Fellowship 2008). Especially in a country like India, with its rapid economic growth, its dynamic processes of social mobility, and its fast rates of urbanisation, much of the future (urban) infrastructures with all involved path dependencies still need to be built. This opportunity needs to be considered now. Unsustainable practices (identified key points of intervention) that have the potential for high levels of future dissemination (and future GHG emissions), such as car driving or air travel, need to be well evaluated in terms of planning-related opportunities and costs (e.g. objective relevance, effects on other practices, reduction potential) but also in terms of "softer" criteria of function, meaning, and knowledge (e.g. symbolic meanings, prestige effects, comfort, knowledge on climate-related impacts). To arrive at this knowledge, it requires further transdisciplinary research with a focus on most relevant areas of consumption.

Chapter 8
Final Conclusions: Understanding Inequalities in Consumption-Based, Personal-Level GHG Emissions

Keywords Income · Carbon intensity of practices · Wealth · Investive consumption · Carbon footprints · Key-points of intervention · Lifestyle policies · Mitigation policies

8.1 Taking an Income Perspective: Identifying Key Sectors Most Relevant in Terms of Income Dynamics

An economic perspective on climate change concerning its drivers, effects, and solutions is common and dominates the research agenda. Many of these approaches are well established, straightforward, and able to provide valuable evidence and proof for issues especially relating to climate change mitigation. However, most of the economically framed trajectories are largely understood. This is especially true for average effects of income on consumption and related GHG emissions. Income can be mainly seen as a conditio sine qua non, both for modern consumption and related GHG emissions. Income enables consumption, and therefore, income will remain the most critical determinant for differences in personal carbon footprints. The results of this study show that income, apart from personal and household wealth, has the most significant impact on individual carbon footprints. This is especially true for all those sectors involving domains with the highest carbon intensities, namely, electricity, individual motorised transport, and air travel.

It is definitely true that income dynamics and their effects on consumption and related carbon emissions are especially relevant in a transitional country such as India. An assessment of these effects enables emission projections that inform national climate policies and facilitates identification of the most relevant sectors for climate change mitigation. In particular, income-specific assessments are relevant concerning a social-economically differentiated view on the carbon footprints of urban areas. However, so far, the approach lacks scope and viability, due to lack of availability of reliable data. Surveying income data, as discussed in Sect. 2.1.2.4, bears huge challenges of reliability. This is the reason why many economic studies build on income approximations based on consumption expenditure data. For consumption-based GHG emission accounting, this approach is problematic, because income is proxied by total household expenditure. At the same time, household consumption expenditure is used to estimate household-based emissions, which involves a tautology to a certain degree. Consumption expenditure reflects actual levels of consumption directly. Based on differing emission intensities of different consumption categories, consumption is translated into associated amounts of GHG emissions. However, consumption expenditure does not adequately reflect differing levels of income, because such an approach blinds out varying levels of savings as well as differing patterns of expenditure in terms of quality (e.g. organic food items vs. conventionally produced goods) and in terms of the structure of the consumption basket (Sect. 2.1.2.4). An explorative multilinear regression analysis has indicated this problem: investive consumption measured by the wealth index has a stronger effect on personal carbon footprints than has income. This study has therefore built the analysis on actually reported levels of income.

The results of this analysis have revealed interesting structural differences in personal carbon footprints based on income differences. The findings indicate that electricity consumption, IMT, air travel, meat, and dairy consumption are the most dynamically changing emission sectors due to income dynamics. Surprisingly, meat consumption shows reverse effects on rising incomes, presumably because of religious traditional reasons. Cooking fuel, public transport, bus and train travel, and rice consumption show minor or no income effects.

Such an income-oriented approach to carbon footprint analysis is straightforward and does not involve huge challenges in terms of measurement. However, it is limited and one-dimensional in terms of scope, differentiation, and explanatory power. Especially studies that approximate income through consumption data largely remain descriptive. As shown in this study, an extended differentiation based on intra-sector disaggregation is an important step forward. However, it does not allow conclusions to be drawn on the average GHG emission effects of certain practice patterns, associated levels of diffusion, and the underlying motivations of individuals and other structural reasons. The practice and lifestyle-oriented approaches therefore deliver deeper, more differentiated insights, and they are more promising for further methodological development.

8.2 Practice-Specific Analysis: Key Points of Intervention and Related Potential for Understanding Practice-Specific Motives, Meanings, and Functions

The practice-specific analysis of personal-level GHG emissions has shown relevant and new results. It indicates at the average emission effects of specific consumption practices and shows how these add to the overall personal carbon footprint. The approach enables one to identify and quantify the typically most carbon-intensive consumption practices (key points), such as the use of cars, or international air travel. This knowledge can serve as basis to inform individual consumer decisions with estimations of average long-term effects on their overall personal carbon footprint.

In addition to these emission effect calculations, the study provides estimates on the dissemination of practices, i.e. the share of people regularly following a certain practice. Emission effects and dissemination levels together offer valuable insights about the different relevance of practices: practices with low carbon intensity and low levels of dissemination are negligible in the first place. Such practices do not offer a key point of intervention. In contrast, practices with high carbon intensity and high prevalence have the highest reduction potential, in terms of both scale effects and direct effects. A practice with low carbon intensity but high prevalence has huge reduction potential due to the scale effect of even minor reductions (e.g. the use of autorickshaws or incandescent light bulbs). Lastly, there are practices with higher carbon intensity, but low levels of dissemination. Here, it is the dynamic potential of the practice which is most relevant for climate mitigation. In the Indian context, air travel and the use of air-conditioners serve as good examples for this category. Mitigation policies can address both efficiency in terms of carbon emissions and the expected dynamics of growth of the practice. The research framework therefore allows all those consumption practices to be highlighted that are most relevant in terms of their overall as well as their personal-level reduction potential, i.e. the most relevant key points for climate policy intervention.

Moreover, the identification of intervention key points in terms of specific social practices can serve as a vantage point for a more targeted, systematic, and qualitative analysis of social practices. Such an approach can draw on methodology and findings from social practice theory that in particular aims to shed light on the different meanings and related motives of practices and thereby reveal the different functions of certain practices for different people. As shown in this study, for instance, Elisabeth Shove (2003) has delivered highly valuable accounts on the issues of comfort, cleanliness, and convenience and the related dynamics of habits and routines. She examines the coevolution of technology and practices, concerning, for instance, air-conditioning, laundry, bathing, and showering. She is able to carve out the related meanings and functions of practices and analyses how these relate to the organisation of everyday life. Such practice-specific knowledge is fundamental for framing environmental policies as well as studies that deal with the question of how to guide consumption towards more sustainability. It helps to understand the structural and motivational barriers of, e.g., more sustainable practices as well as the inherent conflicts and ambivalences that people may have with them. Based on the identification of key points and following social practice theory, more targeted research can be initiated to improve the understanding on the meanings and functions of relevant consumption practices to better inform mitigation and lifestyle policies.

The theoretical chapter (Sect. 2.2.2.1) has shown that lifestyles incorporate some practices while dismissing others, in a way that resembles a patchwork pattern. The question of how certain practices relate to each other in terms of meanings and functions and how certain practices lead to a coevolution of other practices may reveal new insights for the understanding of lifestyles. These dynamic interrelation patterns between different practices provide constructive insights to the question how lifestyles evolve and unfold in relation to the underlying value orientations.

Apart from these valuable insights on the practice-related meanings and functions, social practice theory has highlighted the role of things and consumption infrastructure, i.e. the material basis of practices. The approach of this thesis concerning practice-related emission effects has highlighted the fact that most climate-relevant practices involve a technological or at least a material basis. The author has emphasised this material aspect of social practices also in respect to investive consumption: how certain investments into, e.g., household infrastructure lay the foundations for trajectories of certain patterns of consumption, for everyday routines, and therefore also for lifestyle. Hence, such investments tend to create path dependencies. With this knowledge, estimating practice-related GHG emissions is an approach that can shed light on such trajectories. They are the result of specific investments (e.g. AC, car) or long-term consumption decisions (vegetarian diet, monthly ticket for public transport) and can be expressed in terms of GHG emission averages.

The wealth index supports this view aggregated to an index that reflects the level of investive consumption.[1] The effects on personal carbon footprints have been measured: investive consumption, as conceptualised in this study, was even found to be a better predictor for personal GHG emissions than income.

Based on these considerations and based on the informative power of insights from environment-related social practice theory, the author argues that social practice theory needs to be considered as a complement to lifestyle research and not as being at variance to it. The methodological approach of estimating practice-related GHG emissions can serve as a link between environment-related lifestyle research and social practice theory. Overall, among the greatest advances of this methodical approach is the straightforwardness. The results have an explanatory power for the communication of lifestyle policies. They can also inform consumers and their decisions, as they can easily be explained and communicated.

8.3 Lifestyle-Specific Analysis: Identifying Effects Based on Ideal Typical Value Orientation Patterns

A lifestyle perspective as conceptualised in this study addresses consumption inequalities from a different angle. It draws on the underlying configurations of basic value orientations that are assumed to structure behaviour and investive consumption within the individual scope of action. In the light of these theoretical considerations and insights from other environment-related lifestyle research, the author decided to make a segmentation based on respondents' value orientations. Building on these value segments, the study aims to arrive at a comprehensive

---

[1] In contrast to consumer expenditure, investive consumption reflects investments in durable goods that remain with the person or household for some time. To a varying degree, these investments tend to have a long-term impact on the individuals' conduct of life and his or her lifestyle ("path dependencies"), as explained in Sect. 4.1.3.2.

typology of lifestyles by means of describing the segments in terms of typical patterns of behaviour and consumption as well as social-demographic factors.

With this framework, the study was able to differentiate and interpret the value clusters meaningfully. The cluster-specific descriptions based on behavioural patterns and social demography revealed patterns that largely cohere with the underlying values. Hence, the objective of constructing a meaningful lifestyle typology has been achieved. However, the study was conceptualised on new theoretical grounds and in a completely under-researched context. Therefore, the newly conceptualised components of lifestyle and the findings should be seen as a basis for further research (see Sect. 7.3).

The decision to take values as the single basis for the segmentation was informed by theoretical and empirical evidence as outlined in the theoretical chapter. This fundamental step was fruitful in different regards. *First*, explorative multiple regression analysis has shown that the attitudinal factors as single dimensions are not well able to predict behaviour, consumption, and associated GHG emissions. However, in their configuration with each other (as depicted by the value clusters), values in fact explain variance, but in a non-linear way. A value-based cluster analysis deals with this problem as it builds on an ideal typical interdependent structure or combination of different dimensions and their non-linear cluster-specific effects. In contrast, a post hoc description of behavioural clusters on the basis of value orientations would reduce the possibility to reveal non-linear ideal typical structural patterns of values.

Second, a cluster analysis based on other dimensions than values would have meant a fundamental compromise for the objective of this study. The study aims to get hold of groups that are similar in terms of their values. Patterns of behaviour do not always cohere so well with the underlying values. Depending on the situation and depending on ambivalent value priorities, behaviour can deviate from an expected trajectory, as has been shown empirically and explained with the concept of the knowledge-action gap. It shows that in spite of the growing levels of awareness on environment and climate, an equally growing level of behavioural change towards sustainability cannot be observed: people know of certain impacts of their actions and behavioural patterns, but they often fail to respond accordingly. And even having a normative attitude towards a certain problem does not necessarily lead to consistent reactions.

In this light, it is in fact valid to look at the specific configurations of values. This contributes to better understand the underlying ambivalences, contradictions, and conflicts between different values, as shown in this study. With this perspective, the study is able to deliver valuable insights into underlying motivational drivers of different behaviour and consumption patterns. It was also possible to figure out the group-specific social-economic and external determinants that tend to structure behaviour and that inhibit certain patterns of behaviour contrary to a particular value orientation.

Moreover, it has been proved valid to emphasise general values and not merely environment-related values – as many other studies do. These theme-specific values are deemed most relevant for environment-related behaviour, but they are not necessary. While environmental values may be relevant for some people, holding up these values does not necessarily mean that people are bound to adopt more sustainable practices. Other values and their related life goals may override the environment-related value orientations in some situations. Also, external and social-economic factors may determine certain behavioural patterns. It is therefore constructive to include all those dimensions that may have an impact on situational evaluations. This argument has been shown to be duly important as such conceptualisation helps to get hold of the underlying contradictions and ambivalences.

Based on explorative PCA, the author identified five relevant dimensions which proved to have the greatest differential power: the two traditional value dimensions, frugality, hedonism, and the paradigm of social-ecological consumption. Apart from this statistical evidence, also theoretical considerations suggested this selection. The author has stated in the theoretical framework that especially consumption-related values are most relevant for the constitution of lifestyles. Apart from that, it is promising to figure out which themes of (e-)valuation are more strongly and more widely contested between different social-cultural groups. Tradition and aspects of modern ways of life touch these contested societal themes most directly. Again, as part of these realms, consumption aspects can serve as an anchor to connect with broadly relevant areas of conduct of life. This conceptualisation has proven to work out well. Thereby, in addition to the other mentioned advantages, the lifestyle concept is not bound to be restricted to application in an environment-related topic, but can be employed much more broadly.

The most relevant criterion of validity for this study is the explanatory power of the concept concerning personal GHG emissions. The identified segments have in part shown significant differences in average carbon footprints; however, as expected, these differences were only relatively small compared to those identified between income groups. The greatest variances were observed between cluster 1 (aspiring involuntary economisers), cluster 2 (liberal pragmatists), cluster 3 (well-established traditionalists), and cluster 4 (aspiring modern achievers). These four clusters have also shown the greatest differences in terms of behavioural and consumption patterns as well as social demography. Cluster 5 (hedonic ostentatious consumers) and cluster 6 (religious and nature-oriented traditionalists) are very interesting in terms of their value structure, but in respect to behaviour and related emissions, both clusters largely range along the average.

The process of analysis and interpretation of the value orientation clusters has turned out to be a time-consuming and sometimes challenging task. As it is based on group-specific averages (mean/median), the research can only get hold of tendencies and patterns that often tend to be vague and subtle. At the same time, this undertaking was able to reveal and illuminate very interesting group-specific and non-linear features and relationships between variables that would have not been possible with most other approaches to analysis. For instance, the cluster solution has exposed evidence of the fact that social-economic and objective social-cultural factors apart from income can have a significant effect on the structure of values and thereby on behaviour and emissions. The findings in this study show that social-demographic features are interrelated closely with value orientations and they can influence these to some extent. Hence, attitudes and values tend to respond to, e.g., external factors (as shown with the example of expected dowry), but values are yet more stable than, e.g., social-demographic or behavioural aspects. This stable character of values, and the fact that certain values compete with each other depending on the issue and the situation, is another valid argument to take values as the basis for segmentation.

In conclusion, this study has delivered highly valuable insights into applied lifestyle methodology. Based on the findings of this explorative study, the framework can be more straightforwardly employed in other research contexts in urban India, either with a focus on environment and climate-related issues or with a focus on other issues such as health or consumption. The identified dimensions of general value orientations are found to be valid and highly distinctive. Some refinements of the underlying set of attitudinal items (see Sect. 7.3) would certainly contribute to an improved factor solution, which in turn would allow the value clusters to be analysed and interpreted more concretely, clearly, and meaningfully.

While the approach may not be able to precisely predict group-specific levels of personal carbon footprints, it is still able to shed light on the complex interrelations between a variety of different factors. This characteristic capability is a considerable strength that helps to reveal and explain those interrelations between external determinants, motivations and attitudes, and behaviour and action.

References

Ahmad S, Baiocchi G, Creutzig F (2015) CO_2 emissions from direct energy use of urban households in India. Environ Sci Technol 49:11312–11320
Ajzen I (1991) Theories of cognitive self-regulation: the theory of planned behavior. Organ Behav Hum Decis Process 50:179–211
Ajzen I, Fishbein M (1980) Understanding attitudes and predicting social behavior, Pbk. edn. Prentice-Hall, Englewood Cliffs
Alkire S, Santos ME (2014) Measuring acute poverty in the developing world: robustness and scope of the multidimensional poverty index. World Dev 59:251–274
Allen J (2006) Ambient power: Berlin's Potsdamer Platz and the seductive logic of public spaces. Urban Stud 43:441–455
Ananthapadmanabhan G, Srinivas K, Gopal V (2007) Hiding behind the poor: a report by Greenpeace on climate injustice. Greenpace India Society, Bangalore
Appadurai A (1996) Modernity at large: cultural dimensions of globalization, public worlds. University of Minnesota Press, Minneapolis
Athukorala P, Sen K (2004) The determinants of private saving in India. World Dev 32:491–503
Backhaus K (2011) Multivariate Analysemethoden: Eine anwendungsorientierte Einführung, 13th edn. Springer, Berlin
Banerjee AV, Duflo E (2007) What is middle class about the middle classes around the world? (No. 07/29), MIT Department of Economics Working Paper. Massachusetts Institute of Technology (MIT) – Department of Economics
Baud ISA, Sridharan N, Pfeffer K (2008) Mapping urban poverty for local governance in an Indian mega-city: the case of Delhi. Urban Stud 45:1385–1412
Baynes TM, Wiedmann T (2012) General approaches for assessing urban environmental sustainability. Curr Opin Environ Sustain 4:458–464
BBC (2015) Modi hails social media power at Facebook HQ. BBC News. URL: http://www.bbc.com/news/world-asia-india-34376778. Accessed 4 Mar 2016
Beck U (1998) World risk society. Polity Press, Cambridge
Belk R (2014) You are what you can access: sharing and collaborative consumption online. J Bus Res 67:1595–1600
Benbabaali, D., 2009. Importing new cultures into the city: the role of Kamma migrants in the development of Andhra culture in Hyderabad, Reddy Anant, G., Emerging urban transformations: multilayered cities and urban systems. IGU Urban Geography Commission, Osmania University and Hyderabad Metropolitan Development Authority, Hyderabad, 689–700
Bendix R, Lipset SM (1967) Class, status, and power: social stratification in comparative perspective. Routledge & Kegan Paul, London

Bernard HR (1994) Research methods in anthropology. Qualitative and quantitative approaches, 2nd edn. Sage, Thousand Oaks

Betz J, Destradi S, Neff D, Sen K, Kling V (2013) Indien unter Premierminister Modi: Wandel mit Hindernissen. Polic Soc 32:319–331

Bhoyar SP, Dusad S, Shrivastava R, Mishra S, Gupta N, Rao AB (2014) Understanding the impact of lifestyle on individual carbon-footprint. Procedia Soc Behav Sci 133:47–60

Bilharz M (2009) "Key Points" nachhaltigen Konsums: ein strukturpolitisch fundierter Strategieansatz für die Nachhaltigkeitskommunikation im Kontext aktivierender Verbraucherpolitik, Wirtschaftswissenschaftliche Nachhaltigkeitsforschung, 2nd edn. Metropolis-Verl, Marburg

Bilharz M, Cerny L (2012) Big points of sustainable consumption and lifestyle orientation: How do they fit together? In Beyond consumption. Presented at the beyond consumption: pathways to responsible living, 2nd PERL international conference, 19–20 March 2012, Technical University Berlin, p 54

Bin S, Dowlatabadi H (2005) Consumer lifestyle approach to US energy use and the related CO2 emissions. Energy Policy 33:197–208

Birdsall N (2010) The (indispensable) middle class in developing countries; or, the rich and the rest, not the poor and the rest, Working Paper No. 207. Center for Global Development, Washington, DC

Blasius J (2001) Korrespondenzanalyse, Internationale Standardlehrbücher der Wirtschafts- und Sozialwissenschaften. Oldenbourg, München

Bortz J, Schuster C (2010) Statistik für Human- und Sozialwissenschaftler: mit 163 Tabellen, 7., vollst. überarb. und erw. Aufl. ed, Springer-Lehrbuch. Springer, Berlin

Bourdieu P (1983) Ökonomisches Kapital, kulturelles Kapital, soziales Kapital. In: Kreckel R (ed) Soziale Ungleichheiten. Schwartz, Göttingen, pp 183–198

Bourdieu P (1987) Die feinen Unterschiede. Kritik der gesellschaftlichen Urteilskraft. Suhrkamp, Frankfurt am Main

Brander M, Sood A, Wylie C, Haughton A, Lovel J (2011) Electricity-specific emission factors for grid electricity (Technical Paper). Ecometrica

Bronger D (1996) Indien, 1. Aufl. ed. Perthes-Länderprofile. Perthes

Brosius C (2010) India's middle class: new forms of urban leisure, consumption and prosperity, cities and the urban imperative. Routledge, London

BSI (2013) PAS 2070: 2013 specification for the assessment of greenhouse gas emissions of a city – direct plus supply chain and consumption-based methodologies. British Standards Institute, London

Bühl A (2012) SPSS 20: Einführung in die moderne Datenanalyse. Pearson, München

Bunnell T, Das D (2010) Urban pulse. A geography of serial seduction: urban policy transfer from Kuala Lumpur to Hyderabad. Urban Geogr 31:277–284

Burke J (2010) Avoid US route: Indian minister says American way is "recipe for disaster." The Guardian

Butsch C (2011) Zugang zu Gesundheitsdienstleistungen: Barrieren und Anreize in Pune, Indien, Geographie. Steiner, Stuttgart

cBalance (2010) GHG inventory report for electricity generation and consumption in India. Pune

Chakravarty S, Ahuja DS (2016) Bridging the gap between intentions and contributions requires determined effort. Curr Sci 110:475–476

Chakravarty S, Chikkatur A, de Coninck H, Pacala S, Socolow R, Tavoni M (2009) Sharing global CO2 emission reductions among one billion high emitters. Proc Natl Acad Sci 106:11884–11888

Chan S, van Asselt H, Hale T, Abbott KW, Beisheim M, Hoffmann M, Guy B, Höhne N, Hsu A, Pattberg P, Pauw P, Ramstein C, Widerberg O (2015) Reinvigorating international climate policy: a comprehensive framework for effective nonstate action. Global Policy 6:466–473

Chancel L, Piketty T (2015) Carbon and inequality: from Kyoto to Paris. Paris School of Economics, Paris

Chemnitz C, Becheva S (2014) Meat atlas – facts and figures about the animals we eat. Heinrich Boell Foundation; Friends of the Earth, Berlin

Climate Change Leadership Fellowship (2008) Transitions in practice. Climate Change Leadership Fellowship and Economic and Social Research Council (ESRC). URL: http://www.lancaster.ac.uk/staff/shove/transitionsinpractice/tip.htm. Accessed 25 July 2016

Creswell JW (2003) Research design: qualitative, quantitative, and mixed method approaches, 2nd edn. Sage Publications, Thousand Oaks

CSE (2009) Richest Indians emit less than poorest Americans [WWW Document]. URL: http://www.cseindia.org/equitywatch/pdf/richest_poorest_emissions.pdf. Accessed 12 Nov 2015

Dangschat JS (1996) Raum als Dimension sozialer Ungleichheit und Ort als Bühne der Lebensstilisierung? Zum Raumbezug sozialer Ungleichheit und von Lebensstilen. In: Schwenk O (ed) Lebensstil Zwischen Sozialstrukturanalyse Und Kulturwissenschaft. Leske + Budrich, Opladen, pp 99–135

Das D (2015) Hyderabad: visioning, restructuring and making of a high-tech city. Cities 43:48–58

Datt G, Ravallion M, Murgai R (2016) Growth, urbanization and poverty reduction in India. National Bureau of Economic Research, Washington, DC

Datta A (2015) New urban utopias of postcolonial India: "entrepreneurial urbanization" in Dholera smart city, Gujarat. Dialogues Hum Geogr 5:3–22

Davis SJ, Caldeira K (2010) Consumption-based accounting of CO_2 emissions. Proc Natl Acad Sci 107:5687–5692

de Haan G, Lantermann ED, Linneweber V, Reusswig F (2001) Typenbildung in der sozialwissenschaftlichen Umweltforschung. VS Verlag für Sozialwissenschaften, Wiesbaden

Deaton A, Kozel V (2005) Data and dogma: the great Indian poverty debate. World Bank Res Obs 20:177–199

Deshpande O, Reid MC, Rao AS (2005) Attitudes of Asian-Indian Hindus toward end-of-life care. J Am Geriatr Soc 53:131–135

Devi SM, Balachandar V, Lee SI, Kim IH (2014) An outline of meat consumption in the Indian population – a pilot review. Korean J Food Sci Anim Res 34:507–515

Dodman D (2009) Blaming cities for climate change? An analysis of urban greenhouse gas emissions inventories. Environ Urban 21:185–201

Donner H (2008) Domestic goddesses: maternity, globalization and middle-class identity in contemporary India. Ashgate, Aldershot

Drews S (2011) Aviation and environment. Centre for Science and Environment, Delhi

Etheridge DM, Steele LP, Langenfelds RL, France RJ, Barnola J-M, Morgan VI (1996) Natural and anthropogenic changes in atmospheric CO_2 over the last 1000 years from air in Antarctic ice and firn. J Geophys Res 101:4115–4128

Evans G (1992) Testing the validity of the Goldthorpe class schema. Eur Sociol Rev 8:211–232

Fazio RH (1990) Multiple processes by which attitudes guide behavior: the mode model as an integrative framework. In: Zanna MP (ed) Advances in experimental social psychology. Academic, San Diego, pp 75–109

Fernandes L (2006) India's new middle class: democratic politics in an era of economic reform. Oxford University Press, New Delhi

Fernandes L (2009) The political economy of lifestyle: the political economy of lifestyle: consumption, India's new middle class and state-led development. In: Lange H, Meier L (eds) The new middle classes. Globalizing lifestyles consumerism and environmental concern. Springer, Berlin, pp 219–236

Ferreira FH, Messina J, Rigolini J, López-Calva L-F, Lugo MA, Vakis R, Ló LF (2012) Economic mobility and the rise of the Latin American middle class. World Bank Publications, Washington, DC

Ferreira FH, Chen S, Dikhanov Y, Hamadeh N, Jolliffe D, Narayan A, Prydz EB, Revenga A, Sangraula P, Serajuddin U (2015) A global count of the extreme poor in 2012: data issues, methodology and initial results. World Bank Policy Research Working Paper

Field A (2011) Discovering statistics using SPSS: (and sex and drugs and rock "n" roll), 3rd edn., reprinted ed. Sage, Los Angeles

Filmer D, Pritchett LH (2001) Estimating wealth effects without expenditure data – or tears: an application to educational enrollments in states of India. Demography 38:115–132

Fleurbaey M, Kartha S, Bolwig S, Chee YL, Chen Y, Corbera E, Lecocq F, Lutz W, Muylaert MS, Norgaard RB, Okereke C, Sagar AD (2014) Sustainable development and equity. In: Edenhofer O, Pichs-Madruga R, Sokona Y, Farahani E, Kadner S, Seyboth K, Adler A, Baum I, Brunner S, Eickemeier P, Kriemann B, Savolainen J, Schlömer S, von Stechow C, Zwickel T, Minx JC (eds) Climate change 2014: mitigation of climate change. Contribution of working group III to the fifth assessment report of the Intergovernmental Panel on Climate Change. Cambridge University Press, Cambridge

Flick U (2006) An introduction to qualitative research, 3rd edn. SAGE, London

Fraser I (2015) "Selfie diplomacy": how India's prime minister Modi became such a hit on social media. The Telegraph. URL: http://www.telegraph.co.uk/news/worldnews/asia/india/11986187/How-Indias-prime-minister-Narendra-Modi-became-such-a-hit-on-social-media-selfie.html. Accessed 4 Mar 2016

Freytag T, Gebhardt H, Gerhard U, Wastl-Walter D (eds) (2016) Humangeographie kompakt. Springer, Berlin/Heidelberg

Fromm S (2010) Datenanalyse mit SPSS für Fortgeschrittene 2: Multivariate Verfahren für Querschnittsdaten. VS Verlag für Sozialwissenschaften/GWV Fachverlage GmbH Wiesbaden, Wiesbaden

Fromm S, Baur N (2008) Datenanalyse mit SPSS für Fortgeschrittene: Ein Arbeitsbuch, 2. überarbeitete und erweiterte Auflage. VS Verlag für Sozialwissenschaften, Wiesbaden

Fuchs RJ, Brennan E, Chamie J, Lo FC, Uitto JI (1994) Mega-city growth and the future. United Nations University Press, Tokyo

Gadgil M, Guha R (1995) Ecology and equity: the use and abuse of nature in contemporary India. Penguin Books, New Delhi

Ganesan AV (2014) Come, make In India. The Hindu

Geißler R (2002) Die Sozialstruktur Deutschlands. Bundeszentrale für politische Bildung, Bonn

Giddens A (2015) Preface. In: Werlen B (ed) Global sustainability, cultural perspectives and challenges for transdisciplinary integrated research. Springer, Cham, pp v–viii

Gilani V (2010) Emission factors ready reckoner for India 2010. URL: http://www.no2co2.in/admin/utils/internalresource/intresourceupload/EF_ready_reckoner_india_2010_CC.pdf. Accessed 25 May 2010

Gilani V (2012) Emission factors ready reckoner for India 2012. URL: http://www.no2co2.in/admin/utils/internalresource/intresourceupload/EF_ready_reckoner_india_Mar2012_CC.pdf. Accessed 21 Dec 2015

Girod B, de Haan P (2009) GHG reduction potential of changes in consumption patterns and higher quality levels: evidence from Swiss household consumption survey. Energy Policy 37:5650–5661

GIZ (2015) Non-motorised transport policy in India. The need for a reform agenda (policy brief no. 2), sustainable urban transport project. GIZ, Eschborn

GoI (2011) Census 2011: population enumeration data. URL: http://www.censusindia.gov.in/2011census/C-series/DDWCT-0000C-14.xls. Accessed 24 Sept 2015

GoI (2013) Poverty estimates for 2011–12. Planning Commission, Government of India, New Delhi

GoI (2014a) Employment across various sectors. NSSO 61st and 66th round survey (2009–10). Planning Commission, New Delhi

GoI (2014b) Make in India – new initiatives. URL: http://www.makeinindia.com/policy/new-initiatives. Accessed 3 Mar 2016

GoI (2015) India's intended nationally determined contribution: working towards climate justice

Government of Telangana (2015) Reinventing Telangana – the first steps. Socio-economic outlook 2015. Government of Telangana, Planning Department, Hyderabad

Gregory D (ed) (2009) The dictionary of human geography, 5th edn. Blackwell, Malden

Grießhammer R, Bleher D, Dehoust G, Gensch C-O, Harves K, Hochfeld C, Groß R, Möller M, Seum S (2009) Memorandum product carbon footprint: Positionen zur Erfassung und Kommunikation des Product Carbon Footprint für die internationale Standardisierung und Harmonisierung. BMU, UBA, Öko-Institut e.V, Berlin

Grunewald N, Harteisen M, Lay J, Minx J, Renner S (2012) The carbon footprint of Indian households. Presented at the 32nd General Conference of The International Association for Research in Income and Wealth
Hammond G (2007) Time to give due weight to the "carbon footprint" issue. Nature 445:256–256
Herendeen R, Tanaka J (1976) Energy cost of living. Energy 1:165–178. https://doi.org/10.1016/0360-5442(76)90015-3
Hermann D (2004) Bilanz der empirischen Lebensstilforschung. Kölner Z Soziol Sozialpsychol 56:153–179
Hertwich EG, Peters GP (2009) Carbon footprint of nations: a global trade-linked analysis. Environ Sci Technol 43:6414–6420
Howe LD, Hargreaves JR, Huttly SRA (2008) Issues in the construction of wealth indices for the measurement of socio-economic position in low-income countries. Emerg Themes Epidemiol 5:1–14
Hradil S (1987) Sozialstrukturanalyse in einer fortgeschrittenen Gesellschaft. Von Klassen, Schichten zu Lagen und Milieus. Leske+Budrich Verlag, Opladen
Hradil S (1996) Sozialstruktur und Kultur: Fragen und Antworten zu einem schwierigen Verhältnis. In: Schwenk O (ed) Lebensstil Zwischen Sozialstrukturanalyse Und Kulturwissenschaft. Leske + Budrich, Opladen, pp 13–30
Hradil S (2001a) Soziale Ungleichheit in Deutschland. VS Verlag für Sozialwissenschaften, Wiesbaden
Hradil S (2001b) Eine Alternative? Einige Anmerkungen zu Thomas Meyers Aufsatz "Das Konzept der Lebensstile in der Sozialstrukturforschung". Soziale Welt 52:273–282
IATA (2014) IATA air passenger forecasts global report
Ibrahim N, Sugar L, Hoornweg D, Kennedy C (2012) Greenhouse gas emissions from cities: comparison of international inventory frameworks. Local Environ 17:223–241
ICLEI (2009) International local government GHG emissions analysis protocol (IEAP). Version 1.0. Local Governments for Sustainability (ICLEI), Bonn
IEA (2010) CO_2 emissions from fuel combustion. International Energy Agency, Paris.
IEA (2015) World energy outlook 2015. International Energy Agency. URL: http://www.iea.org/bookshop/700-World_Energy_Outlook_2015. Accessed 1 Mar 2016
Inglehart R (1990) Culture shift in advanced industrial society. Princeton University Press, Princeton
IPCC (2006) 2006 IPCC guidelines for national greenhouse gas inventories. URL: http://www.ipcc-nggip.iges.or.jp/public/2006gl/index.html. Accessed 18 Dec 2015
ISSC, UNESCO, OECD (2013) World social science report 2013. Changing global environments. OECD Publishing; UNESCO Publishing, Paris
Iyer NK, Kulkarni S, Raghavaswamy V (2007) Economy, population and urban sprawl a comparative study of urban agglomerations of Bangalore and Hyderabad, India using remote sensing and GIS techniques. Paper presented at PRIPODE workshop on Urban Population, Development and Environment Dynamics in Developing Countries, Nairobi
Jain P (2011) Dharma and ecology of Hindu communities: sustenance and sustainability, Ashgate new critical thinking in religion, theology and biblical studies. Ashgate, Farnham
Jaiswal T (2014) Indian arranged marriages. In: A social psychological perspective. Routledge, Oxford
Jeuland MA, Pattanayak SK (2012) Benefits and costs of improved cookstoves: assessing the implications of variability in health, forest and climate impacts. PLoS One 7:e30338
Jiang L, O'Neil BC (2004) The energy transition in rural China. Int J Global Energy Issues 21:2
Johannson A, Guillemette Y, Murtin F, Turner D, Nicoletti G, de la Maisonneuve C, Bagnoli P, Bousquet G, Spinelli F (2012) Looking to 2060: long-term global growth prospects: a going for growth report (No. 3), OECD Economic Policy Papers. OECD, Paris
Kaiser HF (1974) An index of factorial simplicity. Psychometrika 39:31–36
Kennedy L (2007) Regional industrial policies driving peri-urban dynamics in Hyderabad. India Cities 24:95–109

Kennedy L, Zérah M-H (2008) The shift to city-centric growth strategies: perspectives from Hyderabad and Mumbai. Econ Polit Wkly:110–117

Kennedy C, Steinberger J, Gasson B, Hansen Y, Hillman T, Havránek M, Pataki D, Phdungsilp A, Ramaswami A, Mendez GV (2009) Greenhouse gas emissions from global cities. Environ Sci Technol 43:7297–7302

Kennedy C, Steinberger J, Gasson B, Hansen Y, Hillman T, Havránek M, Pataki D, Phdungsilp A, Ramaswami A, Mendez GV (2010) Methodology for inventorying greenhouse gas emissions from global cities. Energy Policy 38:4828–4837

Kharas H (2010) The emerging middle class in developing countries, OECD Working Paper No. 285. OECD Development Centre, Paris

Kimmich C, Janetschek H, Meyer-Ohlendorf L, Meyer-Ueding J, Sagebiel J, Reusswig F, Rommel K, Hanisch M (2012) Methods for stakeholder analysis: exploring actor constellations in transition and change processes towards sustainable resource use and the case of Hyderabad, India, emerging megacities. Europäischer Hochschulverlag, Bremen

Kleinhückelkotten S (2005) Suffizienz und Lebensstile. Ansätze für eine milieuorientierte Nachhaltigkeitskommunikation. Berliner Wissenschaftsverlag GmbH, Berlin

Kluckhohn C (1944) Mirror for man: a survey of human behavior and social attitudes. Fawcett, Greenwich

Kluckhohn FR, Strodtbeck FL (1961) Variations in value orientations. Row, Peterson and Company, Evanston

Koehler G (2015) Seven decades of "development", and now what?: the SDGs and transformational change 2015. J Int Dev 27:733–751

Kraas F (2007) Megacities and global change: key priorities. Geogr J 173:79–82

Kraas F, Mertins G (2014) Megacities and global change. In: Kraas F, Aggarwal S, Coy M, Mertins G (eds) Megacities: our global urban future, International Year of Planet Earth. Springer, Dordrecht, pp 1–6

Kraas F, Nitschke U (2006) Megastädte als Motoren globalen Wandels. Int Polit 61:18–28

Krause RM (2013) The motivations behind municipal climate engagement: an empirical assessment of how local objectives shape the production of a public good. Cityscape 15:125–141

Kundu A (2014) India's sluggish urbanization and its exclusionary development. In: Gordon M, George M (eds) Urban growth in emerging economies: lessons from the BRICS. Routledge, London, pp 191–232

Lange H, Meier L, Anuradha NS (2009) Highly qualified employees in Bangalore, India: consumerist predators? In: Lange H, Meier L (eds) The new middle classes. Globalizing lifestyles consumerism and environmental concern. Springer, Berlin, pp 281–298

Latour B (2000) When things strike back: a possible contribution of 'science studies' to the social sciences. Br J Sociol 51:107–123

Leiserowitz AA, Kates RW, Parris TM (2006) Sustainability values, attitudes, and behaviours: a review of multinational and global trends. Annu Rev Environ Resour 31:413–444

Lenzen M, Wier M, Cohen C, Hayami H, Pachauri S, Schaeffer R (2006) A comparative multivariate analysis of household energy requirements in Australia, Brazil, Denmark, India and Japan. Energy 31:181–207

Lin J, Hu Y, Cui S, Kang J, Ramaswami A (2015) Tracking urban carbon footprints from production and consumption perspectives. Environ Res Lett 10:54001

Lorenzoni I, Nicholson-Cole S, Whitmarsh L (2007) Barriers perceived to engaging with climate change among the UK public and their policy implications. Glob Environ Chang 17:445–459

Lüdtke H (1989) Expressive Ungleichheit. Zur Soziologie der Lebensstile. Leske + Budrich, Opladen

Luther N (2008) Hyderabad through foreign eyes. In: Imam S (ed) The untold charminar – writings on Hyderabad. Penguin Books, New Delhi, pp 1–19

Madan TN (1989) Religion in India. Daedalus 118:114–146

Mathur N (2010) Shopping malls, credit cards and global brands: consumer culture and lifestyle of India's new middle class. South Asia Res 30:211–231

Mawdsley E (2004) India's middle classes and the environment. Dev Chang 35:79–103
Mawdsley E (2006) Hindu nationalism, neo-traditionalism and environmental discourses in India. Geoforum 37:380–390
Mayring P (2002) Einführung in die qualitative Sozialforschung. Eine Anleitung zu qualitativem Denken. Beltz, Weinheim
McGranahan G, Schensul D, Singh G (2016) Inclusive urbanization: can the 2030 agenda be delivered without it? Environ Urban 28:13–34
MCH (2005) Hyderabad city development plan. Municipal Corporation of Hyderabad, Hyderabad
McKinsey Global Institute (2007). The "bird of gold": the rise of India's consumer market
Meadows DH, Meadows DL, Randers J, Behrens WW III (eds) (1974) The limits to growth: a report for the club of Rome's project on the predicament of mankind, 2nd edn. Universe Books, New York
Meyer T (2001) Das Konzept der Lebensstile in der Sozialstrukturforschung – eine kritische Bilanz. Soziale Welt 255–272
Meyer C, Birdsall N (2012) New estimates of India's middle class (technical note). Center for Global Development, Washington, DC
Minx JC, Wiedmann T, Wood R, Peters GP, Lenzen M, Owen A, Scott K, Barrett J, Hubacek K, Baiocchi G, Paul A, Dawkins E, Briggs J, Guan D, Suh S, Ackerman F (2009) Input-output analysis and carbon footprinting: an overview of applications. Econ Syst Res 21:187–216
Mitchell A (1983) The nine American lifestyles. Who we are and where we are going. Warner Books, New York
Mitchell D (1995) There's no such thing as culture: towards a reconceptualization of the idea of culture in geography. Trans Inst Br Geogr 20:102
Modi N (2015) The rich world must take greater responsibility for climate change. Financial Times
MoEF (2010) Climate change and India: a 4x4 assessment. A secoral and regional analysis for 2030 (no. 2), Indian Network for Climate Change Assessment (INCCA). Ministry of Environment & Forests, Government of India, New Delhi
Morris MD (1967) Values as an obstacle to economic growth in South Asia: an historical survey. J Econ Hist 27:588–607
Motiram S, Vakulabharanam V (2012) Understanding poverty and inequality in urban India since reforms. Bringing quantitative and qualitative approaches together. Econ Polit Wkly 47:44–52
Mukherjee A (2013) The service sector in India. Economics Working Paper Series. Asian Development Bank
Myers N, Kent J (2003) New consumers: the influence of affluence on the environment. Proc Natl Acad Sci 100:4963–4968
Narayanan V (1999) Y51K and still counting: some Hindu views of time. J Hindu-Christian Stud 12:17–18
National Sample Survey Organisation (2012) Energy sources of Indian households for cooking and lighting 2009–10, NSSO 66th Round
Nelson LE (2000) Purifying the earthly body of God: religion and ecology in Hindu India. D. K. Printworld, New Delhi
Noé21 (2014) Fairconditioning 2014–2015–2016: cooling India efficiently and sustainably. Noé21, Geneva
Nollmann G (2008) Klassen. In: Baur N, Korte H, Löw M, Schroer M (eds) Handbuch Soziologie. Verlag für Sozialwissenschaften, Wiesbaden
NSSO (2001) Concepts and definitions used in NSS. National Sample Survey Organisation, Government of India, New Delhi
O'Brien K, Leichenko R, Kelkar U, Venema H, Aandahl U, Tompkins H, Javed A, Bhadwal S, Barg S, Nygaard L, West J (2004) Mapping vulnerability to multiple stressors: climate change and globalization in India. Glob Environ Chang 14(4):303–313
OPHI (2015) OPHI country briefing 2015: India, OPHI Country Briefing. Oxford Department of International Development, University of Oxford
Otte G (1997) Lebensstile versus Klassen – welche Sozialstrukturkonzeption kann die individuelle Parteipräferenz besser erklären? In: Müller W (ed) Soziale Ungleichheit. Neue Befunde Zu Strukturen, Bewußtsein Und Politik. Leske + Budrich, Opladen, pp 303–346

Otte G (2004) Sozialstrukturanalyse mit Lebensstilen. Eine Studie zur theoretischen und methodischen Neuorientierung der Lebensstilforschung, 1st edn. Verlag für Sozialwissenschaften, Wiesbaden

Otte G (2005) Hat die Lebensstilforschung eine Zukunft? Eine Auseinandersetzung mit aktuellen Bilanzierungsversuchen. Kölner Z Soziol Sozialpsychol 57:1–31

Otte G, Rössel J (2011) Lebensstile in der Soziologie. In: Otte G, Rössel J (eds) Lebensstilforschung. Sonderheft 51 Der Kölner Zeitschrift Für Soziologie Und Sozialpsychologie. Verlag für Sozialwissenschaften, Wiesbaden, pp 7–34

Pachauri S (2004) An analysis of cross-sectional variations in total household energy requirements in India using micro survey data. Energy Policy 32:1723–1735

Pachauri S (2007) An energy analysis of household consumption changing patterns of direct and indirect use in India. Springer, Dordrecht

Pachauri S, Jiang L (2008) The household energy transition in India and China. IIASA, Laxenburg

Pachauri S, Spreng D (2002) Direct and indirect energy requirements of households in India. Energy Policy 30:511–523

Pandey D, Agrawal M, Pandey JS (2011) Carbon footprint: current methods of estimation. Environ Monit Assess 178(1–4):135–160

Parikh J, Panda M, Murthy NS (1997) Consumption patterns by income classes and carbon-dioxide implications for India: 1990–2010. Int J Global Energy Issues 9:4–5

Parikh J, Panda M, Ganesh-Kumar A, Singh V (2009) CO_2 emissions structure of Indian economy. Energy 34:1024–1031

Parnell S (2016) Defining a global urban development agenda. World Dev 78:529–540

Peoples J, Bailey G (1999) Humanity. An introduction to cultural anthropology, 5th edn. Wadsworth, Belmont

Peters GP (2010) Carbon footprints and embodied carbon at multiple scales. Curr Opin Environ Sustain 2:245–250

Peters V, Reusswig F, Altenburg C (2013) European citizens, carbon footprints and their determinants – lifestyles and urban form. In: Khare A, Beckmann T (eds) Mitigating climate change. Springer-Verlag, Berlin, pp 223–245

Pieper J (1984) Hyderabad: a Qur'anic paradise in architectural metaphors. J Islam Environ Des Res Centre:46–51

Planton S (2013) Annex III: glossary. In: Stocker TF, Qin D, Plattner G-K, Tignor M, Allen SK, Boschung J, Nauels A, Xia Y, Bex V, Midgley PM (eds) Climate change 2013: the physical science basis. Contribution of working group I to the fifth assessment report of the Intergovernmental Panel on Climate Change. Cambridge University Press, Cambridge, pp 1447–1466

Poferl A (1998) Wer viel konsumiert, ist reich. Wer nicht konsumiert, ist arm. Ökologische Risikoerfahrung, soziale Ungleichheiten und kulturelle Politik. In: Berger PA, Vester M (eds) Alte Ungleichheiten Neue Spaltungen. VS Verlag für Sozialwissenschaften, Wiesbaden, pp 297–329

Rahmstorf S, Schellnhuber HJ (2007) Der Klimawandel: Diagnose, Prognose, Therapie, 6th edn. Beck'sche Reihe C.-H.-Beck-Wissen, Beck

Ravallion M (2009) The developing world's bulging (but vulnerable) "middle class", Policy Research Working Paper No. WPS4816. World Bank, Washington, DC

Razzaque A, Alam N, Wai L, Foster A (1990) Sustained effects of the 1974–75 famine on infant and child mortality in a rural area of Bangladesh. Popul Stud 44:145–154

Reckien D, Hofmann S, Kit O (2009) Qualitative climate change impact networks for Hyderabad, India (Background study). Sustainable Hyderabad Project

Reckwitz A (2002) Toward a theory of social practices a development in culturalist theorizing. Eur J Soc Theory 5:243–263

ReportLinker (2015) India air conditioners market forecast and opportunities, 2020 [WWW Document]. URL: http://www.reportlinker.com/p0881690-summary/India-Air-Conditioners-Market-Forecast-Opportunities.html. Accessed 7 Oct 2015

Reusswig F (1994a) Lebensstile und Ökologie: Gesellschaftliche Pluralisierung und alltagsökologische Entwicklung unter besonderer Berücksichtigung des Energiebereichs, Sozialökologische Arbeitspapiere. Verlag für Interkulturelle Kommunikation, Frankfurt am Main

Reusswig F (1994b) Lebensstile und Ökologie. In: Dangschat JS (ed) Lebensstile in den Städten: Konzepte und Methoden. Leske + Budrich, Opladen, pp 91–103

Reusswig F, Meyer-Ohlendorf L (2010) Social representation of climate change: a case study from Hyderabad (India), emerging megacities. Europäischer Hochschulverlag GmbH & Co. KG, Bremen

Reusswig F, Meyer-Ohlendorf L (2012) Adapting to what? Climate change impacts on Indian megacities and the local Indian climate change discourse. In: Holt WG (ed) Urban areas and global climate change, research in urban sociology. Emerald Group Publishing Limited, Bradford, pp 197–220

Reusswig F, Otto A, Meyer-Ohlendorf L, Anders U (2009) Climate change discourse in India. An analysis of press articles. Sustainable Hyderabad Project, Berlin

Reusswig F, Meyer-Ohlendorf L, Anders U (2012) Partners for a low-carbon Hyderabad: a stakeholder analysis with respect to lifestyle dynamics and climate change, emerging megacities. Europäischer Hochschulverlag, Bremen

Reusswig F, Hirschl B, Lass W, Becker C, Bölling L, Clausen W, Haag L, Hahmann H, Heiduk P, Hendzlik M, Henze A, Hollandt F, Hunsicker F, Lange C, Meyer-Ohlendorf L, Neumann A, Rupp J, Schiefelbein S, Schwarz U, Weyer G, Wieler U (2014) Klimaneutrales Berlin 2050: Machbarkeitsstudie. Senatsverwaltung für Stadtentwicklung und Umwelt, Berlin

Rieger HC (1995) Die Liberalisierung der Wirtschaft. In: Rothermund D (ed) Indien: Kultur, Geschichte, Politik, Wirtschaft, Umwelt; Ein Handbuch. Beck, München

Rink D (2002) Lebensweise, Lebensstile und Lebensführung. Soziologische Konzepte zur Untersuchung von nachhaltigem Leben. In: Rink D (ed) Lebensstil Und Nachhaltigkeit. Konzepte, Befunde Und Potentiale, Soziologie Und Ökologie. Springer Fachmedien, Wiesbaden, pp 27–52

Rokeach M (1973) The nature of human values. The Free Press, New York

Rosa H, Strecker D, Kottmann A (2007) Soziologische Theorien, UTB. UVK-Verl.-Ges, Konstanz

Rössel J (2005) Plurale Sozialstrukturanalyse. Eine handlungstheoretische Rekonstruktion der Grundbegriffe der Sozialstrukturanalyse. Verlag für Sozialwissenschaften, Wiesbaden

Rössel J (2011) Soziologische Theorien in der Lebensstilforschung. In: Otte G, Rössel J (eds) Lebensstilforschung. Sonderheft 51 Der Kölner Zeitschrift Für Soziologie Und Sozialpsychologie. Verlag für Sozialwissenschaften, Wiesbaden, pp 7–34

Rutstein SO, Johnson K (2004) The DHS wealth index, DHS Comparative Reports, vol 6. ORC Macro, Calverton

Säävälä M (2010) Middle-class moralities: everyday struggle over belonging and prestige in India. Orient Blackswan, New Delhi

Satterthwaite D (2003) The links between poverty and the environment in urban areas of Africa, Asia, and Latin America. Ann Am Acad Polit Soc Sci 590:73–92

Satterthwaite D (2008) Cities' contribution to global warming: notes on the allocation of greenhouse gas emissions. Environ Urban 20:539–549

Schatzki TR, Knorr-Cetina K, von Savigny E (eds) (2001) The practice turn in contemporary theory. Routledge, New York

Schendera CFG (2010) Clusteranalyse mit SPSS: Mit Faktorenanalyse. Oldenbourg Wissenschaftsverlag, München

Schultz I, Weller I (1996) Nachhaltige Konsummuster und postmaterielle Lebensstile. Eine Vorstudie im Auftrag des Umweltbundesamtes. Institut für sozial-ökologische Forschung (ISOE), Frankfurt am Main

Schulze G (1992) Die Erlebnisgesellschaft. Campus Verlag, Frankfurt am Main

Schwartz SH (1992) Universals in the content and structure of values: theoretical advances and empirical tests in 20 countries. In: Zanna MP (ed) Advances in experimental social psychology. Academic, San Diego, pp 1–65

Schwartz SH (1994) Are there universal aspects in the structure and contents of human values? J Soc Issues 50:19–45

Schwartz SH, Melech G, Lehmann A, Burgess S, Harris M, Owens V (2001) Extending the cross-cultural validity of the theory of basic human values with a different method of measurement. J Cross-Cult Psychol 32:519–542

Sen A (1999) Development as Freedom. Anchor Books, New York

Sherwell P (2015) Top Indian artists and scientists return awards in protest at alleged "climate of intolerance" under Narendra Modi. The Telegraph. URL: http://www.telegraph.co.uk/news/worldnews/asia/india/11963542/Top-Indian-artists-and-scientists-return-awards-in-protest-at-alleged-climate-of-intolerance-under-Narendra-Modi.html. Accessed 4 Mar 2016

Shove E (2003) Comfort, cleanliness and convenience: the social organization of normality, new technologies/new cultures. Berg, Oxford

Shove E (ed) (2007) The design of everyday life, cultures of consumption series. Berg, New York

Shove E (2010) Beyond the ABC: climate change policy and theories of social change. Environ Plan A 42:1273–1285

Shove E, Pantzar M, Watson M (2012) The dynamics of social practice: everyday life and how it changes. Sage, Los Angeles

Shukla RK (2010) How India earns, spends and saves: unmasking the real India. SAGE, New Delhi

Shukla RK Dwivedi SK, Sharma A, Jain S (2004) The great Indian middle class: results from the NCAER market information survey of households [WWW Document]. URL: www.ncaer.org/Downloads/PublicationsCatalog.pdf. Accessed 23 Mar 2014

Simmel G (1907) Philosophie des Geldes, 2nd edn. Duncker & Humblot, Leipzig

Simon D, Arfvidsson H, Anand G, Bazaz A, Fenna G, Foster K, Jain G, Hansson S, Evans LM, Moodley N, Nyambuga C, Oloko M, Ombara DC, Patel Z, Perry B, Primo N, Revi A, Van Niekerk B, Wharton A, Wright C (2016) Developing and testing the urban sustainable development goal's targets and indicators – a five-city study. Environ Urban 28:49–63

Singh N (2009) Exploring socially responsible behaviour of Indian consumers: an empirical investigation. Soc Responsib J 5:200–211

Singh DP, Dalei NN, Raju TB (2016) Forecasting investment and capacity addition in Indian airport infrastructure: analysis from post-privatization and post-economic regulation era. J Air Transp Manag 53:218–225

SINUS (2015) Information on Sinus-Milieus [WWW Document]. URL: http://www.sinus-institut.de/fileadmin/user_data/sinus-institut/Dokumente/downloadcenter/Sinus_Milieus/2015-11-10_Information_on_Sinus-Milieus.pdf. Accessed 11 Feb 2016

Sobel ME (1981) Lifestyle and social structure: concepts, definitions, analyses, quantitative studies in social relations. Academic, New York

Sovani NV (1978) The social milieu in India and development: part I. Indian Philisophical Q 3:387–408

Spaargaren G (1997) The ecological modernisation of production and consumption: essays in environmental sociology. Landbouw Universitiet Wageningen, Wageningen

Spellerberg A (1996) Soziale Differenzierung durch Lebensstile: Eine empirische Untersuchung zur Lebensqualität in West- und Ostdeutschland. Sigma, Berlin

Stadtmüller S, Klocke A, Lipsmeier G (2013) Lebensstile im Lebenslauf – Eine Längsschnittanalyse des Freizeitverhaltens verschiedener Geburtskohorten im SOEP. Z Soziol 42:262–290

Stang F (2002) Indien – Geographie, Geschichte, Wirtschaft, Politik. Wissenschaftliche Buchgesellschaft, Darmstadt

Strauss AL (1987) Qualitative analysis for social scientists. Cambridge University Press, Cambridge

Swain G (2014) Environmental movements in India. Global J Multidiscip Stud 4:210–228

Thite M (2014) Commentary – India's Narendra Modi: beacon of hope for political, economic and human resource development in South Asia? South Asian J Hum Res Manag 1:289–292

Thøgersen J (2012) The importance of timing for breaking commuters' car driving habits. In: Warde A, Southerton D (eds) The habits of consumption, Helsinki collegium for advanced studies. Helsinki, pp 130–140

Tiwari G (2011) Key mobility challenges in Indian cities. International Transport Forum Discussion Paper, New Delhi

Tiwari G, Jain D (2013) NMT infrastructure in India: investment, policy and design. UNEP, Roskilde

Triandis S (2002) Cultural influences on personality. Annu Rev Psychol 53:133–160

Trivedi LN (2003) Visually mapping the "nation": swadeshi politics in nationalist India, 1920–1930. J Asian Stud 62:11–41

UNFCC (2015) Wake up call ahead of Paris 2015–400 ppm CO_2 level breached. UN Climate Newsroom. URL: http://newsroom.unfccc.int/unfccc-newsroom/wake-up-call-ahead-of-paris-2015-400ppm-co2-level-breached/. Accessed 2 Dec 2015

United Nations (2014a) World urbanization prospects: the 2014 revision report (No. ST/ESA/SER.A/366). Department of Economic and Social Affairs, Population Division

United Nations (2014b) World urbanization prospects: the 2014 revision. United Nations Department of Economic and Social Affairs, Population Division

Upadhya C (2009) India's new middle classes and the globalising city: software professionals in Bangalore, India. In: Lange H, Meier L (eds) The new middle classes. Springer, Berlin, pp 253–268

Urry J (2011) Climate change and society. Polity Press, Cambridge

van Wessel M (2001) Modernity and identity: an ethnography of moral ambiguity and negotiations in an Indian middle class. University of Amsterdam, Amsterdam

van Wessel M (2004) Talking about consumption: how an Indian middle class dissociates from middle-class life. Cult Dyn 16:93–116

Varma PK (1998) The great Indian middle class. Penguin Books, Delhi

Veblen T (1997) Theorie der feinen Leute. Eine ökonomische Untersuchung der Institutionen, 5th edn. Fischer Taschenbuch Verlag, Frankfurt/Main

Vedwan N (2007) Pesticides in Coca-Cola and Pepsi: consumerism, brand image, and public interest in a globalizing India. Cult Anthropol 22:659–684

Velders GJ, Fahey DW, Daniel JS, McFarland M, Andersen SO (2009) The large contribution of projected HFC emissions to future climate forcing. Proc Natl Acad Sci 106:10949–10954

Venkataraman C, Habib G, Eiguren-Fernandez A, Miguel AH, Friedlander SK (2005) Residential biofuels in South Asia: carbonaceous aerosol emissions and climate impacts. Science 307:1454–1456

Verma A, Velmurugan S, Singh S, Gurtoo A, Ramanayya TV, Dixit M (2015) Urban mobility trends in Indian cities and its implications. In: Gurtoo A, Williams C (eds) Developing country perspectives on public service delivery. Springer India, New Delhi, pp 95–116

Vester M, von Oertzen P, Geiling H, Hermann T, Müller D (2001) Soziale Milieus im gesellschaftlichen Strukturwandel - Zwischen Integration und Ausgrenzung. Suhrkamp, Frankfurt am Main

Von Grebmer K, Ringler C, Rosegrant MW, Olofinbiyi T, Wiesmann D, Fritschel H, Badiane O, Torero M, Yohannes Y (2012) 2012 Global Hunger Index: the challenge of hunger: ensuring sustainable food security under land, water, and energy stresses, Global Hunger Index. International Food Policy Research Institute (IFPRI), Washington, DC

von Stietencron H (1995) Die Erscheinungsformen des Hinduismus. In: Rothermund D (ed) Indien. Beck, München, pp 143–166

Vringer K (2005) Analysis of the energy requirement for household consumption. University of Utrecht, Utrecht

Vyas S, Kumaranayake L (2006) Constructing socio-economic status indices: how to use principal components analysis. Health Policy Plan 21:459–468

Wagner C (2010) India: a difficult partner in international climate policy. In: Dröge S (ed) International climate policy: priorities of key negotiating parties. Stiftung Wissenschaft und Politik, Berlin, pp 63–73

Walker G, Shove E, Brown S (2014) How does air conditioning become "needed"? A case study of routes, rationales and dynamics. Energy Res Soc Sci 4:1–9

Warde A (2014) After taste: culture, consumption and theories of practice. J Consum Cult 14:279–303

WBGU – Wissenschaftlicher Beirat der Bundesregierung Globale Umweltveränderungen (ed) (2016) Der Umzug der Menschheit: Die transformative Kraft der Städte: Hauptgutachten, 1. Auflage. ed. Wissenschaftlicher Beirat der Bundesregierung Globale Umweltveränderungen, Berlin

Weber M (1922) Wirtschaft und Gesellschaft, Grundriss der Sozialökonomik, 1st edn. J. C. B. Mohr (Paul Siebeck), Tübingen

Weber M (1986) Gesammelte Aufsätze zur Religionssoziologie, Bd. II Hinduismus und Buddhismus, 8th edn. Mohr Siebeck, Tübingen

Weiss MJ (1988) The clustering of America. Harpercollins, New York

Werlen B (ed) (2015a) Global sustainability, cultural perspectives and challenges for transdisciplinary integrated research. Springer, Cham

Werlen B (2015b) From local to global sustainability: transdisciplinary integrated research in the digital age. In: Werlen B (ed) Global sustainability, cultural perspectives and challenges for transdisciplinary integrated research. Springer, Cham, pp 3–16

Wiedmann TO, Chen G, Barrett J (2015) The concept of City Carbon Maps: a case study of Melbourne, Australia. J Ind Ecol:1–16

Williams I, Kemp S, Coello J, Turner DA, Wright LA (2012) A beginner's guide to carbon footprinting. Carbon Manag 3:55–67

World Bank (2012) World development indicators. URL: http://databank.worldbank.org/ddp/editReport?REQUEST_SOURCE=search&CNO=2&country=&series=NY.GDP.MKTP.KD.ZG&period=. Accessed 28 Feb 2012

World Bank (2014) India – data. URL: http://data.worldbank.org/country/india#cp_wdi. Accessed 1 Mar 2016

World Bank (2015a) World development report 2015: mind, society, and behavior. Main messages. The World Bank

World Bank (2015b) FAQs: global poverty line update. World Bank. URL: http://www.worldbank.org/en/topic/poverty/brief/global-poverty-line-faq. Accessed 16 Mar 16

World Bank (2016a) World Bank national accounts data and OECD National Accounts data files. World Development Indicators. URL: http://databank.worldbank.org/data/reports.aspx?source=2&country=IND&series=&period=#. Accessed 3 Mar 2016

World Bank (2016b) World development indicators. URL: http://databank.worldbank.org/data/reports.aspx?source=2&country=IND&series=&period=. Accessed 16 Mar 2016

WRI (2015) CAIT climate data explorer. World Resources Institute. URL: http://cait.wri.org. Accessed 29 Mar 2016

WRI, C40, ICLEI (2014) Global protocol for community-scale greenhouse gas emission inventories (GPC) – an accounting and reporting standard for cities. World Resources Institute, C40 Cities Climate Leadership Group, ICLEI Local Governments for Sustainability

WRI/WBCSD (2004) The greenhouse gas protocol: a corporate accounting and reporting standard; revised edition. World Resources Institute and World Business Council for Sustainable Development

Wright LA, Coello J, Kemp S, Williams I (2011) Carbon footprinting for climate change management in cities. Carbon Manag 2:49–60

Xu Y, Zaelke D, Velders GJ, Ramanathan V (2013) The role of HFCs in mitigating 21st century climate change. Atmos Chem Phys 13:6083–6089

Index

A
ABC paradigm, 34
Accounting approaches, 10
Action theory perspective, 26
Aesthetic attitudes, 42
Aestheticisation, 29
Air-conditioners (AC), 189, 227, 229, 230
Air travel, 223, 224
Amartya Sen, 61
American lifestyle, 16
Analysis of variance (ANOVA), 131
Anthropogenic climate change, 1, 57, 68
Appadurai, A., 97
Arranged marriages, 164, 165
Asceticism, 71, 102
Atman, 104
Average carbon footprint, 219

B
Bartlett test of sphericity, 130
Black carbon, 223
Bourdieu, P., 5, 25, 26, 28, 31, 34, 41–44
Brahman, 103
Buddhism, 23

C
Carbon calculator, 123, 127, 187
Carbon dioxide equivalents, 9
Carbon footprint calculations, 175
Carbon footprinting, 15, 121, 130
Carbon footprints, 6, 11, 12, 40, 43, 122, 219, 245, 246, 250, 251
Carbon intensities, 245, 247
Carbon-intensive consumption practices, 246
Carbon-intensive practice, 229
Carbon map approach, 13
Carbon taxing, 13
Caste, 23, 115, 139, 207
Caste endogamy, 164
Caste system, 157
Census 2011, 141, 143, 144
Chandrababu Naidu, 77
City emission inventories, 13
City primacy, 55
Class, 18
Climate change mitigation, 13, 15
Climate change perceptions, 156
Cluster analysis, 37, 38, 129, 160, 165, 166, 170, 173, 232
Cluster sampling approach, 129
CO_2 concentrations, 9
Colonial history, 42
Communalities, 130, 138
Complex patterns, 231
Components of lifestyle, 89
Comte, A., 18
Concept of lifestyle, 29, 139, 231, 239
Conceptualisation of lifestyle, 231, 237
Conduct of life, 22, 23, 27, 31, 88
Conference of Parties (COP13), 69
Configurations of values, 249
Conservative values, 235
Conspicuous consumption, 23, 96, 161–163, 165, 191, 196, 199
Constraint-choice approach, 32
Consumer class, 14
Consumer culture, 208, 210, 211, 213, 217
Consumer culture orientation, 234
Consumer culture paradigm, 234
Consumer decisions, 246

Consumer expenditure, 16, 20, 41, 64
Consumer expenditure data, 59, 60
Consumer lifestyle approach, 24
Consumer orientation, 199
Consumer responsibility, 161
Consumer's decision making, 24
Consumption-based approach, 12
Consumption-based emissions, 10, 11
Consumption-based framework, 10
Consumption-based GHG accounting, 20
Consumption-based inventory, 13
Consumption effects, 189
Consumption elasticity, 17
Consumption expenditure, 14, 40, 226, 246
Consumption expenditure data, 226
Consumption of electricity, 220
Consumption of necessity, 44
Consumption patterns, 1, 4
Consumption practices, 187, 189, 190, 229
Content-based interpretation, 232
Contextual factors, 30
Cultural, 31
Cultural and social capital, 41
Cultural turn, 85
Culture of necessity, 97, 158, 159
Cumulative GHG emissions, 69

D

Definition of lifestyle, 32
Definition of the middle class, 65
Determinants of lifestyle, 18
Developing countries, 15
Development ladder, 16
Dharma, 104, 157, 165
Differentiation, 23
Dimensions and components of lifestyle, 28
Dimensions of lifestyle, 89
Direct and indirect emissions, 10
Distinction, 29
Domestic air travel, 190
Domestication, 23
Double counting, 13
Dowry, 192, 193, 216
Drivers of climate change, 4
Durkheim, E., 18
Dynamics of social mobility, 42, 58
Dynamics of urbanisation, 42

E

Ecological limits, 2
Economic growth, 49
Economic liberalisation, 50

Economic perspective, 245
Economics, 39
Educational level, 144
Eigenart, 5
Eigenart of a city, 3
Eigenvalues, 130, 153
Electricity consumption, 18
Emerging economies, 15, 67
Emerging middle classes, 15
Emerging middle class in India, 63
Emission accounting, 10
Emission effect calculations, 247
Emission factors (EF), 132, 220
Emission inequalities, 15
Emission-intensive consumption practices, 18
Emission inventories, 10
Employment, 145
Enculturation, 97
Endogamous connubial rules, 23
Endogamy, 157
Enhancement, 111
Enlightenment, 18
Environmental concern, 233
Environmental crisis, 39
Environmental friendliness of products, 161
Environmentalism in India, 100
Environmental movements in India, 100
Environmental policies, 247
Environmental values, 250
Environment-related (consumption) practices, 39
Environment-related lifestyle research, 18, 248
Environment-related social science theory, 40
Environment-related social sciences, 39
Environment-related values, 34
Erlebnis, 97, 162
Ethical consumption, 99, 100
Ethnographic research methods, 20
Everyday routines, 248
Expectancy effect, 238
Explorative multiple regression, 249
Expressive inequality, 34
Extreme poverty, 59
Extreme weather events, 68

F

Factor analysis, 129, 155
Family tradition, 106, 163–165, 191, 196, 199, 202, 204, 206, 208, 212, 233
Fatalism, 200, 206, 233
Fatalistic paradigm, 213
FCKW, 9
Financial, 31

Flânerie, 96
Focus group discussions, 239
Food-related emissions, 222
Fossil fuels, 9
Foundations of sociology, 18
Frugality, 159, 160, 165, 192, 193, 195, 196, 199, 202, 203, 206, 208, 210–213, 233, 234
Frugality and thrift, 97, 158–160, 191–193, 232, 233

G
Gandhi, 71, 94, 99
Gandhian ideals of simplicity, 91, 160
Gated communities, 86, 88
Geography, 19, 39
Georg F.W. Hegel, 18
German Advisory Council on Global Change, 2
Ghandi, 102
Giddens, A., 19
GHG emissions, 57, 74
Global emissions, 15
Global environmental change, 19
Global North, 15
Global poverty, 59
Global Protocol for Community-Scale Greenhouse Gas Emission Inventories (GPC), 11
Global sustainability, 19
Global warming potential (GWP), 9, 73, 223
Goldthorpe classification, 145
Great transformation, 1, 2
Green growth paradigm, 39
Greenhouse effect, 9
Greenhouse gases (GHGs), 9
Greenhouse gas inventories, 11
Greenpeace India, 14, 15
Ground truthing, 241

H
Habitat III, 2
Habitus, 31, 41, 43
Hedonic, 211
Hedonic consumer culture paradigm, 213
Hedonic consumer values, 210, 234
Hedonism, 91, 113, 160, 163, 191
Hedonistic consumer, 206
Hedonistic consumer values, 203
Hiding behind the poor, 4, 74
Hierarchical cluster analysis, 131
Highest income class, 224

Highest income group, 224
Hindu caste system, 23
Hindu castes, 23
Hinduism, 23, 102
Hindu mythology, 106, 156
Historical emissions, 15
Historical materialism, 21
Holidays, 159, 205, 207
Horizontal differences, 36
Household consumption, 14
Household-level emissions, 14
Household survey, 20
H-test, 172, 179
Human activities, 9
Humanities, 39
Hyderabad, 75–79, 84, 123

I
Identity, 97, 106, 160
Identity management, 162, 208
Impacts of climate change, 57, 68
Inclusive development, 2
Income, 14, 146, 179, 181, 183, 185–187, 192, 193, 200, 204, 207, 215, 216, 245
Income classes, 14
Income dynamics, 187, 189
Income elasticity, 17
Income elasticity of emissions, 14
Income groups, 14
Income-oriented approach, 246
Income-related GHG accounting, 187
Indian consumer culture, 162, 163
Indian middle classes, 63
Indian monsoon, 68
Indian Network on Climate Change Assessment (INCCA), 71
Indirect energy consumption, 17
Indirect GHG emissions, 17
Individual carbon footprints, 125
Individualisation, 23
Individual-level carbon footprints, 20
Individual scope of action, 6, 31
Input-output (IO), 14
Intended Nationally Determined Contributions (INDCs), 71
Internal locus of control, 204
International accounting guidelines, 11
International air travel, 189, 223, 229
International development policies, 4
International negotiations on climate change, 15
International Social Science Council (ISSC), 4
Intra-caste marriage, 157

Intra-cluster homogeneity, 232
Investive consumption, 16, 116–119, 122, 123, 130, 135, 138, 148, 160, 187, 202, 203, 206, 216, 217, 228, 229, 233, 242, 246, 248

J
Jairam Ramesh, 16

K
Kaiser-Meyer-Olkin (KMO), 130
Kali Yuga, 104–106, 156
Karma, 156, 210
Karman, 102, 103
Key points, 230, 231, 242, 247
Key points of intervention, 243
Kin-related hierarchies, 163
Kluckhohn-Strodtbeck framework, 90
K-means algorithm, 131, 232
Kruskal-Wallis, 179
Kruskal-Wallis test, 131, 170
Kyoto Protocol, 69

L
Latour, B., 35
Legally binding commitment, 15
Leisure, 162
Liberalisation, 49, 50
Life cycle assessment (LCA), 12
Lifestyle, 6
Lifestyle analysis, 135, 175
Lifestyle concept, 21, 25, 26, 30, 33, 34, 36
Lifestyle dynamics, 84
Lifestyle policies, 12
Lifestyle-related environmental research, 42
Lifestyle segmentations, 27, 142, 144
Lifestyle sociology, 36
Lifestyle survey, 127
Lifestyle typology, 139, 170
Lifeworld, 27
Locus of control, 191
Lower income classes, 224
Lower middle-class, 94

M
Mahabharata, 104
Make in India, 51, 100
Manifest variables, 129, 130

Marx, K., 18
Mass consumer culture, 161
Material configuration of urban areas, 87
McKinsey Global Institute (MGI), 63
Meanings, 35
Measure of sampling adequacy (MSA), 130
Meat consumption, 177, 222, 226
Media analysis, 124
Megacities, 53, 159
Megacities in India, 58
Mega-urban regions, 54
Mentalities, 29, 88, 89
Methane, 9, 17
Middle-class definition, 65
Middle classes, 5, 15, 49, 63, 64, 68, 159
Middle classes in India, 91
Middle-class households, 64
Middle-class identity, 100
Milieus, 27
Milk consumption, 221, 225
Millennium Development Goals (MDGs), 2, 61
Mitigating global climate change, 10
Modernisation, 23, 101
Modernity, 25, 106
Modern urban lifestyle, 108, 162–163
Morality, 94, 104, 165, 202
Motivation, 28
Multidimensional measures of poverty, 63
Multidimensional Poverty Index (MPI), 61
Multimodal Transport System (MMTS), 177
Muslims, 142, 201, 203, 233

N
Narendra Modi, 16, 51, 52, 71, 100
National Action Plan on Climate Change (NAPCC), 71
National Council of Applied Economic Research (NCAER), 63, 64
National Family Health Survey (NFHS-3), 62
National inventories, 12
National Tendulkar poverty line, 62
Natural sciences, 39
Necessity, 44
Nehru socialism, 42
Nehruvian era of socialism, 92
Nehruvian socialism, 160
Neolithic revolution, 56
New Economic Policy (NEP), 55, 58, 72, 77, 92, 100
New middle class, 85

Index

The new middle class, 66
New middle class rhetoric, 66
New middle-class, 66
New middle-class narrative, 66
New Telangana government, 78
New (urban) middle classes, 42
no2co2, 120, 121, 132
Non-parametric tests, 179
Normative compass, 2–3, 5
Nuclear family, 108

O
Obedience, 165
Objective living conditions, 24

P
P. V. Narasimha Rao, 50
Paradigm of ABC, 119
Paradigm of continuous growth, 39
Paris Agreement, 2
Passive variables, 171
Path dependencies, 1, 187, 229, 243, 248
Patriotic consumption, 100
Patterns of everyday life, 6
People-oriented approach, 2
Per capita emission effects, 179
Per capita GHG emissions, 74
Performance, 28, 29
Personal carbon footprints, 171, 246
Personal-level GHG accounting, 81
Personal-level GHG emissions, 246
Piketty, T., 15, 17
Pluralisation, 23
6-point Likert scale, 155
Poor, 44
Poor people, 43
Portrait Values Questionnaire (PVQ), 90, 91
Post-independence, 76
Post-Kyoto, 69
Potential of social mobility, 67
Poverty, 43, 58, 61, 62, 238
Poverty in India, 74
Poverty lines, 15, 59, 60, 62
Practical habitus, 34
Practice, 35
Practice-related emission effects, 248
Practice-related GHG emissions, 248
Practice theory, 33, 35
Predatory consumerism, 63
Principal component analysis (PCA), 88, 107, 130, 131, 136, 138, 151–153, 155, 158, 163, 170, 171, 240

Problem-oriented approach, 4
Product carbon footprint (PCF), 12
Production-based approaches, 10
Pro-environmental consumption, 191
Proxy to income, 16
Psychological aspects of behaviour, 3
Public transport, 177

Q
Qualitative research, 238
Qualitative social research, 241
Qualitative survey, 138
Quality of life, 5, 39–40, 57
Quantitative research, 238
Questionnaire, 127–129

R
Rational choice-based approach, 41
Rational choice models, 3
Rationalisation, 23
Religious tradition, 190–192, 196, 202, 204, 206, 208, 210, 211, 213, 233, 234
Religious traditional values, 204
Residential welfare associations (RWA), 86
Rice consumption, 221
Routines, 117–119, 228
Routinised behaviours, 35
Routinised pattern of consumption, 187
Rural-to-urban migration, 49

S
Saint-Simon, C.H., 18
Sampling, 127
Samsara, 102
Savings, 16
Scale effect, 247
Schwartz Portrait Value Questionnaire (PVQ), 110, 128
Schwartz value theory, 90, 111
Scope 1, 11
Scope 2, 11
Scope 3, 11
Scope 3 emissions, 13
Scope of action, 33, 43, 248
Scopes, 10
Scree plot, 130
Secularism, 163
Seductive spaces, 87
Self-efficacy, 99, 161, 197, 204, 211, 233
Self-enhancement, 110
Self-expression, 162

Self-fulfillment, 235
Self-realisation, 162
Self-transcendence, 91, 110, 111
Simmel, G., 18, 25
Simplicity, 159
Simplicity and thrift, 94, 102
SINUS-Milieus, 27
Situative contexts, 29
Spencer, H., 18
Social accounting matrix (SAM), 14
Social and cultural geography, 19, 39
Social behaviour, 5
Social capital, 31
Social differentiation, 18
Social disparities, 49
Social distinction, 27–28, 41
Social expectancy effect, 226
Social identity, 22, 27
Social inequalities, 21, 36
Social mobility, 42, 193, 230
Social organisation, 21
Social position, 18, 29, 88
Social practice, 18, 24, 35, 108, 247
Social practice theory, 242
Social science environmental research, 20
Social science research, 25, 35
Social sciences, 3–5, 26, 39
Social status, 22, 193
Social status group, 23
Social stratification, 157
Social structure, 18, 23, 45
Social structure analysis, 23, 37–38
Social values, 34, 45, 101
Social-cultural inequalities, 18
Social-ecological consumption, 161
Social-ecological orientation, 161, 204, 236
Social-ecological paradigm, 208, 234
Social-ecological values, 195, 206, 211, 233, 235
Social-ecologically conscious consumption, 210
Social-ecologically responsible consumption, 204
Social-economic constraints, 45
Social-economic position, 6
Social-economic situation, 28
Sociological lifestyle research, 26
Sociology of consumption, 85
SPSS, 136
Standard guideline for GHG emission accounting in cities, 13
Status groups, 22

Structuration theory, 34
Stylisation, 29, 32, 43, 44
Stylisation of living, 22
Stylisation options, 43
Stylisations of life, 22
Subjective lifeworld, 25
Subjective ways of living, 24
Subordination, 163
Sufficiency, 39, 113
Sustainability, 39, 57
Sustainable consumption, 161, 189, 242
Sustainable development, 1
Sustainable Development Goals (SDGs), 2
Sustainable lifestyles, 58
Sustainable products, 161
Sustainable urban development, 2, 3
Symbolic capital, 43

T
Targeted interventions, 231
Taste, 41
Technocratic, 4
Technology ladder, 43
Telangana, 75, 76, 78
Tendulkar poverty line, 62
Territorial approach to carbon footprinting, 12
Theodicy, 102
Theory of planned action, 29
Thrift and frugality, 195
Tradition, 101, 106
Traditional values, 106–108, 163, 165, 236
Tradition and modernity, 107, 109
Transdisciplinarity, 4
Transdisciplinary, 19
Transdisciplinary perspective, 4
Transdisciplinary social-ecological research, 4
Transformation, 1, 2, 101
Transformation processes, 49
Typification of lifestyle, 240
Typifications, 20, 36

U
Universalism, 111
Upper middle class, 107, 159
Urban India, 42
Urban middle classes, 101
Urban population in India, 55
Urban transformation pathways, 5
Urban transformation towards sustainability, 3

Urbanisation, 49, 50, 52–58, 69
Urbanisation in India, 54, 55, 58

V
Value orientation, 29
Value-action gap, 30
Value-based cluster analysis, 249
Values, 89, 109
Veblen, T., 25
Vertical approach, 36
Virtues of frugality and thrift, 41
Vulnerability, 62, 65

Vulnerability to climate change, 68
Vulnerable to climate change, 72

W
Ward method, 131
Way of life, 30
Wealth, 200, 205, 209
Wealth index, 20, 148, 183, 216, 242
Weber, M., 18, 19, 21, 25, 101–103
Werlen, B., 19
World Development Report, 3
World risks, 1

Printed by Printforce, the Netherlands